Biopolymers as Therapeutic Adjuvants: Innovations and Advancement

Edited by

Sudhanshu Mishra
Faculty of Pharmaceutical Sciences Mahayogi Gorakhnath University
Gorakhpur 273007, Uttar Pradesh
India

Smriti Ojha
Department of Pharmaceutical Science & Technology
Madan Mohan Malaviya University of Technology
Gorakhpur, Uttar Pradesh
India

Shashi Kant Singh
Faculty of Pharmaceutical Sciences
Mahayogi Gorakhnath University
Gorakhpur 273007, Uttar Pradesh
India

Rishabha Malviya
Department of Pharmacy
Galgotias University
Greater Noida, Uttar Pradesh
India

&

Saurabh Kumar Gupta
Rameshwaram Institute of Technology & Management
Lucknow, Uttar Pradesh
India

Biopolymers as Therapeutic Adjuvants: Innovations and Advancements

Editors: Sudhanshu Mishra, Smriti Ojha, Shashi Kant Singh, Rishabha Malviya and Saurabh Kumar Gupta

ISBN (Online): 979-8-89881-141-9

ISBN (Print): 979-8-89881-142-6

ISBN (Paperback): 979-8-89881-143-3

Published by Bentham Science Publishers Pte. Ltd. Singapore,

in collaboration with Eureka Conferences, USA. All Rights Reserved.

First published in 2025.

need for a court order if at any point you breach any terms of this License Agreement. In no event will any delay or failure by Bentham Science Publishers in enforcing your compliance with this License Agreement constitute a waiver of any of its rights.

3. You acknowledge that you have read this License Agreement, and agree to be bound by its terms and conditions. To the extent that any other terms and conditions presented on any website of Bentham Science Publishers conflict with, or are inconsistent with, the terms and conditions set out in this License Agreement, you acknowledge that the terms and conditions set out in this License Agreement shall prevail.

Bentham Science Publishers Pte. Ltd.
No. 9 Raffles Place
Office No. 26-01
Singapore 048619
Singapore
Email: subscriptions@benthamscience.net

BENTHAM SCIENCE

CONTENTS

FOREWORD

The field of biopolymers represents a remarkable confluence of biology, chemistry, and material science, offering innovative solutions to some of the most pressing challenges in medicine and therapeutics. From their humble beginnings as naturally occurring substances to their sophisticated modern-day applications, biopolymers have continuously evolved, providing the scientific community with tools for creating safer, more effective, and personalized healthcare solutions.

The book ***Biopolymers as Therapeutic Adjuvants: Innovations and Advancemet*** arrives at a pivotal moment in this field, bridging the gap between foundational knowledge and the latest advancements. It captures the spirit of interdisciplinary collaboration that defines modern science, bringing together contributions from seasoned researchers and emerging scholars. Their collective effort paints a detailed picture of the transformative potential of biopolymers across diverse domains, including drug delivery, tissue engineering, cancer therapy, and beyond.

The book invites readers to not only absorb its wealth of knowledge but to actively participate in shaping the future of biopolymer applications. By fostering a community of inquiry and shared purpose, it becomes more than a resource; it is a catalyst for advancement in therapeutic science.

I commend Mr. Sudhanshu Mishra and his dedicated team of contributors for producing a publication that is both scientifically rigorous and forward-looking. This work serves as a beacon of inspiration, guiding researchers and practitioners toward the shared goal of improving human health through biopolymer innovation.

With best wishes.

Pranesh Kumar
Institute of Pharmacy
University of Lucknow
Lucknow, Uttar Pradesh
India

PREFACE

Biopolymers, derived from natural sources or synthesized to mimic biological molecules, have emerged as revolutionary tools in therapeutics and drug delivery systems. The unique combination of biocompatibility, biodegradability, and tunable properties positions biopolymers as pivotal components in advancing healthcare technologies. This book, ***Biopolymers as Therapeutic Adjuvants: Innovations and Advancemet***, delves into the multidisciplinary realm of biopolymer science, encompassing its historical evolution, scientific fundamentals, and cutting-edge applications. The chapters are thoughtfully curated to provide a comprehensive understanding of the subject. The book begins with an exploration of the historical context and foundational principles of biopolymers, tracing their development from early discoveries to their contemporary therapeutic relevance. Subsequent chapters address the diverse biological sources, chemical characteristics, and extraction methods of biopolymers, providing readers with a solid scientific foundation.

A detailed discussion on synthetic methodologies and the physicochemical properties of biopolymers sets the stage for understanding their pharmacokinetics and pharmacodynamics. Moving beyond the basics, the text explores the role of biopolymers in modern therapeutics, including their use in chemotherapeutic regimens, synergistic drug combinations, and disease-targeting strategies through nano-based systems. Special emphasis is given to advancements in the field, such as the integration of artificial intelligence in biopolymer engineering and the development of stimuli-responsive and enzyme-activated drug delivery systems. Readers will also discover applications in tissue engineering, bone regeneration, and autoimmune disease immunotherapy, highlighting the transformative potential of biopolymers in addressing complex medical challenges.

Authored by a diverse group of experts and researchers, this book serves as a vital resource for professionals, academics, and students involved in pharmaceutical sciences, biomaterials research, and biomedical engineering. By merging theoretical insights with practical applications, it aspires to inspire further innovation and exploration in the burgeoning field of biopolymer therapeutics.

We hope this book enriches your understanding of biopolymers and motivates you to contribute to this fascinating and impactful area of research.

Sudhanshu Mishra
Faculty of Pharmaceutical Sciences Mahayogi Gorakhnath University
Gorakhpur 273007, Uttar Pradesh
India

Smriti Ojha
Department of Pharmaceutical Science & Technology
Madan Mohan Malaviya University of Technology
Gorakhpur, Uttar Pradesh
India

Shashi Kant Singh
Faculty of Pharmaceutical Sciences
Mahayogi Gorakhnath University
Gorakhpur 273007, Uttar Pradesh
India

Rishabha Malviya
Department of Pharmacy
Galgotias University
Greater Noida, Uttar Pradesh
India

&

Saurabh Kumar Gupta
Rameshwaram Institute of Technology & Management
Lucknow, Uttar Pradesh
India

List of Contributors

Ajay Pandey	Department of Pharmaceutical Engineering and Technology, Indian Institute of Technology (Banaras Hindu University), Varanasi, India
Aamir Anwar	Department of Pharmacy, Integral University, Lucknow, Uttar Pradesh, India
Asad Ahmad	Department of Pharmacy, Integral University, Lucknow, Uttar Pradesh, India
Anindita De	Department of Pharmaceutics, College of Pharmacy, JSS University, Noida 201301, India
Anubhav Anand	Shri Ramswaroop College of Engineering and Management (Pharmacy), Lucknow, Uttar Pradesh, India
Aarti Tiwari	Institute of Pharmacy, Dr Rammanohar Lohia Avadh University, Ayodhya, Uttar Pradesh, India
Ajay Kumar Shukla	Institute of Pharmacy, Dr Rammanohar Lohia Avadh University, Ayodhya, Uttar Pradesh, India
Bharat Mishra	Institute of Pharmacy, DR. Shakuntala Misra National Rehabilitation University, Lucknow, 226017, India
Bhaveshwari Wagh	Institute of Pharmaceutical Sciences, Faculty of Pharmacy, Parul University, Vadodara, Gujarat, 391760, India
Bishambar Singh	PHTI Department, SMS Medical College and Hospital, Jaipur, Rajasthan, 302017, India
Deepak Kumar	Faculty of Pharmaceutical Sciences, Mahayogi Gorakhnath University, Gorakhpur, Uttar Pradesh, 273007, India
Ganesh Lal	KJ College of Pharmacy, Babatpur, 221006, Varanasi, India
Gaurish Narayan Singh	Institute of Pharmacy, Deen Dayal Upadhyaya Gorakhpur University, Gorakhpur, 273009, Uttar Pradesh, India
Gowthamarajan Kuppusamy	Department of Pharmaceutics, JSS College of Pharmacy, JSS Academy of Higher Education & Research, Ooty 643001, Tamil Nadu, India
Juhi Tiwari	Faculty of Pharmaceutical Sciences, Mahayogi Gorakhnath University, Gorakhpur, Uttar Pradesh, 273007, India
Kunal Agam Kanujia	Institute of Pharmacy, Dr Rammanohar Lohia Avadh University, Ayodhya, Uttar Pradesh, India
Manoj Kumar Mishra	Shambhunath Institute of Engineering and Technology, Prayagraj, Uttar Pradesh, India
Manya Modi	Department of Pharmacy, Banasthali Vidyapith, Rajasthan, 304022, India
Nancy Gupta	Department of Pharmacy, Banasthali Vidyapith, Rajasthan, 304022, India
Nandani Jayaswal	Faculty of Pharmaceutical Sciences, Mahayogi Gorakhnath University, Gorakhpur, 273007, Uttar Pradesh, India
Piyush Anand	Faculty of Pharmaceutical Sciences, Mahayogi Gorakhnath University, Gorakhpur, Uttar Pradesh, 273007, India

Pooja Pooja	Department of Pharmacy, Banasthali Vidyapith, Rajasthan, 304022, India
Pooja Jaiswal	Faculty of Pharmaceutical Sciences, Mahayogi Gorakhnath University, Gorakhpur, 273007, Uttar Pradesh, India
Piyush Anand	Faculty of Pharmaceutical Sciences, Mahayogi Gorakhnath University, Gorakhpur, Uttar Pradesh, 273007, India
Rufaida Wasim	Department of Pharmacy, Integral University, Lucknow, Uttar Pradesh, India
Rama Sankar Dubey	Department of Pharmacy, MMM University, Gorakhpur, Uttar Pradesh, India
Sharda Sambhakar	Department of Pharmacy, Banasthali Vidyapith, Rajasthan, 304022, India
Shreya Maddesiya	Faculty of Pharmaceutical Sciences, Mahayogi Gorakhnath University, Gorakhpur, Uttar Pradesh, 273007, India
Taufik Mulla	Institute of Pharmaceutical Sciences, Faculty of Pharmacy, Parul University, Vadodara, Gujarat, 391760, India
Tarique Mahmood	Department of Pharmacy, Integral University, Lucknow, Uttar Pradesh, India
Srishti Verma	Department of Pharmacy, Banasthali Vidyapith, Rajasthan, 304022, India
Saba Parveen	Department of Pharmacy, Madan Mohan Malaviya University of Technology, Gorakhpur, Uttar Pradesh, India
Sonali Jayronia	Department of Pharmaceutics, College of Pharmacy, JSS University, Noida 201301, India
Young Joon Park	Department of Formulation and Drug Delivery, College of Pharmacy, Ajou University, 206 Worldcup-ro, Yeongtong-gu, Suwon-si 16499, Republic of Korea
Surbhi Gupta	Ashoka Institute of Technology and Management, Varanasi, Uttar Pradesh, India
Vimal Kumar Yadav	Institute of Pharmacy, Dr Rammanohar Lohia Avadh University, Ayodhya, Uttar Pradesh, India
Vishnu Prasad Yadav	Institute of Pharmacy, Dr Rammanohar Lohia Avadh University, Ayodhya, Uttar Pradesh, India

CHAPTER 1

Introduction and Historical Overview of Biopolymers as Therapeutics

Bhaveshwari Wagh[1,*] and **Taufik Mulla**[1]

1 Institute of Pharmaceutical Sciences, Faculty of Pharmacy, Parul University, Vadodara, Gujarat 391760, India

Abstract: Biopolymers are naturally occurring polymers that are produced by living organisms. They include proteins, nucleic acids, polysaccharides, and other biomolecules. Due to their biocompatibility, biodegradability, and low toxicity, biopolymers have gained significant attention in medicine, particularly as therapeutic agents. As a class of materials, biopolymers offer unique advantages over synthetic polymers, including the ability to interact with biological systems more naturally. Their applications span drug delivery, tissue engineering, wound healing, and gene therapy, making them essential to modern biomedical research. The historical development of biopolymers as therapeutics spans from ancient uses in traditional medicine to modern biotechnology advancements. In the 19th century, the foundation was laid with the discovery of proteins and nucleic acids. In the early 20th century, therapeutic use of proteins such as insulin and polysaccharides like heparin emerged. The mid-20th century marked the rise of nucleic acids as therapeutic agents, while the late 20th century introduced biotechnology, enabling large-scale production of biopolymer-based drugs. In the 21st century, innovations in drug delivery, gene therapy, and regenerative medicine have further advanced the use of biopolymers in treating diseases.

Keywords: Biocompatibility, Biodegradability, Biopolymers, Drug delivery, Polysaccharides, Therapeutic agents, Tissue engineering.

INTRODUCTION TO BIOPOLYMERS IN THERAPEUTICS

Biopolymers are a remarkable class of naturally occurring polymers produced by living organisms, including plants, animals, and microbes. These polymers differ fundamentally from synthetic ones, as they are derived from renewable biological sources rather than petroleum-based resources [1]. Due to their inherent biodegradability and biocompatibility, biopolymers have garnered increasing attention in the medical field, where there is a constant need for materials that can

---

* **Corresponding author Bhaveshwari Wagh:** Institute of Pharmaceutical Sciences, Faculty of Pharmacy, Parul University, Vadodara, Gujarat 391760, India; E-mail: bhaveshwari.wagh35656@paruluniversity.ac.in

Sudhanshu Mishra, Smriti Ojha, Shashi Kant Singh, Rishabha Malviya & Saurabh Kumar Gupta (Eds.)

integrate seamlessly with human biology. This introduction will explore the unique qualities that set biopolymers apart, their essential role in therapeutic applications, and a high-level overview of the primary types of biopolymers in use today [2].

Definition of Biopolymers

Biopolymers are large, chain-like molecules composed of repeating subunits identified as monomers. Monomers, which are covalently bonded, make a long chain that can exhibit a diverse array of structures and functionalities. The variability in structure among different types of biopolymers contributes to their unique properties and functions, making each type suitable for specific applications [3]. For instance, the structural organization of protein-based biopolymers allows for complex three-dimensional forms, enabling them to perform precise biological functions. Nucleic acids, on the other hand, encode genetic information, while polysaccharides offer structural and energy storage solutions. This structural and functional diversity makes biopolymers indispensable in therapeutic applications [4].

Importance of Biopolymers in Therapeutics

The therapeutic potential of biopolymers can be attributed to several critical properties that make them well-suited for interaction with biological systems.

Biocompatibility

One of the most important features of biopolymers is their compatibility with biological tissues. Because biopolymers are typically well-tolerated by the body, they pose a lower risk of immune reactions, making them ideal to apply in drug delivery, tissue engineering, and wound care [5].

Biodegradability

Unlike synthetic polymers, which often persist in the body and environment, biopolymers can be broken down by natural enzymatic or hydrolytic processes. This eliminates concerns related to long-term accumulation and reduces potential complications. This property is advantageous in applications requiring a temporary scaffold or carrier, such as drug delivery systems or tissue engineering [6].

Customizability with Interactive Capabilities

Biopolymers can be modified or engineered to exhibit specific interactions with biological targets, increasing their efficacy and versatility in therapeutic applications. For example, biopolymers can be designed for precise drug release, to target specific cells, or to facilitate tissue regeneration by promoting cell adhesion and growth [7].

Overview of Biopolymer Types

Biopolymers used in therapeutics can be classified into three main types: proteins, nucleic acids, and polysaccharides. Each class has unique properties and applications that make it suitable for specific therapeutic purposes [8].

Proteins

Proteins are perhaps the most versatile type of biopolymer, with roles that range from structural support to enzymatic catalysis. Composed of amino acids linked with peptide bonds, proteins adopt intricate three-dimensional shapes that determine their specific functions [9]. Within therapeutics, various types of protein-based biopolymers have specific applications.

Enzymes

As natural catalysts, enzymes accelerate biochemical reactions in the body. Therapeutic enzymes are used in enzyme-replacement therapy to treat metabolic disorders or as catalysts in drug synthesis, offering a biologically safe and efficient alternative to chemical catalysts [10].

Antibodies

Antibodies are crucial components of the immune system, identifying and neutralizing pathogens. Therapeutically, monoclonal antibodies are used in cancer immunotherapy and autoimmune disease treatments due to their ability to specifically target disease-causing cells or molecules [11].

Hormones

Hormones are regulatory proteins that modulate physiological processes. Synthetic hormone analogs, such as insulin for diabetes and growth hormone for growth disorders, are widely used in medicine [12].

Nucleic Acids

Examples include DNA and RNA, which store and transport genetic information, playing an essential role in cell functioning and gene expression. Advances in genetic engineering have expanded their potential in medicine.

DNA

Deoxyribonucleic acid (DNA) serves as the genetic blueprint for cellular functions [13]. In gene therapy, therapeutic DNA can be delivered to cells to correct genetic mutations, offering a promising avenue for the treatment of inherited diseases.

RNA

Ribonucleic acid (RNA) facilitates protein synthesis and regulates gene expression. Recently, mRNA-based vaccines have been developed to guide cells in producing antigens that activate an immune response, providing a new tool in vaccine development [14].

Polysaccharides

Polysaccharides are carbohydrate-based polymers composed of monosaccharide units. They play diverse roles in structural support, cellular recognition, and energy storage, and have unique therapeutic applications due to their bioactivity and biodegradability.

Cellulose

This abundant polysaccharide is used as an excipient in drug formulations and as a component in controlled-release drug delivery systems [15]. It provides structural support and stability to formulations, enhancing their efficacy.

Chitosan

Chitosan, derived from chitin, is well known for its biocompatibility and its ability to enhance drug absorption across biological membranes [16]. Its applications span from wound dressings to drug and gene delivery vehicles.

Hyaluronic Acid

Found in connective tissues, hyaluronic acid has hydrating and viscoelastic properties, making it valuable in orthopedic and cosmetic applications, such as joint lubricants and dermal fillers for skin rejuvenation [17].

ANCIENT AND TRADITIONAL USES OF BIOPOLYMERS IN MEDICINE

Throughout history, biopolymers have served a pivotal role in healing practices for ancient civilizations. Early cultures developed a deep understanding of their natural environment and utilized various natural substances, many of which were rich in biopolymers, to treat ailments and improve overall health. Biopolymers are naturally occurring macromolecules produced by living organisms and include materials such as polysaccharides, proteins, and complex carbohydrates [18]. These biopolymers were often derived from plants, animals, and microorganisms, and their use in traditional medicine paved the way for many modern therapeutic practices, as shown in Table **1**. This exploration will delve into the different ways ancient civilizations used biopolymeric substances, detailing how Egyptians, Chinese, and other cultures developed early medical applications based on these naturally occurring polymers [19].

Table 1. Ancient and traditional uses of biopolymers in medicine.

Culture	Biopolymer	Use(s)
Egyptian	Honey	Wound treatment, antimicrobial
Egyptian	Tree Resin	Wound protection, preservative
Various	Aloe Vera	Wound healing, skin treatment, and inflammation reduction
Chinese	Ginseng	Immunity boost, vitality enhancement
Various	Silk	Sutures, wound closure
Chinese	Chitosan	Wound healing, infection control
Various	Honey	Wound treatment, antimicrobial, preservative
Various	Resin	Wound protection, antimicrobial, preservative
Various	Propolis	Wound healing, antimicrobial, and anti-inflammatory
Various	Gums and Mucilages	Wound healing, inflammation reduction, and digestive aid

Early Medicinal Practices with Natural Substances

Ancient societies developed extensive medicinal knowledge from their surroundings, using natural substances for their therapeutic effects. These cultures employed empirical methods to observe and pass down the healing properties of specific plants and animal-derived substances, many of which contained biopolymers that supported wound healing, infection prevention, and overall wellness [20]. The application of biopolymeric materials, such as plant extracts, honey, silk, chitosan, and others, became embedded in these early medicinal systems and was documented in ancient texts that informed subsequent medical practices for centuries [21].

Egyptian Medicine

The Egyptians were pioneers in using natural substances with biopolymeric properties for medicinal purposes. Medical papyri dating back to 1500 BCE reveal that the Egyptians utilized substances like honey and tree resin in wound treatment [22]. Honey, for instance, was applied to wounds for its high sugar content and antimicrobial properties, which created a moist environment that promoted tissue repair and regeneration. Honey's enzymatic activity produced hydrogen peroxide, providing further antibacterial effects. Tree resin, a naturally occurring biopolymer, was also applied to wounds to form a protective barrier against pathogens [23]. The Egyptians not only valued these materials for their medicinal properties but also used them in mummification practices due to their preservative qualities. This empirical understanding of biopolymers as protective and healing agents laid a foundation for their extensive use in wound management [24].

Plant Extracts

Plants provided a rich source of biopolymers that ancient cultures harnessed for healing. *Aloe vera*, a widely used plant in ancient medicine, was especially valued for treating burns, wounds, and skin conditions. The polysaccharides in aloe vera gel, particularly Ace Mannan, are known to stimulate cell proliferation, reduce inflammation, and form a barrier over wounds, facilitating a moist healing environment [25]. The gel's high water content and polysaccharide composition not only provided immediate relief from burns but also promoted long-term tissue repair. Beyond Egypt, aloe vera was also used by the Greeks and Romans, who recognized its soothing and healing properties. This early reliance on plant-based biopolymers contributed significantly to the development of herbal medicine and phytotherapy [26].

Chinese Medicine

In Traditional Chinese Medicine (TCM), biopolymer-rich substances like ginseng and various herbal formulations have been used for centuries. Ginseng, a root that contains polysaccharides and ginsenosides, was traditionally employed to boost immunity, enhance vitality, and improve overall health [27]. The biopolymers in ginseng were believed to support the body's natural defenses and maintain balance, a core principle in TCM. Additionally, herbal preparations rich in biopolymers were used to treat respiratory, digestive, and skin disorders [28]. These natural formulations were often prepared as teas, tinctures, or topical ointments, and their continued use in modern TCM demonstrates the lasting impact of ancient biopolymer applications.

Traditional Healing Methods Using Biopolymers

Biopolymers have been integral to traditional healing practices worldwide. The following sections explore the use of various natural biopolymers across different cultural contexts and highlight their distinct healing properties [29].

Aloe Vera

Aloe vera has been one of the most valued plants in traditional medicine, used by ancient cultures across the globe. The gel extracted from the aloe vera leaf is rich in polysaccharides such as acemannan, which plays a key role in wound healing by stimulating fibroblast activity, promoting collagen synthesis, and reducing inflammation [30]. Ancient Egyptians referred to it as the "plant of immortality" and used it extensively for wound care and skin treatments. The gel formed a protective coating over wounds, which helped prevent infection and kept the area moist, accelerating the healing process [31]. *Aloe vera's* therapeutic use also extended to the ancient Greeks, Romans, and Chinese, who appreciated its cooling and restorative properties for skin ailments. Its ability to support tissue repair has made aloe vera a lasting natural remedy in both traditional and modern herbal medicine [32].

Silk

Silk, produced by silkworms, has been used in traditional wound care practices for thousands of years, particularly in East Asia. The primary component of silk, fibroin, is a protein with excellent biocompatibility and durability, making it suitable for sutures and wound closures. Ancient Chinese and Japanese medical records detail the use of silk sutures to close wounds, benefiting from silk's tensile strength and low tendency to cause immune reactions. These sutures supported the wound healing process by minimizing inflammation and infection risks. Silk sutures have continued to be used in modern medicine due to their biocompatibility and biodegradability, and their long history highlights the ingenuity of ancient medical practitioners who leveraged natural biopolymers for surgical applications [33].

Chitosan

It is taken *via* chitin originates within the exoskeletons of crustaceans, and has a long history of use in traditional Chinese medicine. It is remembered for its functions against microbes as well as its capability to form a gel-like matrix. Chitosan was used to treat wounds and prevent infection [34]. Gel form allowed it to cover wounds, creating a barrier that prevented bacterial entry and promoted the formation of new tissue. Chitosan's wound-healing effects are due to its capacity to stimulate the growth of granulation tissue, a crucial component of the healing process. Today, chitosan is recognized for its potential in drug delivery systems due to its biodegradable and biocompatible nature, providing a sustained release of therapeutic agents at wound sites [35].

Honey

Honey's therapeutic properties have been celebrated for centuries, with documented use in ancient Egyptian, Greek, and Roman medicine. Known for its antimicrobial and protective qualities, honey was applied to wounds to prevent infection and maintain a moist environment favorable for tissue regeneration [36]. Its high sugar content created an osmotic effect that drew moisture out of the wound, effectively cleaning it and aiding in debris removal. Honey's enzymes produce hydrogen peroxide, further preventing microbial growth. In addition to its use in wound care, honey served as a remedy for sore throats, digestive issues, and other ailments. Honey remains a popular natural remedy [37], and its long-standing use illustrates the enduring relevance of biopolymer-rich substances in medicine.

Resin

Resin, a sticky biopolymer secreted by trees, particularly conifers, was used by various ancient cultures for its medicinal properties. The Egyptians used resin in wound treatment and their mummification process, capitalizing on its antimicrobial and preservative qualities [38]. The resin forms a protective coating over wounds, safeguarding them from infection while promoting tissue regeneration. Similarly, Native American tribes used tree resin to treat skin injuries, burns, and infections. The antiseptic and anti-inflammatory effects of resin made it a valuable resource in traditional medicine, and it remains a component in various modern natural remedies [39].

Propolis

Propolis, a resinous substance produced by bees, has been used in traditional medicine for its antimicrobial and wound-healing properties [40, 41]. Rich in flavonoids and other biopolymers, propolis was employed by ancient Greeks and Egyptians to treat wounds and infections. Its sticky texture allowed it to form a barrier over wounds, preventing microbial infiltration and facilitating tissue repair [42]. Propolis also exhibited anti-inflammatory effects, reducing pain and swelling. In traditional healing, propolis was used not only for skin injuries but also as a treatment for respiratory and gastrointestinal ailments. Today, propolis is valued for its antioxidant properties and is included in many natural health products [43].

Gums and Mucilages

Plant-derived gums and mucilages have been utilized in traditional medicine systems such as Ayurveda and Unani. These biopolymers, extracted from plants like Plantago and fenugreek, form viscous solutions that provide a protective layer over mucous membranes and wounds [44]. In ancient practices, these substances were used to soothe inflamed tissues, promote wound healing, and treat gastrointestinal issues. The mucilage from Plantago seeds, for example, has anti-inflammatory properties that help reduce irritation in the digestive tract [45]. These natural biopolymers create a moist environment that supports tissue repair and relieves inflammation, making them indispensable in traditional healing practices.

INFLUENCE OF ANCIENT BIOPOLYMER USE ON MODERN THERAPEUTICS

The traditional uses of biopolymers in medicine have laid the groundwork for numerous modern therapeutic applications [46]. From wound healing and antimicrobial formulations to drug delivery systems, the knowledge derived from ancient practices has been invaluable. The continued interest in biopolymers for their biodegradability, biocompatibility, and bioactivity underscores their lasting relevance in medical science [47]. The historical use of biopolymeric substances demonstrates the longstanding relationship between humans and nature, and it serves as a testament to the efficacy of traditional healing practices in addressing health concerns [48].

The study of ancient biopolymer applications has inspired new research into how these natural compounds can be modified and optimized for modern medical use. Scientists are now exploring innovative ways to harness and enhance the properties of biopolymers to create advanced biomaterials, contributing to advancements in regenerative medicine, pharmacology, and biomedical engineering [49]. The legacy of traditional biopolymer use continues to shape contemporary medicine, emphasizing the importance of preserving and building upon this ancient wisdom.

19th Century: Foundation of Biopolymer Science

The 19th century was an era of scientific awakening and monumental progress in the biological sciences, particularly in the realm of biopolymers [50]. During this period, foundational discoveries about proteins and nucleic acids were made, which ultimately paved the way for modern biopolymer science. These early explorations and breakthroughs began to reveal the molecular basis of life, sparking curiosity and research that would expand throughout the next century

[51]. This exploration of the intricate molecules within living organisms not only revolutionized biological understanding but also set the stage for groundbreaking advancements in medicine, genetics, and biochemistry.

Discovery of Proteins and Nucleic Acids

Proteins

The 19th century marked the beginning of our understanding of proteins, which are among the most fundamental molecules in biological systems [52]. In 1838, Dutch chemist Gerardus Johannes Mulder introduced the word "protein," which originated in Greek, *i.e.*, "proteins," which means "primary" or "of first importance." Mulder had conducted elemental analyses of animal and plant substances and observed a common chemical composition in many of these samples, which he identified as protein [53]. He speculated that this essential component might serve as a fundamental building block of life.

Initially, the structure and function of proteins remained a mystery. Although Mulder's studies revealed their elemental makeup, the complexity and size of protein molecules were not yet understood [54]. It was not until later in the 19th century that scientists began to comprehend proteins as polymers composed of smaller subunits known as amino acids. A significant breakthrough occurred in 1806, when French druggist Louis-Nicolas Vauquelin, along with his colleague Pierre Jean Robiquet, separated asparagine from asparagus, marking the first discovery of an amino acid [55]. This was an unprecedented achievement that directed isolation for other amino acids over the following decades, with chemists slowly identifying and cataloging the building blocks that make up proteins [56].

Despite the progress, the exact structure and arrangement of amino acids in proteins were still elusive at this time. The realization that proteins were high-molecular-weight polymers of amino acids was a crucial discovery that opened new avenues for scientific inquiry [57]. At the termination of the 19th century, a conceptual foundation had been laid, and scientists were beginning to suspect that proteins had complex, yet highly organized structures [58]. This understanding laid the groundwork for 20th century advances, such as X-ray crystallography, which would later enable scientists to visualize the three-dimensional structures of proteins and understand their roles in biological functions [59].

Nucleic Acids

Alongside the discovery of proteins, the identification of nucleic acids in the 19th century was equally transformative. In 1869, Swiss chemist Friedrich Miescher isolated something from the nuclei of WBCs, which was initially known as "nuclein" [60]. Miescher was the first to discover that cell nuclei contained a unique, phosphorus-rich substance that differed from proteins in its chemical composition. This substance, later named nucleic acid, turned out to be a mixture of deoxyribonucleic acid (DNA) and associated proteins [61].

Though Miescher had identified DNA as a component of the nucleus, its functional significance remained unclear. At the time, the role of nucleic acids in heredity and cellular function was not yet recognized, and it would take several more decades of research to establish the connection between DNA and genetic information [62]. However, Miescher's discovery was monumental because it highlighted the existence of another class of biopolymers within cells, separate from proteins, which hinted at the complexity and diversity of molecular structures within organisms [63].

The discovery of nuclein prompted further research into the components of nucleic acids. Between the 19th and 20th centuries, scientists identified individual building blocks of nucleic acids—purines and pyrimidines—which eventually led to the structural elucidation of DNA. This early work contributed to the foundation of the remarkable finding of a double-helix structure through Watson and Crick, an achievement that fundamentally changed our understanding of genetics and molecular biology [64].

Early Understanding of Biopolymer Structures and Functions

With the discovery of proteins and nucleic acids, scientists in the 19th century embarked on a quest to unravel their structures and biological roles. This period marked the dawn of structural biology, as researchers sought to understand how the arrangement of atoms within these molecules influenced their functions in living organisms [65]. By investigating the relationships between molecular structure and biological activity, scientists of this era began to develop the earliest models of biopolymer behavior, which informed the future of biochemical and medical research.

Proteins

A key area of focus for 19th century scientists was the structure and function of proteins, particularly their role as enzymes. In 1873, German chemist Emil Fischer projected a "lock and key" prototype for enzyme and substrate interaction, introducing the concept that enzymes' functions are dictated by their structures [66]. Fischer theorized that enzymes, which are proteins, had specific sites that could bind substrates with a high degree of specificity, much like a lock and key. These revolutionary ideas highlighted a relationship between a protein's structure and its function, emphasizing that the sequence and arrangement of amino acids within a protein molecule directly affect its biological role.

Fischer's work extended to the synthesis and analysis of peptides, called short chains of amino acids. He demonstrated that they could link with each other in a specific manner to make larger protein structures, a finding that further supported the understanding of proteins as polymers of amino acids [67]. His insights into enzyme-substrate specificity were instrumental in the study of biochemistry and molecular biology, and they laid the groundwork for understanding protein functionality at the molecular level.

By the close of the 19th century, it had become evident that proteins were not merely simple compounds but rather complex, high-molecular-weight molecules. This realization represented a paradigm shift, encouraging scientists to delve deeper into the structural intricacies of proteins and consider the potential applications of this knowledge in areas such as medicine and agriculture. Fischer's contributions, in particular, spurred the advancement of protein chemistry, inspiring 20th century breakthroughs that included the three-dimensional visualization of proteins using emerging techniques like X-ray crystallography [68].

Nucleic Acids

In parallel with protein research, scientists began to investigate the structure and function of nucleic acids, albeit with a more limited understanding initially. Miescher's discovery of nuclein introduced a new class of biomolecules, yet the biological significance of DNA and RNA remained unknown throughout much of the 19th century. However, the basic components of nucleic acids, namely, purines and pyrimidines, were identified during this period. This work on nucleotide bases established the groundwork for understanding how nucleic acids encode genetic information [69].

The understanding of nucleic acids as molecules essential for heredity was not fully appreciated until the 20th century. The eventual recognition that DNA stores and transmits genetic information marked a pivotal moment in biopolymer science, as it established a direct link between chemical structure and biological inheritance. The concept that nucleic acids carry genetic information reshaped biology and medicine, providing new insights into how traits are passed from one generation to the next [70]. This revelation inspired numerous applications, from genetic engineering to forensic science, and continues to be a driving force in biotechnology and genomics.

The early studies on nucleic acids underscored the significance of chemical structure in determining biological function. While the 19th century provided only a preliminary understanding of DNA and RNA, it laid the essential groundwork for the discovery of the DNA double helix. The connection between DNA's structure and its role in heredity has since enabled scientists to manipulate genetic material, advancing fields like synthetic biology and personalized medicine. These discoveries ultimately affirmed the importance of nucleic acids as central to life's blueprint [71].

Legacy of 19th Century Biopolymer Research

The biopolymer research conducted in the 19th century marked the fields of biology, chemistry, and medicine. Foundational understanding of proteins and nucleic acids, along with their structural and functional properties, catalyzed scientific advances that are still unfolding today. The work of pioneers like Gerardus Mulder, Friedrich Miescher, and Emil Fischer demonstrated that complex molecules are central to biological processes and that their structural configurations underpin diverse cellular functions [72].

The discovery of biopolymers has influenced countless research areas, leading to innovations in drug design, agricultural biotechnology, and disease treatment. Scientists today continue to build upon the knowledge established by 19th century researchers, developing sophisticated techniques to study proteins and nucleic acids at the atomic and molecular levels. Advances in technologies like X-ray crystallography, NMR, as well as cryo-electron microscopy now allow for unprecedented insights into the intricate structures and dynamics of these biomolecules.

The legacy of 19th century biopolymer science is a testament to the profound impact that foundational research can have on future scientific developments. By uncovering the nature of proteins and nucleic acids, early biochemists illuminated the complex mechanisms underlying life itself. Their contributions have laid a foundation for modern scientific pursuits, including genetic engineering, synthetic biology, and nanotechnology [73]. As research into biopolymers continues to evolve, the 19th century stands as a period of remarkable discovery that forever changed our understanding of biology and opened new avenues for innovation and exploration in science.

EARLY 20TH CENTURY: EMERGENCE OF PROTEIN AND POLYSACCHARIDE THERAPEUTICS

The early 20th century was a transformative period in the field of medical science, particularly in the therapeutic applications of biopolymers like proteins and polysaccharides. This era marked the beginning of modern biochemistry and pharmacology, leading to life-saving treatments that are still foundational in today's medical landscape. With landmark discoveries in the isolation and therapeutic use of insulin and the development of polysaccharides like heparin, the early 1900s set the stage for the integration of complex biomolecules in clinical practice [74].

Discovery and Therapeutic Use of Insulin

The discovery of insulin and its therapeutic applications in managing diabetes represent one of the most significant achievements in 20th century medicine. Before insulin's availability, diabetes was often a fatal disease, especially for individuals with Type 1 diabetes, who could only survive through a strict starvation diet. However, the collaborative work of scientists in the early 1920s led to the breakthrough that changed the prognosis for diabetic patients worldwide [75].

Discovery of Insulin

The groundwork for understanding insulin's role in blood sugar regulation was laid in the late 19th century when scientists first identified the pancreas as central to diabetes. Experiments revealed that damage to the pancreas resulted in diabetes symptoms, sparking interest in uncovering the specific pancreatic substances involved. Yet, it was not until 1921 that Frederick Banting, a Canadian physician, and his assistant Charles Best succeeded in isolating the hormone responsible for blood sugar regulation. Working in J.J.R. Macleod's lab at the University of Toronto, Banting and Best managed to extract insulin from the pancreas of dogs, an arduous process that required them to carefully isolate the pancreatic islets of Langerhans, where insulin is produced. This research, followed by further refinement by biochemist James Collip, allowed them to purify insulin for clinical use [76].

The team's research provided the first concrete evidence that insulin could control blood sugar levels in diabetic patients. Insulin was identified as a protein hormone, making it one of the first proteins recognized for its pivotal role in human health. This understanding was groundbreaking, as it demonstrated how hormones function as molecular messengers and laid the foundation for hormone replacement therapies [77].

Therapeutic Use of Insulin

In January 1922, it was applied therapeutically for the first time. Leonard Thompson, a boy with severe diabetes, was the recipient of this experimental treatment. At the time, Thompson was nearing death from diabetic ketoacidosis, but following an injection of insulin, his blood glucose levels dramatically decreased, and his symptoms improved significantly [78]. This first treatment marked the start of an era where Type 1 diabetes could be managed effectively rather than being a terminal condition.

The introduction of insulin transformed diabetes management, shifting it from a fatal disease to a chronic, manageable condition. Following Banting and Best's success, insulin production became a priority for pharmaceutical companies. Early insulin was extracted from animal pancreas, primarily cows and pigs, which allowed for the mass production of the hormone [79]. Over time, the field saw tremendous advances in insulin refinement, including the development of synthetic and recombinant DNA-derived insulin. These modern insulins offer precise dosing and tailored formulations, leading to improved glycemic control and fewer side effects for patients.

The legacy of insulin's discovery extends beyond diabetes management; it has had a profound impact on biotechnology, inspiring researchers to pursue hormone therapy for various conditions [80]. Today, insulin remains indispensable in diabetes care, with innovations such as insulin analogs and automated insulin delivery systems continually improving patient outcomes. The discovery of insulin highlighted the therapeutic potential of proteins, paving the way for advancements in protein-based therapies and expanding the understanding of hormone regulation.

Development of Polysaccharides like Heparin as Medicinal Agents

Alongside protein-based therapies, the early 20th century witnessed groundbreaking advancements in polysaccharide therapeutics, particularly with the development of heparin, an anticoagulant essential in preventing blood clots. Polysaccharides, which are complex carbohydrates, play a variety of roles in biological systems, including structural support and immune responses. Heparin's discovery and subsequent use in clinical settings revolutionized the management of thrombotic disorders and facilitated the progress of surgical and cardiovascular medicine [81].

Discovery of Heparin

Heparin was discovered in 1916 when Jay McLean, a medical student working under Dr. William Howell at Johns Hopkins University, accidentally stumbled upon a substance with strong anticoagulant properties. Initially, McLean was researching procoagulant agents—compounds that promote blood clotting—when he identified an unexpected anticoagulant effect in certain tissue extracts. This substance, which would later be named heparin, was found to possess remarkable properties for preventing blood coagulation [82].

Heparin's molecular structure, characterized by its sulfated glycosaminoglycan chains, was only partially understood in its early days. However, subsequent research in the 1920s and 1930s clarified its anticoagulant properties, unveiling its role in inhibiting thrombin and other enzymes critical for clot formation. Early methods of heparin extraction were inefficient and yielded variable potency, but ongoing research efforts led to standardized production techniques, enabling heparin to be used therapeutically on a larger scale by the 1940s [83].

Therapeutic Use of Heparin

Heparin's anticoagulant properties have made it an indispensable drug in various clinical applications, particularly in preventing and treating thromboembolic diseases like deep vein thrombosis (DVT), pulmonary embolism (PE), and arterial thromboembolism. Heparin works by binding to antithrombin, a natural inhibitor of blood coagulation, thereby inactivating thrombin and other clotting factors. This mechanism prevents the formation of blood clots, making heparin highly effective for use in medical situations where blood clot prevention is critical [84].

The therapeutic use of heparin has transformed cardiovascular medicine and surgical procedures. During cardiopulmonary bypass surgeries, where blood is circulated outside the body, heparin prevents clot formation in the external machinery, allowing surgeons to operate safely on the heart. Additionally, heparin is widely used in dialysis to prevent clotting in the blood-filtering apparatus, essential for patients with kidney failure. The anticoagulant is also commonly used in maintaining the patency of intravenous lines and catheters, reducing the risk of blockages and complications [85].

Beyond surgical and procedural applications, heparin has also played a critical role in the long-term management of blood clotting disorders. In cases of DVT and PE, where patients are at high risk of recurrent clots, heparin provides an effective solution to reduce complications and improve survival rates. For patients with conditions requiring anticoagulation therapy, heparin has become a trusted, life-saving treatment that has been refined over the decades, with formulations now including low-molecular-weight heparin (LMWH), which offers greater stability, reduced bleeding risk, and the possibility of at-home administration.

The discovery and use of heparin underscored the therapeutic potential of polysaccharides, illustrating their versatility in biological and clinical applications. Heparin remains a vital component of modern medicine, and its introduction has facilitated the development of other anticoagulants, expanding treatment options for patients with clotting disorders. The use of polysaccharides in therapy has inspired ongoing research into biopolymer-based treatments, fueling innovations in drug delivery and targeted therapeutics [86].

Broader Implications and Lasting Impact of Early 20th Century Biopolymer Therapeutics

The discovery and application of insulin and heparin exemplify the early 20th century shift towards using biopolymers for therapeutic purposes, showcasing how naturally occurring molecules can be harnessed for medical interventions. These achievements laid the groundwork for biopolymer research and inspired new approaches to treating chronic and acute conditions, thus shaping the future of biopharmaceuticals and molecular medicine.

Advances in Protein-based Therapies

The success of insulin as a therapeutic protein highlighted the vast potential for protein-based treatments. This realization prompted further exploration into other protein hormones, enzymes, and antibodies that could be developed into drugs. Protein therapeutics have since expanded to include monoclonal antibodies, enzyme replacement therapies, and growth factors, which have applications in cancer treatment, genetic disorders, and immunology. Today, the field of protein therapeutics is a booming industry with a diverse portfolio of drugs that aim to treat illnesses on the molecular level, offering accuracy and efficacy that traditional drugs cannot achieve [87].

One significant advancement in protein therapeutics was the development of monoclonal antibodies in the 1970s, which became a staple in cancer treatment and autoimmune disease management. The ability to create antibodies that specifically target disease-causing molecules allowed for treatments that minimize harm to healthy cells, a breakthrough that would not have been possible without the foundational understanding of proteins as therapeutic agents. These advancements illustrate how insulin's initial success opened doors to a wide range of protein-based therapies, influencing fields as diverse as oncology, hematology, and infectious diseases.

Polysaccharide-based Therapeutics and Drug Delivery

The success of heparin also illustrated the therapeutic potential of polysaccharides, leading to research on other carbohydrate-based molecules with pharmacological effects. Beyond heparin, polysaccharides like chitosan and alginate have been explored for their applications in wound healing, drug delivery, and tissue engineering. These molecules possess characteristic abilities, like biocompatibility and biodegradability, which make them suitable for sustained drug release and localized delivery systems. Polysaccharide-based drug delivery systems are being cast off to improve the efficacy of medications by targeting specific sites within the body, enhancing absorption, and minimizing side effects [88].

One notable application of polysaccharides is in the field of wound healing, where materials like chitosan have shown promise in promoting tissue regeneration and preventing infections. Chitosan-based dressings are now commonly used for wound care.

Mid-20th Century: Rise of Nucleic Acids as Therapeutics

The Discovery of DNA Structure

The mid-20th century was a revolutionary time for genetics and molecular biology, primarily owing to the landmark finding of the double helix structure of DNA in 1953 by James Watson and Francis Crick. It was monumental, revealing that DNA, with its paired nucleotide bases and helical shape, is the molecule responsible for storing and transmitting genetic information. This understanding laid the foundation for modern genetic research, illuminating the molecular mechanisms of heredity and opening avenues for numerous scientific advancements.

The double helix model also explains how DNA replicates and how genetic information is passed from one generation to the next. It illustrated how the sequence of bases (adenine, thymine, cytosine, and guanine) encoded commands essential to the development, function, and reproduction of living organisms [89]. This insight transformed biology and medicine, influencing research directions and leading to the development of various biotechnological applications.

Early Research into Nucleic Acid-based Therapeutics

The latter half of the 20th century witnessed burgeoning interest in nucleic acids as potential therapeutic agents. Researchers began to explore the possibilities of manipulating genetic material to treat diseases, marking the advent of gene therapy. This approach involves correcting defective genes responsible for disease development by introducing healthy copies of the gene into the patient's cells.

One of the pioneering efforts in this field was the expansion of recombinant DNA technology, which permitted inventors to cut and recombine DNA sequences from different sources. This technology paved the way for producing therapeutic proteins, like insulin, as well as growth hormones, by inserting human genes into bacterial plasmids [90].

Moreover, antisense oligonucleotides emerged as a novel class of therapeutic agents. These small, synthetic strands of DNA or RNA are designed to bind with specific messenger RNA (mRNA) molecules, blocking the production of proteins that contribute to disease. This concept demonstrated the potential of nucleic acids in maintaining gene expression as well as combating different diseases at the molecular level.

Overall, the mid-20th century laid the groundwork for modern nucleic acid therapeutics, highlighting the potential of DNA and RNA in revolutionizing medicine. Advances in this era have shaped the development of cutting-edge therapies, including mRNA vaccines and CRISPR-based gene editing, offering new hope for treating previously intractable diseases.

MID-20TH CENTURY: RISE OF NUCLEIC ACIDS AS THERAPEUTICS

The mid-20th century heralded a transformative period in genetics and molecular biology, with the invention of the structure of DNA at the forefront. This era laid the groundwork for a new understanding of heredity and molecular medicine, eventually leading to revolutionary advances in nucleic acid-based therapeutics. It was during this period that scientists began to recognize DNA and RNA as not just molecules central to life, but also as potential tools for manipulating genetic information to combat diseases. These developments set the stage for today's innovations in gene therapy, CRISPR-based gene editing, and mRNA-based vaccines, fundamentally altering the landscape of modern medicine [91]. A timeline of nucleic acid therapeutics is shown in Table **2**.

Table 2. A timeline of nucleic acid therapeutics.

Era	Key Development	Significance
Mid-20th Century	Discovery of DNA Structure	Unraveling the genetic code and understanding the mechanisms of heredity.
Mid-20th Century	Early Research into Nucleic Acid-based Therapeutics	Pioneering efforts in gene therapy, recombinant DNA technology, and antisense oligonucleotide therapy.
Late 20th Century	Recombinant DNA Technology	Enabling the production of therapeutic proteins like insulin and growth hormones.
21st Century	mRNA Vaccines	Revolutionizing vaccine development with rapid and effective responses to infectious diseases.
21st Century	CRISPR-Cas9 Gene Editing	Precise genome editing for potential treatments of genetic diseases and other medical conditions.

The Discovery of DNA Structure

James Watson and Francis Crick, along with the findings from Rosalind Franklin's X-ray crystallography work as well as Maurice Wilkins' studies, unveiled the double helix structure of DNA. This discovery was groundbreaking, providing insights into how genetic information is stored, replicated, and transmitted from one generation to the next. The double helix model revealed that DNA is made with two complementary strands wound about each other, with pairs of nucleotide bases (adenine with thymine, and cytosine with guanine) acting as rungs of a twisted ladder. This pairing explained how genetic information could be precisely copied and passed down [92].

Significance of the Double Helix Discovery

The double helix model illustrated not only how genetic information is stored in the sequence of nucleotide bases but also how DNA replication occurs, ensuring accurate genetic inheritance. Each strand of the DNA molecule serves as a template for the synthesis of a new, complementary strand, a process essential to cell division and growth. Understanding the molecular structure of DNA unlocked mysteries of heredity that had puzzled scientists for centuries and laid the scientific basis for modern genetics.

Watson and Crick's model also clarified how mutations could lead to genetic diversity and, at times, disease. Since each base pair sequence encodes specific instructions for protein synthesis, any alteration (mutation) in the sequence could lead to a change in protein structure and function. This understanding of mutations helped scientists connect genetic errors with hereditary diseases, sparking the development of genetic screening, diagnostic tools, and therapeutic interventions aimed at correcting or compensating for these errors.

The impact of the DNA discovery extended beyond biology and medicine, as it sparked public interest in genetics, raising questions about the ethical implications of genetic engineering. This societal interest influenced policies and discussions around genetic privacy, testing, and research ethics. DNA's discovery was monumental not only for its immediate implications but also for its influence on both scientific research and societal perspectives on genetic intervention [93].

Early Research into Nucleic Acid-based Therapeutics

The discovery of DNA's structure opened a new frontier for exploring nucleic acids as therapeutic agents. Scientists began to investigate the potential for manipulating DNA and RNA to treat diseases at their genetic roots. Early research in this field led to groundbreaking advances, including gene therapy, recombinant

DNA technology, and antisense oligonucleotide therapy—each offering novel ways to combat diseases by targeting their molecular causes rather than merely addressing symptoms.

Gene Therapy: Correcting Genetic Defects

One of the earliest concepts to emerge from nucleic acid research was gene therapy, an approach aimed at treating diseases by correcting or compensating for defective genes. Gene therapy involves inserting a healthy copy of a gene into a patient's cells to replace or supplement a faulty gene responsible for a disease. The development of gene therapy began in the latter half of the 20th century, with early studies focused on hereditary disorders, such as cystic fibrosis and muscular dystrophy, where single-gene mutations cause significant health issues.

While initial experiments were met with challenges—such as difficulties in delivering genes to the right cells and concerns about potential immune responses—gene therapy has since evolved into a sophisticated field. Today, viral vectors and other gene delivery systems have been refined to improve the safety and efficacy of gene therapy treatments. Clinical trials have shown promise in treating previously untreatable genetic diseases, including certain forms of blindness and spinal muscular atrophy. These advancements demonstrate the potential of gene therapy to provide long-lasting or even curative treatments by addressing the root genetic causes of disease [94].

Accurate Gene Delivery to Target Cells

One of the major challenges in gene therapy is delivering therapeutic genes precisely to the target cells. Successful gene delivery requires an efficient carrier, often referred to as a vector, to transport the genetic material into the patient's cells.

Delivery Methods

1. **Viral Vectors:** Modified viruses, such as adenoviruses, lentiviruses, and adeno-associated viruses (AAVs), are commonly used due to their natural ability to enter cells and deliver genetic material efficiently. AAVs are particularly preferred for their low immunogenicity and ability to target specific tissues.
2. **Non-viral Methods:** These include lipid nanoparticles (LNPs), electroporation, and gene guns that facilitate gene transfer without using viruses, reducing potential immune responses.
3. **Targeted Approaches:** Advances in molecular biology have enabled cell-specific targeting, such as using tissue-specific promoters or receptor-ligand

interactions to ensure that only the intended cells receive the therapeutic gene.

Integration of Genes into the Genome

Once delivered, the new gene must be either integrated into the host genome for long-term expression or remain as an episomal element for transient effects, depending on the treatment requirements.

Integration Strategies

1. **Random Integration:** Some viral vectors, such as retroviruses, integrate randomly into the genome. However, this carries a risk of insertional mutagenesis, potentially leading to cancer if the gene disrupts a critical regulatory sequence.
2. **Site-specific Integration:** Gene-editing tools like CRISPR-Cas9, TALENs, and zinc-finger nucleases (ZFNs) allow for precise integration at safe loci, minimizing unintended effects [95].
3. **Extrachromosomal Persistence:** Some approaches, such as AAV-based therapies, allow the gene to remain outside the genome (episomal expression), reducing the risk of mutation but requiring periodic re-administration.

Possible Immune Rejection

The immune system poses a significant challenge to gene therapy by recognizing foreign genetic material or viral vectors as threats, leading to potential rejection.

Immune Challenges and Mitigation Strategies

1. **Immune Response to Viral Vectors:** Some patients may have pre-existing immunity to commonly used vectors like AAV, reducing therapy efficacy. Strategies to overcome this include.
   - Using novel serotypes of viruses to bypass existing immunity.
   - Administering immunosuppressive drugs to dampen the immune response.
   - Developing non-viral vectors to avoid immune recognition altogether.
2. **Innate and Adaptive Immune Responses:** Inflammatory responses can lead to the clearance of transduced cells. Researchers are exploring ways to.
   - Modify viral vectors to be less immunogenic.
   - Use stealth coatings, such as PEGylation, to shield the vector from immune detection.
   - Optimize dosing regimens to minimize immune activation while maintaining effectiveness.
3. **Long-term Tolerance Strategies:**
   - Engineering immune-privileged gene delivery to avoid activation of immune pathways.

○ Exploring tolerance induction protocols, where immune cells are trained to accept the new gene as self [96].

Recombinant DNA Technology: A New Era in Therapeutic Proteins

The development of recombinant DNA technology in the 1970s was a major milestone in the application of nucleic acids for therapeutic purposes. This technology allows scientists to cut and recombine DNA sequences from different sources, enabling the creation of "recombinant" organisms that can produce therapeutic proteins. By inserting a human gene into a bacterial plasmid, for instance, scientists can program bacteria to produce human proteins, such as insulin or growth hormones, in large quantities.

Recombinant DNA technology was the catalyst for the biotechnology industry, leading to the mass production of therapeutic proteins that are essential for managing various health conditions. Insulin, initially derived from animal sources, could now be produced in a purer, safer form through recombinant DNA technology, providing a significant benefit to patients with diabetes. Similarly, recombinant growth hormones and clotting factors have been developed to treat conditions like growth hormone deficiencies and hemophilia.

The impact of recombinant DNA technology extends beyond protein production. It has facilitated the development of genetically engineered vaccines, such as the hepatitis B vaccine, which has significantly reduced the incidence of hepatitis-related liver disease. This technology has also paved the way for personalized medicine, as researchers can now produce custom proteins tailored to an individual's genetic profile, enhancing treatment effectiveness and minimizing side effects [97].

Antisense Oligonucleotides: Regulating Gene Expression

As researchers explored nucleic acid therapeutics further, they developed antisense oligonucleotides (ASOs)—short, synthetic strands of DNA or RNA designed to bind to specific messenger RNA (mRNA) molecules. By binding to target mRNA, ASOs can effectively block the production of proteins that contribute to disease. This approach has shown promise in regulating gene expression and combating diseases at the molecular level, especially in genetic disorders where an excess or abnormal protein causes symptoms.

The concept of ASO opened up new possibilities for treating diseases with a known genetic basis. For example, antisense therapies have been developed for conditions such as Duchenne muscular dystrophy, where a mutation in the dystrophin gene leads to severe muscle deterioration. By using ASOs to modify the splicing of the dystrophin gene, scientists have been able to partially restore protein function in some patients, demonstrating a new way to approach treatment for genetic disorders. Moreover, ASOs are being explored as treatments for neurodegenerative diseases like Huntington's and Alzheimer's, where they can potentially reduce the production of toxic proteins that drive disease progression [98].

ASOs offer a targeted approach to treating diseases by intervening directly in the gene expression process. This specificity reduces the likelihood of side effects associated with traditional drugs, which may affect multiple pathways in the body. ASOs continue to be a focus of research, with several therapies approved by regulatory agencies for use in clinical settings and many more in development.

Impact of Nucleic Acid Discoveries on Modern Medicine

The mid-20th century advancements in nucleic acid research have fundamentally transformed the landscape of medicine, leading to innovations that continue to shape healthcare today. From gene therapy and recombinant DNA technology to antisense oligonucleotide therapy, these nucleic acid-based approaches represent a shift toward more personalized, targeted treatments. By addressing diseases at the genetic level, nucleic acid therapeutics offer the potential for greater precision, efficacy, and durability than many traditional treatments.

mRNA Vaccines: A Breakthrough in Infectious Disease Prevention

One of the most notable applications of nucleic acid research in recent years has been the development of mRNA vaccines, which became globally recognized with the rollout of COVID-19 vaccines. mRNA vaccines, such as those developed by Pfizer-BioNTech and Moderna, leverage synthetic mRNA to instruct cells to produce a protein that stimulates an immune response, offering protection against the targeted virus.

The success of mRNA vaccines demonstrates the power and flexibility of nucleic acid-based therapeutics. mRNA vaccines can be developed and produced more rapidly than traditional vaccines, a critical advantage in responding to emerging infectious diseases. The technology behind mRNA vaccines holds promise for preventing not only viral infections but also other diseases, including certain cancers. Researchers are exploring mRNA-based vaccines that target specific cancer antigens, potentially leading to personalized cancer immunotherapies.

CRISPR and Genome Editing: Precision Medicine in Action

The discovery of the CRISPR-Cas9 system, a tool for precise genome editing, has revolutionized genetic engineering and holds enormous potential for treating genetic diseases. CRISPR allows scientists to make targeted changes to DNA sequences, correcting or modifying genes with unprecedented accuracy. Since its introduction in 2012, CRISPR has been used in research settings to model diseases, study gene functions, and explore therapeutic possibilities.

CRISPR technology has shown particular promise for treating genetic diseases that arise from single-gene mutations, such as sickle cell anemia and cystic fibrosis. In these cases, researchers can use CRISPR to directly correct the genetic mutation within a patient's cells, potentially offering a permanent cure. In 2020, the first clinical trials using CRISPR to treat sickle cell anemia showed promising results, with patients experiencing significant symptom relief following treatment [99].

Beyond single-gene disorders, CRISPR is being explored for applications in cancer therapy, infectious disease treatment, and organ transplantation. For example, CRISPR has been used to engineer immune cells that can target and destroy cancer cells more effectively, offering hope for new treatments in oncology. As CRISPR technology continues to evolve, it promises to expand the therapeutic potential of gene editing across a wide range of medical applications.

21ST CENTURY: MODERN ADVANCES IN BIOPOLYMER THERAPEUTICS

mRNA Vaccines and Nucleic Acid-based Therapeutics

The 21st century has witnessed a remarkable shift in vaccine technology, driven primarily by the development of mRNA vaccines. This transformative approach has showcased the immense potential of nucleic acid-based therapeutics, particularly evident during the COVID-19 pandemic.

Development of mRNA Vaccines

mRNA vaccines work by introducing a small piece of messenger RNA (mRNA) into the body. This mRNA provides the instructions for cells to produce a protein that triggers an immune response, without using live virus particles. This method is significantly faster and more flexible than traditional vaccine development.

The most notable examples are the COVID-19 vaccines developed by Pfizer-BioNTech and Moderna. Both vaccines use mRNA technology to instruct cells to produce the spike protein found on the surface of the SARS-CoV-2 virus, thereby eliciting an immune response. The rapid development, testing, and distribution of these vaccines have demonstrated the potential of mRNA technology in responding to emerging infectious diseases [100].

Regulation of Protein Synthesis After mRNA Enters the Cell

Once mRNA enters the cell *via* lipid nanoparticles, it must be efficiently translated into the target protein while maintaining stability and avoiding excessive immune detection. This process involves several critical steps.

1. **mRNA Stability and Translation Efficiency**
   - **5' Cap and Poly(A) Tail:** These modifications enhance mRNA stability and ensure efficient ribosome binding for translation.
   - **Codon Optimization:** Using synonymous codons that match the host cell's tRNA availability increases translation speed and protein yield.
   - **Untranslated Regions (UTRs):** Specific sequences in the 5' and 3' UTRs regulate ribosomal recruitment and translation initiation.
2. **Ribosome Recruitment and Translation Initiation**
   - **Eukaryotic Initiation Factors (eIFs):** These proteins assist in ribosome assembly at the start codon, ensuring efficient translation.
   - **mRNA Circularization:** Interaction between the 5' cap and the poly(A) tail enhances translation efficiency.

3. **Protein Folding and Post-translational Modifications**
   - Once synthesized, the protein undergoes proper folding with the help of chaperone proteins to achieve its functional form.
   - **Glycosylation and Phosphorylation:** These modifications ensure the antigen mimics its natural structure, improving immune recognition.

Triggering an Effective Immune Response

A strong and appropriate immune response is essential to developing long-term immunity. This process involves antigen presentation, immune cell activation, and memory formation.

1. **Antigen Presentation by Dendritic Cells**
   - Once the target protein is synthesized, it is processed and presented on the cell surface *via* **Major Histocompatibility Complex (MHC)** molecules [101].
   - **MHC-I Presentation:** Triggers cytotoxic T cell responses (CD8+ T cells) to eliminate infected cells.
   - **MHC-II Presentation:** Stimulates helper T cells (CD4+ T cells), which enhance antibody production by B cells.

2. **Activation of the Innate Immune System**
   - **Pattern Recognition Receptors (PRRs):** Sensors like Toll-like receptors (TLRs) recognize foreign mRNA and activate innate immunity.
   - **Interferon Response:** Type I interferons initiate antiviral defenses and enhance adaptive immunity.
3. **Adaptive Immune Activation and Memory Formation**
   - **B Cell Activation:** This leads to antibody production that neutralizes the pathogen.
   - **T Cell Activation:** CD8+ T cells eliminate infected cells, while CD4+ T cells provide help to B cells and regulate the immune response.
   - **Memory Cell Formation:** Ensures rapid and effective immune responses upon future exposure to the virus.

Potential of Nucleic Acid-based Therapeutics

Beyond vaccines, nucleic acid-based therapeutics encompass a range of treatments involving DNA, RNA, and other nucleic acid molecules. These therapies aim to correct or modulate gene expression to treat diseases at their genetic root. Gene therapy, for example, involves delivering nucleic acids to correct genetic defects, potentially curing inherited disorders.

Antisense oligonucleotides (ASOs) and small interfering RNAs (siRNAs) are other notable examples. ASOs are short DNA or RNA molecules designed to bind to specific mRNA molecules, blocking their translation into proteins. This approach can be used to silence disease-causing genes. Similarly, siRNAs can degrade target mRNA molecules, reducing the production of harmful proteins.

Innovations in Drug Delivery Systems

The effectiveness of biopolymer-based therapeutics relies heavily on the development of advanced drug delivery systems. Innovations in this field have significantly improved the delivery, stability, and efficacy of these therapies.

Hydrogels

Hydrogels are three-dimensional polymer networks that can absorb and retain large amounts of water. They offer a versatile platform for drug delivery due to their biocompatibility, tunable properties, and ability to encapsulate a wide range of therapeutics.

Hydrogels can be engineered to release drugs in a controlled manner, responding to specific stimuli such as pH, temperature, or enzymes. This controlled release ensures a sustained therapeutic effect, reducing the frequency of administration and improving patient compliance. Hydrogels are being explored for various applications, including wound dressings, tissue engineering scaffolds, and delivery vehicles for proteins, peptides, and nucleic acids.

Underlying Mechanisms of Hydrogels in Drug Delivery

Hydrogels function as effective drug delivery systems due to their unique structural and physicochemical properties. The underlying mechanisms that govern their drug encapsulation and release include.

1. **Swelling-controlled Release**
   - Hydrogels absorb water, leading to network expansion and increased pore size.
   - Drug molecules diffuse out as water content increases, allowing for a controlled and sustained release [102].
2. **Diffusion-controlled Release**
   - Hydrophobic or hydrophilic drugs diffuse through the hydrogel matrix at a rate dependent on polymer composition and crosslinking density.
   - The diffusion process can be tailored by modifying polymer properties to achieve desired release kinetics.
3. **Stimuli-responsive Release**
   - **pH-Sensitive Hydrogels:** Designed to release drugs in specific pH environments, such as acidic tumor tissues or the gastrointestinal tract.
   - **Temperature-sensitive Hydrogels:** Undergo phase transitions at predetermined temperatures, leading to controlled drug release.
   - **Enzyme-responsive Hydrogels:** Release drugs in response to specific enzymes found in target tissues or diseased sites.
4. **Biodegradable Hydrogels**
   - Hydrogel degradation occurs through hydrolysis or enzymatic cleavage, releasing encapsulated drugs in a predetermined manner.
   - The degradation rate can be adjusted by modifying polymer composition and crosslinking density.

Potential Challenges of Hydrogel-based Drug Delivery

While hydrogels present promising advantages, several challenges must be addressed to optimize their clinical applications.

1. **Mechanical Strength and Stability**
   - Hydrogels often have weak mechanical properties, limiting their application in load-bearing tissues.
   - Strategies such as composite hydrogels and reinforced crosslinking are being explored to enhance durability.

2. **Controlled and Predictable Drug Release**
   - Achieving consistent and reproducible drug release profiles remains challenging.
   - Variability in polymer degradation rates and drug-polymer interactions can lead to fluctuations in therapeutic efficacy.
3. **Biocompatibility and Immune Response**
   - Some synthetic hydrogels may elicit an immune response, leading to inflammation or foreign body reactions.
   - Using naturally derived polymers (*e.g.*, alginate, hyaluronic acid) can improve biocompatibility [103].
4. **Manufacturing and Scalability**
   - Large-scale production with reproducible quality is complex due to batch-to-batch variations.
   - Advanced fabrication techniques, such as 3D printing and microfluidics, are being developed to enhance scalability.

Nanoparticles

Nanoparticles have revolutionized drug delivery by enabling targeted and efficient delivery of therapeutics to specific tissues or cells. These tiny particles, typically ranging from 1 to 100 nanometers, can be designed to carry drugs, genes, or imaging agents.

Nanoparticles offer several advantages, including improved drug solubility, protection from degradation, and Enhanced Permeability and Retention (EPR) effect, which allows for preferential accumulation in tumor tissues. Functionalizing nanoparticles with targeting ligands further enhances their ability to deliver drugs specifically to diseased cells, minimizing off-target effects.

Underlying Mechanisms of Nanoparticles in Drug Delivery

Nanoparticles offer targeted and efficient drug delivery through several key mechanisms. One of the most significant is the Enhanced Permeability and Retention (EPR) effect, which allows nanoparticles to accumulate preferentially in tumor tissues. This occurs due to the leaky vasculature and poor lymphatic drainage characteristic of tumor microenvironments. By leveraging this passive targeting mechanism, nanoparticles increase drug concentration at the diseased site while minimizing systemic exposure, thereby reducing adverse effects on healthy tissues.

Beyond passive targeting, nanoparticles can be functionalized with targeting ligands such as antibodies, peptides, or small molecules. These surface modifications enable active targeting of specific cell receptors, enhancing drug efficacy and reducing off-target side effects. By binding to overexpressed markers on diseased cells, functionalized nanoparticles improve selectivity and therapeutic outcomes. Additionally, nanoparticles are engineered for controlled drug release, ensuring sustained, triggered, or environmentally responsive drug administration. For example, pH-sensitive nanoparticles are designed to release drugs in acidic tumor environments, enhancing localized therapy and minimizing exposure to healthy tissues. This controlled-release approach optimizes therapeutic efficiency and patient compliance [104].

Potential Challenges of Nanoparticle-based Drug Delivery

Despite their advantages, nanoparticle-based drug delivery systems face several challenges. One primary concern is stability and degradation, as nanoparticles may aggregate or degrade prematurely, affecting drug loading capacity and release kinetics. To address this, researchers incorporate surface coatings and polymer stabilizers to enhance structural integrity and maintain the functional properties of nanoparticles during circulation and storage.

Another critical challenge is immune system recognition, where nanoparticles are often identified and cleared by immune cells before reaching their target tissues. This rapid clearance reduces their therapeutic efficacy. Strategies such as PEGylation and biomimetic coatings help nanoparticles evade immune detection and prolong circulation time, improving drug delivery efficiency. By mimicking natural biological structures, these modifications enhance nanoparticle compatibility within the body.

Manufacturing and reproducibility also pose significant hurdles in nanoparticle-based therapies. Large-scale production with consistent nanoparticle size, shape, and drug loading remains technically complex. Variability in formulation can impact drug efficacy and safety. To overcome this, researchers are exploring advanced nanofabrication techniques, such as microfluidics and high-precision synthesis methods, to improve batch-to-batch consistency and scalability. As these technologies evolve, they hold the potential to enhance the commercial viability and widespread adoption of nanoparticle-based drug delivery systems.

Liposomes

Liposomes are spherical vesicles composed of phospholipid bilayers, capable of encapsulating both hydrophilic and hydrophobic drugs. They have been widely used in drug delivery due to their biocompatibility and ability to protect drugs from degradation.

Liposomes can be modified with polyethylene glycol (PEG) to improve their circulation time in the bloodstream [105]. Additionally, targeting ligands can be attached to the liposome surface to enhance delivery to specific cells or tissues. Liposomal formulations of chemotherapy drugs, such as doxorubicin, have demonstrated improved efficacy and reduced toxicity compared to traditional formulations.

Underlying Mechanisms of Liposomes in Drug Delivery

Liposomes play a crucial role in drug delivery by leveraging multiple mechanisms. They encapsulate both hydrophilic and hydrophobic drugs, ensuring versatile drug delivery applications. The aqueous core houses hydrophilic drugs, while the lipid bilayer integrates hydrophobic compounds. This dual-encapsulation capability enables the controlled and sustained release of therapeutic agents, minimizing fluctuations in drug concentration and reducing potential side effects. Additionally, liposomal encapsulation protects drugs from enzymatic and chemical degradation in the bloodstream, enhancing their stability and bioavailability. By forming a protective barrier around the drug, liposomes prevent premature metabolism and excretion, ensuring efficient delivery to the intended site of action.

To prolong circulation time and target specific tissues, PEGylation prevents immune recognition, while targeting ligands on the liposomal surface facilitate precise drug delivery to desired cells or organs. PEGylation involves the attachment of polyethylene glycol (PEG) molecules to the liposomal surface, reducing interactions with serum proteins and immune cells, thus extending the drug's half-life. Active targeting strategies involve the functionalization of liposomes with ligands such as antibodies, peptides, or aptamers, allowing them to bind selectively to overexpressed receptors on diseased cells. These targeting mechanisms improve drug accumulation at the disease site while minimizing off-target effects, thereby enhancing therapeutic efficacy.

Potential Challenges of Liposome-Based Drug Delivery

Despite their advantages, liposome-based drug delivery systems face several challenges. Stability and storage issues arise as liposomes are prone to leakage and fusion, which may affect drug retention. Optimization of lipid composition and storage conditions can enhance stability, ensuring a prolonged shelf life. Factors such as temperature fluctuations, pH variations, and exposure to mechanical stress can influence liposomal integrity. Researchers are developing advanced formulations, such as freeze-dried liposomes and lyophilized preparations, to improve storage stability.

Another significant hurdle is the rapid clearance by the reticuloendothelial system (RES), particularly by macrophages in the liver and spleen. The immune system recognizes and eliminates foreign particles, leading to a reduced therapeutic window. Surface modifications, such as PEGylation, help evade immune detection and prolong circulation. Additionally, the development of "stealth liposomes" with surface coatings that mimic endogenous cell membranes is being explored to further enhance immune evasion [106].

Manufacturing liposomal drug formulations involves specialized production techniques, leading to high costs. The precise control of particle size, lipid composition, and drug encapsulation efficiency requires sophisticated methodologies such as extrusion, sonication, and microfluidic systems. Scaling up production while maintaining consistency and reproducibility remains a challenge. Current research focuses on developing scalable and cost-effective manufacturing methods, including high-throughput production technologies and automated quality control measures to facilitate commercialization.

Design Principles of Drug Delivery Systems

The design of drug delivery systems, including hydrogels, nanoparticles, and liposomes, is guided by key principles to optimize therapeutic outcomes.

Hydrogels are developed using natural polymers like alginate and hyaluronic acid or synthetic polymers such as PEG and PVA to ensure biocompatibility and controlled drug release. These polymers influence the mechanical properties, degradation rate, and drug-loading capacity of the hydrogel. Crosslinking methods—chemical (covalent) or physical (ionic, hydrogen bonding)—determine their mechanical strength, degradation rate, and drug release kinetics. Hydrogels are also engineered to swell and release drugs in response to environmental stimuli such as pH, temperature, or enzymatic activity. Stimuli-responsive hydrogels have been extensively explored for targeted therapies, such as wound healing applications and localized cancer treatments.

Nanoparticles, typically ranging from 1 to 100 nm, are optimized for size and shape to enhance tissue penetration, cellular uptake, and circulation time. Surface functionalization with targeting ligands, such as antibodies or peptides, enables active targeting of diseased tissues. Nanoparticles are designed with core-shell structures, where hydrophobic drug-loaded cores are protected by hydrophilic shells (*e.g.*, PEGylation) to improve solubility, stability, and immune evasion. Stimuli-responsive nanoparticles release drugs in specific physiological environments, such as the tumor microenvironment, ensuring precise drug delivery while reducing systemic toxicity [107].

Liposomes, another crucial drug delivery system, rely on lipid bilayer composition to influence membrane rigidity, stability, and drug encapsulation efficiency. The encapsulation of hydrophilic and hydrophobic drugs within their respective compartments enhances delivery efficiency. Liposomal formulations can be modified to control the rate of drug release, allowing for sustained and long-term therapeutic effects. Surface modifications, such as PEGylation, extend circulation time by reducing immune clearance, while ligand-functionalized liposomes improve site-specific drug delivery. Recent advancements in lipid-based nanocarriers, such as hybrid liposomes and lipid-polymer conjugates, are being investigated for their enhanced stability and targeted drug delivery capabilities.

Factors Affecting the Performance of Drug Delivery Systems

The efficacy of drug delivery systems is influenced by physiological barriers, stability, drug release control, and immune system recognition.

Physiological barriers include degradation rates and tissue interactions that affect hydrogels, while nanoparticles may be cleared by the liver, spleen, or kidneys, reducing their circulation time. Similarly, liposomes face rapid uptake by the RES, limiting effective drug delivery. Strategies such as surface modification with stealth coatings and biomimetic materials help improve bioavailability and therapeutic efficiency.

Stability and drug release control are also critical concerns; premature drug leakage or inconsistent swelling in hydrogels can alter therapeutic outcomes, while nanoparticles may suffer from aggregation or instability in circulation. Liposomes, if not properly designed, may undergo membrane leakage and fusion, leading to unintended drug release before reaching the target site. Advances in formulation techniques, including the use of stabilizing agents and controlled-release matrices, are helping overcome these limitations.

Immune system recognition further impacts drug delivery efficiency. Some synthetic polymers in hydrogels can induce immune responses or inflammation, while nanoparticles and unmodified liposomes are prone to macrophage uptake and rapid clearance, thereby shortening their circulation time. Researchers are working on incorporating biomimetic coatings derived from cell membranes to create more immune-evasive drug carriers.

Strategies to Overcome Existing Challenges

To address these challenges, various strategies have been implemented. Enhancing stability and bioavailability involves optimizing hydrogel crosslinking density and polymer composition, using polymeric coatings or lipid bilayers in nanoparticles, and incorporating cholesterol or PEGylation in liposomes to improve membrane stability and circulation time. Advanced stabilization techniques, such as the incorporation of nanostructured lipid carriers and solid lipid nanoparticles, are being explored to improve drug retention and release profiles.

Improving targeting efficiency is another critical focus, achieved through bioresponsive hydrogels that release drugs based on local physiological conditions, functionalized nanoparticles with targeting ligands for specific cellular uptake, and liposomes modified with antibodies, peptides, or aptamers for precise tissue targeting. Personalized medicine approaches utilizing patient-specific biomarkers are being integrated into these targeting strategies to enhance therapeutic efficacy.

To reduce immune clearance, naturally derived materials in hydrogels enhance biocompatibility, while PEGylation and stealth coatings in nanoparticles prevent immune system recognition. Similarly, stealth liposomes minimize protein adsorption, allowing for prolonged circulation and improved therapeutic effectiveness. Researchers are also investigating exosome-based drug delivery systems, which mimic natural extracellular vesicles to achieve enhanced biocompatibility and reduced immune clearance.

The continuous evolution of drug delivery technologies is driving the development of next-generation nanocarriers that combine multiple functionalities, such as multi-responsive systems that adapt to changes in pH, temperature, and enzymatic activity. These innovations hold great promise for improving patient outcomes and expanding the potential of precision medicine [108].

Advances in Regenerative Medicine and Tissue Engineering

Biopolymers have become integral to regenerative medicine and tissue engineering, offering solutions for repairing and replacing damaged tissues. These fields leverage the unique properties of biopolymers to create scaffolds, matrices, and other structures that support cell growth and tissue regeneration [19].

Scaffolds for Tissue Engineering

Biopolymer-based scaffolds provide a three-dimensional structure that mimics the extracellular matrix, promoting cell adhesion, proliferation, and differentiation. These scaffolds can be designed to degrade over time, gradually being replaced by new tissue.

Materials such as collagen, chitosan, and hyaluronic acid are commonly used in scaffold fabrication due to their biocompatibility and biodegradability. The development of advanced manufacturing techniques, such as electrospinning and 3D printing, has enabled the creation of complex, customized scaffolds that closely resemble the native tissue architecture.

Hydrogels in Regenerative Medicine

Hydrogels are particularly well-suited for regenerative medicine due to their high water content, which provides a hydrated environment for cells. They can be used as injectable matrices for cell delivery, promoting tissue repair and regeneration.

One notable application is in cartilage regeneration. Hydrogels loaded with chondrocytes (cartilage-forming cells) or stem cells can be injected into cartilage defects, providing a supportive environment for cell growth and matrix production. This approach has the potential to restore damaged cartilage, offering a less invasive alternative to surgical procedures.

Bioprinting

Bioprinting is an emerging technology that combines 3D printing with biomaterials and cells to create complex tissue structures. This technique allows

for precise control over the placement of cells and biomaterials, enabling the fabrication of tissue constructs with defined architectures.

Biopolymers such as alginate, gelatin, and fibrin are commonly used as bioinks in bioprinting. These materials provide a supportive matrix for cell growth and can be tailored to achieve the desired mechanical and biological properties. Bioprinting has the potential to revolutionize regenerative medicine by enabling the production of functional tissues and organs for transplantation.

Advances in Wound Healing

Biopolymers are also playing a crucial role in advancing wound healing technologies. Hydrogels, films, and sponges made from biopolymers can be used as dressings to promote wound healing.

Chitosan-based dressings, for example, have shown excellent hemostatic properties, making them effective in controlling bleeding. Biopolymer dressings can be loaded with growth factors, antimicrobial agents, or stem cells to enhance the healing process.

Gene Therapy and Tissue Engineering

Biopolymers are being explored as carriers for gene therapy in tissue engineering applications. Delivering genes that encode growth factors or extracellular matrix proteins can enhance tissue regeneration. Plasmid DNA or viral vectors encoding vascular endothelial growth factor (VEGF) can be delivered to promote angiogenesis in tissue-engineered constructs. This approach improves vascularization and nutrient supply, enhancing the survival and function of the engineered tissue.

Biopolymers in Personalized Medicine

Personalized medicine is transforming the landscape of modern healthcare by tailoring treatments to individual patients based on their genetic makeup, disease characteristics, and biological responses. Biopolymers play a crucial role in advancing this field by providing highly adaptable and responsive therapeutic solutions.

One of the key applications of biopolymers in personalized medicine is in drug delivery systems, where customized polymeric carriers can be designed to release specific drugs at optimal rates based on a patient's metabolic profile. Stimuli-responsive biopolymers can adjust their release behavior in response to pH changes, temperature variations, or specific biomolecular signals, ensuring precise dosing and reduced side effects. For example, polylactic-co-glycolic acid (PLGA) nanoparticles have been used to deliver cancer drugs in a controlled manner, reducing adverse effects while maintaining therapeutic efficacy [109].

Additionally, biopolymer-based nanocarriers are being developed for personalized cancer therapies. By conjugating targeting ligands or antibodies to polymeric nanoparticles, drug formulations can be directed specifically to cancerous tissues, minimizing harm to healthy cells. For example, HER2-targeted nanoparticles carrying chemotherapy agents have shown significant improvements in breast cancer treatment, reducing toxicity and enhancing drug accumulation at the tumor site.

In the realm of gene therapy, biopolymers enable safe and efficient genetic material delivery tailored to individual patient needs. Synthetic and natural polymer-based vectors offer a non-viral alternative for gene editing technologies like CRISPR, ensuring higher safety and better compatibility with different cell types. These innovations are expanding the potential for precision medicine applications, from correcting genetic disorders to engineering patient-specific regenerative therapies.

Furthermore, the development of bioengineered polymer scaffolds for tissue regeneration is enhancing personalized regenerative medicine approaches. 3D-printed biopolymer scaffolds are customized to match a patient's tissue architecture, allowing for more effective cell growth and tissue repair. This is particularly valuable in areas such as orthopedic and cardiovascular medicine, where personalized implants and grafts significantly improve clinical outcomes. An example of this approach is the use of polycaprolactone (PCL) scaffolds in bone regeneration, where they support new bone growth while gradually degrading over time.

Emerging Trends and Potential Breakthroughs in Biopolymer Therapeutics

Recent advancements in biopolymer therapeutics have been driven by the need for more efficient, targeted, and patient-specific treatments. Innovations in material science, nanotechnology, and biomedical engineering have enabled the development of sophisticated drug-delivery systems and regenerative medicine applications. One of the most significant developments is the emergence of smart and responsive biopolymer systems that can regulate drug release based on environmental stimuli such as pH, temperature, or enzymatic activity. These systems improve therapeutic efficacy while minimizing systemic side effects, making them ideal for controlled and localized treatment approaches.

Another major area of progress is the integration of biopolymer-based nanocarriers in precision medicine. Advances in polymeric nanoparticles have enabled highly specific drug targeting, allowing for the delivery of therapeutic agents directly to diseased cells. The combination of biopolymers with artificial intelligence and machine learning is further refining drug formulation and optimization strategies, making treatments more personalized and effective. Hybrid nanoparticle systems that incorporate lipids, polymers, and inorganic elements are also being explored to enhance the bioavailability and stability of therapeutic compounds [110].

Liposomal and polymeric drug carriers are undergoing significant innovation, particularly in gene therapy and vaccine delivery. The incorporation of DNA- and RNA-loaded liposomes has played a crucial role in the development of mRNA-based vaccines and targeted genetic treatments. Biodegradable and stimuli-responsive liposomal formulations enhance drug release control, reducing toxicity and improving treatment safety profiles. These advancements are paving the way for more sophisticated gene-based therapies that can address previously untreatable conditions.

In the field of tissue engineering and regenerative medicine, biopolymer scaffolds are being integrated into 3D bioprinting techniques to fabricate functional organ and tissue replacements. Bioinspired hydrogels that mimic the extracellular matrix are being developed to support cell growth, differentiation, and tissue regeneration. Additionally, nanostructured polymeric biomaterials are improving wound healing by facilitating sustained drug release and providing antimicrobial protection, which is essential for preventing infections and promoting tissue recovery.

Sustainability is also becoming a key focus in biopolymer research, with an increasing emphasis on biodegradable and eco-friendly materials. The shift toward plant-derived and recombinant biopolymers ensures better biocompatibility while reducing environmental impact. Moreover, antimicrobial coatings made from biopolymers are being explored for medical devices to enhance patient safety and prevent infections in clinical settings.

The future of biopolymer therapeutics lies in the continued integration of emerging technologies such as AI-driven drug design, advanced nanomedicine approaches, and biofabrication techniques. As research progresses, biopolymers will play an essential role in developing safer, more effective, and personalized medical solutions. The convergence of biotechnology, material science, and computational tools is set to revolutionize drug delivery, regenerative medicine, and beyond, making biopolymers a cornerstone of next-generation healthcare advancements [111].

CONCLUSION

The journey of biopolymers from ancient remedies to cutting-edge therapeutics underscores their enduring versatility and significance in medicine. Over the centuries, these naturally occurring polymers have transitioned from simple, traditional uses in early medicinal practices to becoming foundational elements in modern therapeutic approaches. Ancient civilizations harnessed the healing properties of natural substances rich in biopolymers, such as plant extracts and animal products, laying the groundwork for future scientific exploration. The 19th and early 20th centuries saw the identification and understanding of proteins, nucleic acids, and polysaccharides, which paved the way for their therapeutic application. Landmark discoveries, such as the isolation of insulin and the development of heparin, revolutionized the treatment of diabetes and blood clotting disorders, respectively. The latter half of the 20th century witnessed the advent of recombinant DNA technology and PEGylation, which enhanced the production and stability of protein -based drugs, propelling the field into a new era. In the 21st century, the advent of mRNA vaccines has showcased the transformative potential of nucleic acid-based therapeutics. Innovations in drug delivery systems, such as hydrogels and nanoparticles, have further improved the delivery and efficacy of biopolymer-based drugs. Advances in regenerative medicine and tissue engineering have highlighted the role of biopolymers in repairing and replacing damaged tissues. Looking ahead, future research will undoubtedly focus on optimizing the efficacy, safety, and delivery mechanisms of biopolymer therapeutics. The continued exploration of biopolymers holds immense promise for developing innovative treatments for a wide range of diseases, opening new avenues for personalized medicine and improving patient outcomes. The versatility and potential of biopolymers ensure they will remain at the forefront of therapeutic advancements, driving the evolution of medicine and biotechnology.

REFERENCES

[1]　Biswas MC, Jony B, Nandy PK, *et al.* Recent advancement of biopolymers and their potential biomedical applications. J Polym Environ 2022; 30(1): 51-74.
[http://dx.doi.org/10.1007/s10924-021-02199-y]

[2]　Babu RP, O'Connor K, Seeram R. Current progress on bio-based polymers and their future trends. Prog Biomater 2013; 2(1): 8.
[http://dx.doi.org/10.1186/2194-0517-2-8] [PMID: 29470779]

[3]　Opriş O, Mormile C, Lung I, Stegarescu A, Soran M-L, Soran A. An overview of biopolymers for drug delivery applications. Appl Sci (Basel) 2024; 14(4): 1383.
[http://dx.doi.org/10.3390/app14041383]

[4]　Nitta S, Numata K. Biopolymer-based nanoparticles for drug/gene delivery and tissue engineering. Int J Mol Sci 2013; 14(1): 1629-54.
[http://dx.doi.org/10.3390/ijms14011629] [PMID: 23344060]

[5]　Gulati S, Ansari N, Moriya Y, *et al.* Nanobiopolymers in cancer therapeutics: advancing targeted drug delivery through sustainable and controlled release mechanisms. J Mater Chem B Mater Biol Med 2024; 12(46): 11887-915.
[http://dx.doi.org/10.1039/D4TB00599F] [PMID: 39502076]

[6]　Bejenaru C, Radu A, Segneanu AE, *et al.* Pharmaceutical applications of biomass polymers: review of current research and perspectives. Polymers (Basel) 2024; 16(9): 1182.
[http://dx.doi.org/10.3390/polym16091182] [PMID: 38732651]

[7]　Biswas MC, Jony B, Nandy PK, et al. Recent advancement of biopolymers and their potential biomedical applications. J Polym Environ. 2022; 30: 51–74. Available from: https://link.springer.com/article/10.1007/s10924-021-02199-y

[8]　Bacterial enzyme makes new type of biodegradable polymer. ACS Cent Sci 2022. Available from: https://www.acs.org/pressroom/presspacs/2022/acs-presspac-march-16-2022/bacterial-enzyme-makes-new-type-of-biodegradable-polymer.html

[9]　Narayan B, Ed. Introduction to biopolymers and their applications. Biopolymers and Their Industrial Applications. 1st ed. Elsevier 2021; pp. 1-18. Available from: https://www.sciencedirect.com/science/article/abs/pii/B9780323852333000021

[10]　Taib NAAB, Rahman MR, Huda D, *et al.* A review on poly lactic acid (PLA) as a biodegradable polymer. Polym Bull 2023; 80(2): 1179-213.
[http://dx.doi.org/10.1007/s00289-022-04160-y]

[11]   Yadav P, Yadav H, Shah VG, Shah G, Dhaka G. Biomedical biopolymers, their origin and evolution in biomedical sciences: a systematic review. J Clin Diagn Res 2015; 9(9): ZE21-5.
[http://dx.doi.org/10.7860/JCDR/2015/13907.6565] [PMID: 26501034]

[12]   Pardi N, Hogan MJ, Porter FW, Weissman D. mRNA vaccines — a new era in vaccinology. Nat Rev Drug Discov 2018; 17(4): 261-79.
[http://dx.doi.org/10.1038/nrd.2017.243] [PMID: 29326426]

[13]   Corey DR, Damha MJ, Manoharan M. Challenges and opportunities for nucleic acid therapeutics. Nucleic Acid Ther 2022; 32(1): 8-13.
[http://dx.doi.org/10.1089/nat.2021.0085] [PMID: 34931905]

[14]   Swarnalatha KM, Saikiran CH, Mounika B, Swetha B, Ramarao T. A Comprehensive Review on Role of Polymers in Transdermal Drug Delivery System. Int J Pharm Bio-Med Sci. 2023; 3(10): 568-574.
[http://dx.doi.org/10.47191/ijpbms/v3-i10-11]

[15]   Dadwal A, Baldi A, Kumar Narang R. Nanoparticles as carriers for drug delivery in cancer. Artif Cells Nanomed Biotechnol 2018; 46(sup2): 295-305.
[http://dx.doi.org/10.1080/21691401.2018.1457039]

[16]   Oleksy M, Dynarowicz K, Aebisher D. Advances in Biodegradable Polymers and Biomaterials for Medical Applications — A Review. Molecules. 2023; 28(17): 6213.
[http://dx.doi.org/10.3390/molecules28176213]

[17]   Langer R, Tirrell DA. Designing materials for biology and medicine. Nature. 2004; 428(6982): 487-492.
[http://dx.doi.org/10.1038/nature02388]

[18]   Murphy SV, Atala A. 3D bioprinting of tissues and organs. Nat Biotechnol 2014; 32(8): 773-85.
[http://dx.doi.org/10.1038/nbt.2958] [PMID: 25093879]

[19]   Jayakumar R, Prabaharan M, Sudheesh Kumar PT, Nair SV, Tamura H. Biomaterials based on chitin and chitosan in wound dressing applications. Biotechnol Adv 2011; 29(3): 322-37.
[http://dx.doi.org/10.1016/j.biotechadv.2011.01.005] [PMID: 21262336]

[20]   Agarwal S, Wendorff JH, Greiner A. Use of electrospinning technique for biomedical applications. Polymer (Guildf) 2008; 49(26): 5603-21.
[http://dx.doi.org/10.1016/j.polymer.2008.09.014]

[21]   Huynh, C.T., Nguyen, M.K., Huynh, D.P. *et al.* Biodegradable star-shaped poly(ethylene glycol)-poly(β-amino ester) cationic pH/temperature-sensitive copolymer hydrogels. Colloid Polym Sci 2011; 289: 301–308.
[http://dx.doi.org/10.1007/s00396-010-2349-9]

[22]   Soppimath, Kumaresh S., Tejraj Malleshappa Aminabhavi, Anandrao R. Kulkarni and Walter. Rudzinski. Biodegradable polymeric nanoparticles as drug delivery devices. Journal of controlled release : official journal of the Controlled Release Society. 2011; 70: 1-20.

[23]   Anderson JM, Shive MS. Biodegradation and biocompatibility of PLA and PLGA microspheres. Adv Drug Deliv Rev 1997; 28(1): 5-24.
[http://dx.doi.org/10.1016/S0169-409X(97)00048-3] [PMID: 10837562]

[24]   Arthanari Y, Subramani K. Chitosan-based nanomaterials: A review. Int J Biol Macromol 2018; 114: 1025-33.
[http://dx.doi.org/10.1016/j.ijbiomac.2018.03.123]

[25]   Bae YH, Park K. Targeted drug delivery to tumors: Myths, reality and possibility. J Control Release 2011; 153(3): 198-205.
[http://dx.doi.org/10.1016/j.jconrel.2011.06.001] [PMID: 21663778]

[26]   Baldwin AD, Kiick KL. Tunable degradation of maleimide–thiol adducts in reducing environments. Bioconjug Chem 2010; 21(5): 1047-54.

[http://dx.doi.org/10.1021/bc900533y] [PMID: 21863904]

[27]   Kang S, Min H. Ginseng, the 'Immunity Boost': The Effects of Panax on Immune Function. J Ginseng Res. 2012; 36(1): 25-33.
[PMID: 23717137]

[28]   Banik BL, Fattahi P, Brown JL. Polymeric nanoparticles: the future of nanomedicine. Wiley Interdiscip Rev Nanomed Nanobiotechnol 2016; 8(2): 271-99.
[http://dx.doi.org/10.1002/wnan.1364] [PMID: 26314803]

[29]   Zhao, Y., *et al.* Polysaccharides from Traditional Chinese Medicines: Extraction, Purification, Modification, and Biological Activity. [Internet]. 2016. Available from: https://pubmed.ncbi.nlm.nih.gov/27983593/

[30]   Biondi M, Ungaro F, Quaglia F, Netti PA. Controlled drug delivery in tissue engineering. Adv Drug Deliv Rev. 2008; 60(2): 229-42.
[http://dx.doi.org/10.1016/j.addr.2007.08.038]

[31]   Aloe Vera Throughout History. Aloe.se. Available from: https://aloe.se/en/aloe-vera-throughout-history/

[32]   Bourges X, Weiss P, Daculsi G, Legeay G. Synthesis and general properties of silated-hydroxypropyl methylcellulose in prospect of biomedical use. Adv Colloid Interface Sci 2002; 99(3): 215-28.
[http://dx.doi.org/10.1016/S0001-8686(02)00035-0] [PMID: 12509115]

[33]   Brannon-Peppas L, Blanchette JO. Nanoparticle and targeted systems for cancer therapy. Adv Drug Deliv Rev 2004; 56(11): 1649-59.
[http://dx.doi.org/10.1016/j.addr.2004.02.014] [PMID: 15350294]

[34]   Brazel CS, Peppas NA. Modeling of drug release from Swellable polymers. Eur J Pharm Biopharm 2000; 49(1): 47-58.
[http://dx.doi.org/10.1016/S0939-6411(99)00058-2] [PMID: 10613927]

[35]   Buwalda SJ, Boere KWM, Dijkstra PJ, Feijen J, Vermonden T, Hennink WE. Hydrogels in a historical perspective: From simple networks to smart materials. J Control Release 2014; 190: 254-73.
[http://dx.doi.org/10.1016/j.jconrel.2014.03.052] [PMID: 24746623]

[36]   Caló E, Khutoryanskiy VV. Biomedical applications of hydrogels: A review of patents and commercial products. Eur Polym J 2015; 65: 252-67.
[http://dx.doi.org/10.1016/j.eurpolymj.2014.11.024]

[37]   Cao L, Mooney D. Spatiotemporal control over growth factor signaling for therapeutic neovascularization. Adv Drug Deliv Rev 2007; 59(13): 1340-50.
[http://dx.doi.org/10.1016/j.addr.2007.08.012] [PMID: 17868951]

[38]   Chandra R, Rustgi R. Biodegradable polymers. Prog Polym Sci 1998; 23(7): 1273-335.
[http://dx.doi.org/10.1016/S0079-6700(97)00039-7]

[39]   Chen G, Amsden B. Biodegradable hydrogels for tissue engineering. J Biomed Mater Res A 2008; 95A(2): 573-86.
[http://dx.doi.org/10.1002/jbm.a.32894]

[40]   Wagh VD. Propolis: A Wonder Bees Product and Its Pharmacological Potentials. Adv Pharmacol Pharm Sci. 2013; 308249.
[http://dx.doi.org/10.1155/2013/308249]

[41]   Neeraj K Pawaskar, Jan Mohd Muneeb, Akhilesh Kumar, Praveen K Gupta, Shaurya Yadav, Mohini Saini and Mudasir M Rather. Synthesis and characterization of chitosan nanoparticles: Insights from in-vitro analysis. Int. J. Adv. Biochem. Res. 2024; 8(8): 421-426.
[http://dx.doi.org/10.33545/26174693.2024.v8.i8f.1767]

[42]   Propolis: a new frontier for wound healing? Burns & Trauma. 2015; 22;3(1): 9.
[http://dx.doi.org/10.1186/s41038-015-0010-z]

[43] Coviello T, Matricardi P, Marianecci C, Alhaique F. Polysaccharide hydrogels for modified release formulations. J Control Release 2007; 119(1): 5-24.
[http://dx.doi.org/10.1016/j.jconrel.2007.01.004] [PMID: 17382422]

[44] Crommelin DJA, Storm G. Liposomes: from the bench to the bed. J Liposome Res 2003; 13(1): 33-6.
[http://dx.doi.org/10.1081/LPR-120017488] [PMID: 12725726]

[45] Dash TK, Konkimalla VB. Poly-ε-caprolactone based formulations for drug delivery and tissue engineering: A review. J Control Release 2012; 158(1): 15-33.
[http://dx.doi.org/10.1016/j.jconrel.2011.09.064] [PMID: 21963774]

[46] de Jong WH, Borm PJ. Drug delivery and nanoparticles: Applications and hazards. Int J Nanomedicine 2008; 3(2): 133-49.
[http://dx.doi.org/10.2147/IJN.S596] [PMID: 18686775]

[47] Langer R, Tirrell DA. Designing materials for biology and medicine. Nature. 2004; 1;428(6982): 487□492.
[http://dx.doi.org/10.1038/nature02388]

[48] Zhang H, Li X, *et al.* A review on the applications of Traditional Chinese medicine polysaccharides in drug delivery systems. Chin Med (Lond). 2022; 17:12.
[http://dx.doi.org/10.1186/s13020-021-00567-3]

[49] Duncan R. The dawning era of polymer therapeutics. Nat Rev Drug Discov 2003; 2(5): 347-60.
[http://dx.doi.org/10.1038/nrd1088] [PMID: 12750738]

[50] Duncan R, Vicent MJ. Polymer therapeutics-prospects for 21st century: The end of the beginning. Adv Drug Deliv Rev 2013; 65(1): 60-70.
[http://dx.doi.org/10.1016/j.addr.2012.08.012] [PMID: 22981753]

[51] Wikipedia contributors. Protein. Wikipedia [Internet]. 2025. Available from: https://en.wikipedia.org/wiki/Protein

[52] Medium. History of protein requirements. Medium. 2023. Available from: https://medium.com/startoday/history-of-protein-requirements-a2b6261c9479

[53] Bioinformatics Home. History of protein research. Bioinformaticshome.com. 2024. Available from: https://bioinformaticshome.com/bioinformatics_tutorials/history/history_Page_5.html

[54] Gaharwar AK, Peppas NA, Khademhosseini A. Nanocomposite hydrogels for biomedical applications. Biotechnol Bioeng 2014; 111(3): 441-53.
[http://dx.doi.org/10.1002/bit.25160] [PMID: 24264728]

[55] Vauquelin LN, Robiquet PJ. Sur une nouvelle substance extraite de l'asperge. Ann Chim Phys. 1806; 61: 66□79. Available from: https://gallica.bnf.fr/ark:/12148/bpt6k6544251j/f66.item

[56] Gopi S, Amalraj A, Sukumaran NP, Haponiuk JT, Thomas S. Biopolymers and their composites for drug delivery: A brief review. Macromol Symp 2018; 380(1): 1800114.
[http://dx.doi.org/10.1002/masy.201800114]

[57] Hacker MC, Mikos AG. Synthetic polymers. Principles of Regenerative Medicine 2010; 587-622.

[58] Hennink WE, van Nostrum CF. Novel crosslinking methods to design hydrogels. Adv Drug Deliv Rev 2012; 64: 223-36.
[http://dx.doi.org/10.1016/j.addr.2012.09.009] [PMID: 11755704]

[59] Hoffman AS. Hydrogels for biomedical applications. Adv Drug Deliv Rev 2002; 54(1): 3-12.
[http://dx.doi.org/10.1016/S0169-409X(01)00239-3] [PMID: 11755703]

[60] Huang X, Brazel CS. On the importance and mechanisms of burst release in matrix-controlled drug delivery systems. J Control Release 2001; 73(2-3): 121-36.
[http://dx.doi.org/10.1016/S0168-3659(01)00248-6] [PMID: 11516493]

[61] Jain KK. Drug delivery systems - an overview. Methods Mol Biol 2008; 437: 1-50.

[http://dx.doi.org/10.1007/978-1-59745-210-6_1] [PMID: 18369961]

[62]   Jain RA. The manufacturing techniques of various drug loaded biodegradable poly(lactide-co-glycolide) (PLGA) devices. Biomaterials 2000; 21(23): 2475-90.
[http://dx.doi.org/10.1016/S0142-9612(00)00115-0] [PMID: 11055295]

[63]   Jeong B, Bae YH, Kim SW. Drug release from biodegradable injectable thermosensitive hydrogel of PEG–PLGA–PEG triblock copolymers. J Control Release 2000; 63(1-2): 155-63.
[http://dx.doi.org/10.1016/S0168-3659(99)00194-7] [PMID: 10640589]

[64]   Khan A, Alamry KA, Asiri AM. Multifunctional biopolymers-based composite materials for biomedical applications: A systematic review. ChemistrySelect 2021; 6(2): 154-76.
[http://dx.doi.org/10.1002/slct.202003978]

[65]   Watson JD, Crick FH. Molecular structure of nucleic acids: a structure for deoxyribose nucleic acid. Nature. 1953; 25;171(4356): 737□738.
[http://dx.doi.org/10.1038/171737a0]

[66]   Kopeček J. Hydrogel biomaterials: A smart future?. Biomaterials 2007; 28(34): 5185-92.
[http://dx.doi.org/10.1016/j.biomaterials.2007.07.044] [PMID: 17697712]

[67]   Ravi Kumar MNV. A review of chitin and chitosan applications. React Funct Polym 2000; 46(1): 1-27.
[http://dx.doi.org/10.1016/S1381-5148(00)00038-9]

[68]   Langer R, Tirrell DA. Designing materials for biology and medicine. Nature 2004; 428(6982): 487-92.
[http://dx.doi.org/10.1038/nature02388] [PMID: 15057821]

[69]   Lavik E, Langer R. Tissue engineering: current state and perspectives. Appl Microbiol Biotechnol 2004; 65(1): 1-8.
[http://dx.doi.org/10.1007/s00253-004-1580-z] [PMID: 15221227]

[70]   Li J, Mooney DJ. Designing hydrogels for controlled drug delivery. Nat Rev Mater 2016; 1(12): 16071.
[http://dx.doi.org/10.1038/natrevmats.2016.71] [PMID: 29657852]

[71]   Li S, McCarthy S. Further investigations on the hydrolytic degradation of poly (DL-lactide). Biomaterials 1999; 20(1): 35-44.
[http://dx.doi.org/10.1016/S0142-9612(97)00226-3] [PMID: 9916769]

[72]   Liu Y, Chan-Park MB. Hydrogel based on interpenetrating polymer networks of dextran and gelatin for vascular tissue engineering. Biomaterials 2009; 30(2): 196-207.
[http://dx.doi.org/10.1016/j.biomaterials.2008.09.041] [PMID: 18922573]

[73]   Luo Y, Prestwich G. Cancer-targeted polymeric drugs. Curr Cancer Drug Targets 2002; 2(3): 209-26.
[http://dx.doi.org/10.2174/1568009023333836] [PMID: 12188908]

[74]   Ma PX. Biomimetic materials for tissue engineering. Adv Drug Deliv Rev 2008; 60(2): 184-98.
[http://dx.doi.org/10.1016/j.addr.2007.08.041] [PMID: 18045729]

[75]   Malmsten, M. (Ed.). (2002). Surfactants and Polymers in Drug Delivery (1st ed.). CRC Press.
[http://dx.doi.org/10.1201/9780824743758]

[76]   Mao HQ, Leong KW. Design of polyphosphoester-DNA nanoparticles for non-viral gene delivery. Adv Genet 2005; 53: 275-306.
[http://dx.doi.org/10.1016/S0065-2660(05)53011-6] [PMID: 16240998]

[77]   Matsumura Y, Maeda H. A new concept for macromolecular therapeutics in cancer chemotherapy: mechanism of tumoritropic accumulation of proteins and the antitumor agent smancs. Cancer Res 1986; 46(12 Pt 1): 6387-92.
[PMID: 2946403]

[78]   Mikos AG, Temenoff JS. Formation of highly porous biodegradable scaffolds for tissue engineering. Electron J Biotechnol 2000; 3(2): 114-9.

[http://dx.doi.org/10.2225/vol3-issue2-fulltext-5]

[79]   Moghimi SM, Hunter AC. Poloxamers and poloxamines in nanoparticle engineering and experimental
       medicine. Trends Biotechnol 2000; 18(10): 412-20.
       [http://dx.doi.org/10.1016/S0167-7799(00)01485-2] [PMID: 10998507]

[80]   Gote V, McCulloch D, McKinney J. Antisense oligonucleotides and their applications in rare
       neurological diseases. Front Neurosci. 2024; 18: 1414658.
       [http://dx.doi.org/10.3389/fnins.2024.1414658]

[81]   Peppas N, Bures P, Leobandung W, Ichikawa H. Hydrogels in pharmaceutical formulations. Eur J
       Pharm Biopharm 2000; 50(1): 27-46.
       [http://dx.doi.org/10.1016/S0939-6411(00)00090-4] [PMID: 10840191]

[82]   Howlett PC. Heparin: historical perspective and development. Br J Haematol. 1990; 74(1): 1□7.
       [http://dx.doi.org/10.1111/j.1365-2141.1990.tb07514.x]

[83]   Reinbold J, Hierlemann T, Hinkel H, *et al.* Development and *in vitro* characterization of poly(lactide-
       co-glycolide) microspheres loaded with an antibacterial natural drug for the treatment of long-term
       bacterial infections. Drug Des Devel Ther 2016; 10: 2823-32.
       [http://dx.doi.org/10.2147/DDDT.S105367]

[84]   Kim YJ, Choi S, Koh JJ, Lee M, Ko KS, Kim SW. Controlled release of insulin from injectable
       biodegradable triblock copolymer. Pharm Res 2001; 18(4): 548-50.
       [http://dx.doi.org/10.1023/A:1011074915438]

[85]   Park SB, Lih E, Park KS, Joung YK, Han DK. Biopolymer-based functional composites for medical
       applications. Prog Polym Sci 2017; 68: 77-105.
       [http://dx.doi.org/10.1016/j.progpolymsci.2016.12.003]

[86]   Tan H, Marra KG. Injectable, biodegradable hydrogels for tissue engineering applications. Materials
       (Basel) 2010; 3(3): 1746-67.
       [http://dx.doi.org/10.3390/ma3031746]

[87]   Kopeček J, Yang J. Hydrogels as smart biomaterials. Polym Int 2007; 56(9): 1078-98.
       [http://dx.doi.org/10.1002/pi.2253]

[88]   Suhail M, Alamgir , Wahab , *et al.* Magnetically responsive hydrogel systems: fundamental features,
       emerging applications, and future horizons. Coord Chem Rev 2025; 543: 216916.

[89]   Tibbitt MW, Rodell CB, Burdick JA, Anseth KS. Progress in material design for biomedical
       applications. Proc Natl Acad Sci U S A 2015; 112(47): 14444-51.
       [http://dx.doi.org/10.1073/pnas.1516247112]

[90]   Sharma P, Kumar P, Sharma R, Bhatt VD, Dhot PS. Tissue engineering; current status & futuristic
       scope. J Med Life 2019; 12(3): 225-9.
       [http://dx.doi.org/10.25122/jml-2019-0032]

[91]   Dreiss CA. Hydrogel design strategies for drug delivery. Curr Opin Colloid Interface Sci 2020; 48: 1-
       17.
       [http://dx.doi.org/10.1016/j.cocis.2020.02.001]

[92]   Mason NS, Miles CS, Sparks RE. Hydrolytic degradation of poly DL-(lactide). Biomedical and Dental
       Applications of Polymers. Gebelein CG, Koblitz FF, editors. Polymer Science and Technology.
       Boston: Springer 1981; pp. 279-91.
       [http://dx.doi.org/10.1007/978-1-4757-9510-3_20]

[93]   Li L, Ge J, Ma PX, Guo B. Injectable conducting interpenetrating polymer network hydrogels from
       gelatin-graft-polyaniline and oxidized dextran with enhanced mechanical properties. RSC Adv 2015;
       5: 92490-8.
       [http://dx.doi.org/10.1039/C5RA19467A]

[94]   Wang J, Liu X, Sunir M, Zheng G, Zhang A. Targeted delivery and controlled release of polymeric

nanomedicines in tumor therapy. Acta Pharm Sin B 2025; 15(5): 1447-67.
[http://dx.doi.org/10.1016/j.apsb.2024.12.005]

[95]   Kim HD, Amirthalingam S, Kim SL, Lee SS, Rangasamy J, Hwang NS. Biomimetic materials and fabrication approaches for bone tissue engineering. Adv Healthc Mater 2017; 6(23): 1700612.
[http://dx.doi.org/10.1002/adhm.201700612]

[96]   Bjerk TR, Severino P, Jain S, *et al.* Biosurfactants: properties and applications in drug delivery, biotechnology and ecotoxicology. Bioengineering (Basel) 2021; 8(8): 115.
[http://dx.doi.org/10.3390/bioengineering8080115]

[97]   Wang C, Pan C, Yong H, *et al.* Emerging non-viral vectors for gene delivery. J Nanobiotechnol 2023; 21(1): 272.
[http://dx.doi.org/10.1186/s12951-023-02044-5]

[98]   McDowall S, McCulloch D, McKinney J. Antisense oligonucleotides and their applications in rare neurological diseases. Front Neurosci. 2024; 18:1414658.
[http://dx.doi.org/10.3389/fnins.2024.1414658]

[99]   Mikos AG, Temenoff JS. Formation of highly porous biodegradable scaffolds for tissue engineering. Electron J Biotechnol 2000; 3(2): 114-9.
[http://dx.doi.org/10.2225/vol3-issue2-fulltext-5]

[100]   Gote V, Bolla PK, Kommineni N, Butreddy A, Nukala PK, Palakurthi SS, Khan W. A. Comprehensive Review of mRNA Vaccines. Int J Mol Sci. 2023; 24(3): 2700.
[http://dx.doi.org/10.3390/ijms24032700]

[101]   Chu TW, Feng J, Yang J, Kopeček J. Hybrid polymeric hydrogels *via* peptide nucleic acid (PNA)/DNA complexation. J Control Release 2015; 220(Pt B): 608-16.
[http://dx.doi.org/10.1016/j.jconrel.2015.09.035]

[102]   Klouda L, Mikos AG. Thermoresponsive hydrogels in biomedical applications. Eur J Pharm Biopharm 2008; 68(1): 34-45.
[http://dx.doi.org/10.1016/j.ejpb.2007.02.025]

[103]   Peppas NA, Langer R. New challenges in biomaterials. Science 1994; 263(5154): 1715-20.
[http://dx.doi.org/10.1126/science.8134835] [PMID: 8134835]

[104]   Peppas NA, Mikos AG. Preparation methods and structure of hydrogels. Hydrogels Med Pharm 1986; 1: 1-27.

[105]   Peppas NA, Reinhart CT. Solute diffusion in swollen membranes. Part I. A new theory. J Membr Sci 1983; 15(3): 275-87.
[http://dx.doi.org/10.1016/S0376-7388(00)82304-2]

[106]   Peppas NA, Sahlin JJ. Hydrogels as mucoadhesive and bioadhesive materials: a review. Biomaterials 1996; 17(16): 1553-61.
[http://dx.doi.org/10.1016/0142-9612(95)00307-X] [PMID: 8842358]

[107]   Peppas NA, Wong JE. Polymeric systems for oral protein delivery. J Drug Deliv Sci Technol 2000; 10(2): 117-25.

[108]   Jabeen N, Atif M. Polysaccharides based biopolymers for biomedical applications: A review. Polym Adv Technol 2024; 35(1): e6203.
[http://dx.doi.org/10.1002/pat.6203]

[109]   Mitura S, Sionkowska A, Jaiswal A. Biopolymers for hydrogels in cosmetics: review. J Mater Sci Mater Med 2020; 31(6): 50.
[http://dx.doi.org/10.1007/s10856-020-06390-w] [PMID: 32451785]

[110]   Alharbi HM. Exploring the Frontier of Biopolymer-Assisted Drug Delivery: Advancements, Clinical Applications, and Future Perspectives in Cancer Nanomedicine. Drug Des Devel Ther. 2024; 10(18): 2063-2087.

[http://dx.doi.org/10.2147/DDDT.S441325] [PMID: 38882042]

[111]   Kumar A, Negi YS. Chitosan-based hydrogels: Recent advances in design and applications. Carbohydr Polym 2018; 196: 1-14.
[http://dx.doi.org/10.1016/j.carbpol.2018.05.002]

CHAPTER 2

Biological Sources, Chemistry, and Extraction of Biopolymers

Sharda Sambhakar[1,*], Bishambar Singh[2], Srishti Verma[1], Nancy Gupta[1], Manya Modi[1] and Pooja[1]

[1] *Department of Pharmacy, Banasthali Vidyapith, Rajasthan 304022, India*

[2] *PHTI Department, SMS Medical College and Hospital, Jaipur, Rajasthan 302017, India*

Abstract: To foster a green environment, considerable efforts have been made to replace synthetic polymers with biodegradable materials, such as biopolymers, particularly for the development of green drug delivery systems. Biopolymers are a prominent class of functional materials with high-value applications, generated either by biological systems or derived from biological sources. Natural sources of biopolymers include plants, animals, microorganisms, and agricultural wastes. Biopolymers exhibit excellent properties, including flexibility, tensile strength, stability, reusability, and many more. Biopolymers are composed of repetitive monomers bound covalently *via* polymerization reaction or enzyme-catalyzed assemblies of monomeric units that occur in the biosynthetic pathway within biological systems. Biopolymers can be classified based on their source, chemical composition, functional properties, degradability, type of charges, and other factors. The extraction of biopolymers involves a range of chemical and enzymatic processes that vary specifically for each biopolymer. Some of the extraction methods include the use of coagulating agents, hydrolysis, alkali and acid treatments, bleaching, deproteination, and demineralization, among others. Following extraction, purification, and often modification, biopolymers are prepared for potential applications. Due to their renewability, abundance, biodegradability, and unique properties, such as higher absorption capabilities and ease of functionalization, biopolymers have been explored for various industrial applications. This chapter examines the sources, chemistry, and extraction procedures for several important biopolymers, including polyhydroxyalkanoates, polylactic acid, chitosan, alginate, polyesteramide, starch, gelatin, polyglycolic acid, and pectin, as well as their biomedical applications.

Keywords: Biopolymers, Bio-based technology, Biodegradable polymers, Chemistry, Extraction.

---

* **Corresponding author Sharda Sambhakar:** Department of Pharmacy, Banasthali Vidyapith, Rajasthan 302017, India; E-mail: ssambhakar@yahoo.co.in

Sudhanshu Mishra, Smriti Ojha, Shashi Kant Singh, Rishabha Malviya & Saurabh Kumar Gupta (Eds.)

INTRODUCTION

The concept of biopolymers emerged when people became aware of the deleterious effects of synthetic polymers on the environment and human health. The increasing use of synthetic, non-biodegradable polymers in our daily lives has significantly increased the risks of cancers and other health hazards. These are costly and derived from petrochemical sources. Biopolymers are sustainable, biodegradable, and eco-friendly alternatives to synthetic polymers. The term "biopolymer" originates from the Greek prefix *bio* (life) and *polymer* (many parts), reflecting the biological origin and the large molecular structure of these materials, which are composed of repeating monomeric units covalently bonded, forming highly structured macromolecules [1]. Biopolymers are synthesized either through direct biosynthesis, a process involving living organisms, such as microorganisms or plants, which enables the production of polymers with specific structural and functional characteristics, or through chemical processes that utilize biological raw materials. The latter involves the chemical conversion of natural substrates, such as corn starch or cellulose, into polymeric forms. The chemical synthesis approach enables the tailored production of biopolymers with desired molecular properties, thereby facilitating the creation of specialized materials for various applications [2].

SOURCES OF BIOPOLYMERS

Biopolymers are sourced from various origins, including plants, animals, seaweed, mushrooms, and microorganisms like bacteria, fungi, and yeasts (Table **1**).

Natural Sources

These biopolymers are formed organically from living organisms. Examples of polymers obtained from plants include cellulose, starch, and pectin, whereas collagen, chitin, gelatin, and hyaluronic acids are obtained from animals. Biotechnological techniques produce microorganism-based biopolymers (polyhydroxyalkanoates, polylactic acid), whereas polycaprolactone synthesis requires petrochemical sources (Table **1**).

Synthetic Biopolymers

Synthetic biopolymers are produced or chemically modified by chemical reactions, either as biodegradable or non-biodegradable polymers. Synthetic polymers offer both stability and flexibility, making them suitable for a wide range of applications. Synthetic biopolymers are economically produced by enzymes and living cells in a controlled environment, utilizing biopolymer synthesis, and they exhibit improved mechanical characteristics. They are widely used in tissue engineering. Commonly used biodegradable polymers in the pharmaceutical field include Polyglycolic acid (PGA), Polylactic Acid (PLA), Polycaprolactone (PCL), Polyhydroxybutyrate (PHB), Polypropylene Fumarate (PPF), and Polydioxanone (PDS), among others. The non-biodegradable polymers include Polyamide (PA), Polyvinyl Chloride (PVC), Polypropylene (PP), Polyethylene (PE), Polymethyl Methacrylate (PMMA), Polycarbonate (PC), and Polyurethane (PU), among others. They do not get degraded or metabolized inside the body and, therefore, accumulate and sometimes require surgical removal; hence, they are preferred for non-medical purposes [3, 4].

Table 1. Sources of biopolymers.

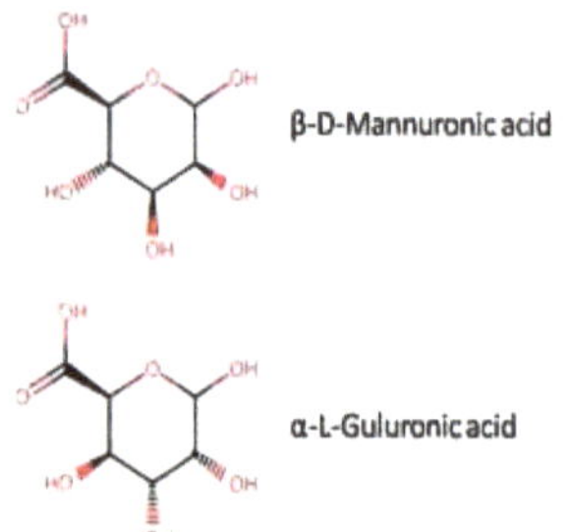

β-D-Mannuronic acid

α-L-Guluronic acid

Alginate(β-D-mannuronic acid and α-L-guluronic acid)
Source: Brown algae and some bacteria

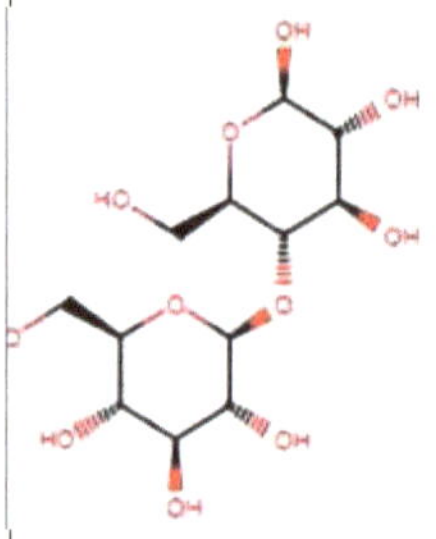

Cellulose(Cellubiose)
Source: Plants and agricultural residues

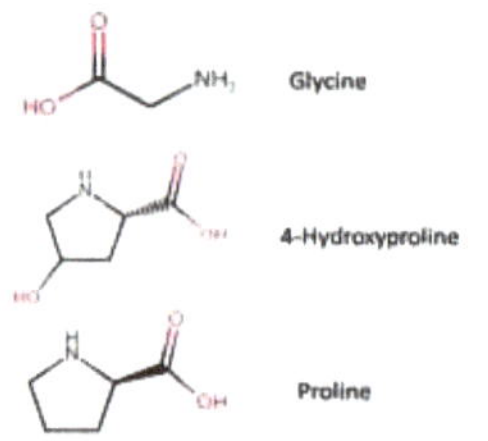

Glycine

4-Hydroxyproline

Proline

Gelatin(Collagen: composed of glycine, proline and hydroxyproline)
Source: Pig skin, bovine bones, tendons and ligaments

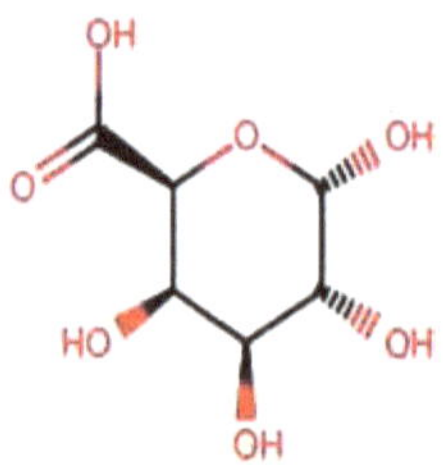

Pectin(α-D-galacturonic acid)
Source: Citrus fruits and their byproducts

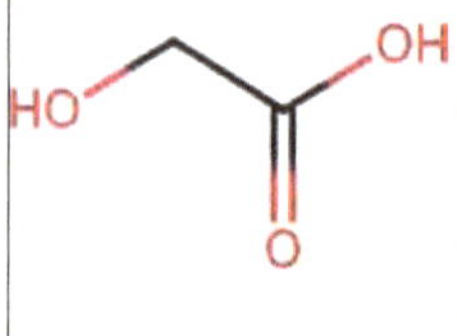

Glycolic acid monomer (PGA)
Source: Glycolic acid present in fruits and vegetables

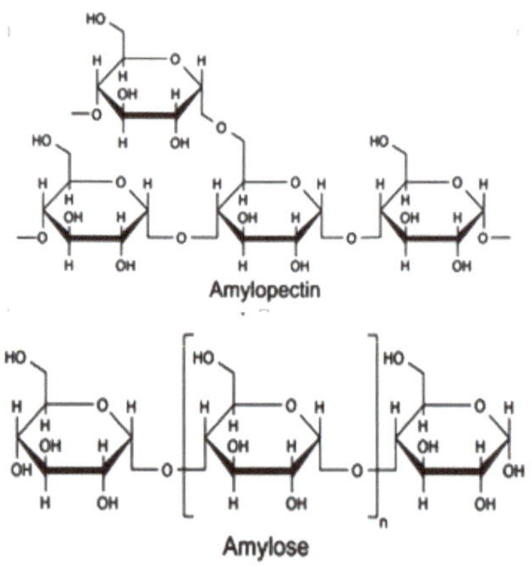

Amylopectin

Amylose

Starch(Amylose and amylopectin)
Source: Potato, maize, rice

(Table 1) cont.....

Polyhydroxyalkanoates(R-hydroxy fatty acids)
Source: Bacteria, Microalgae, waste material

Lactic acid monomers
(PLA)
Source: Fermented resources like corn and cane sugar, milk, and food products

Chitin
Source: Exoskeleton of crustaceans and insects

CLASSIFICATION OF BIOPOLYMERS

Biopolymers (BPs) are classified in various ways, including by chemical composition, source, functional properties, biodegradability, and degree of polymerization (Fig. **1**). Composition-wise, biopolymers can be categorized as polysaccharides, proteins, nucleic acids, and polymeric lipids. Examples of polysaccharides include chitosan, chitin, alginic acid, starch, cellulose, and guar gum, whereas proteins encompass collagen, silk, and fibrinogen, among others. According to the source, they can be obtained from plants, animals, microorganisms, petrochemicals, and biotechnology. Biopolymers can also be categorized into structural biopolymers, energy storage biopolymers, functional biopolymers, and bioactive biopolymers. Another way of classification includes biodegradable and non-biodegradable polymers, and based on the degree of polymerization, they can be categorized as low-molecular-weight and high-molecular-weight BPs [3, 4].

Polysaccharide BPs play important roles in the food and pharmaceutical industries, as well as in tissue engineering. They can further be categorized based on their monosaccharide composition into homopolysaccharides (consisting of a single type of monosaccharide, *e.g.*, starch) and heteropolysaccharides (containing two or more different monomeric units, *e.g.*, pectin). Depending on their charge, polysaccharides can be sub-classified as anionic (hyaluronic acid, alginate, heparin, xanthan gum, xylan, dextran, pectin, gellan, mannan, inulin, carrageenan), cationic (chitosan, chitin), and nonionic (cellulose, starch, dextran, agarose, pullulan). Protein BPs perform many important functions in the body and can further be classified as plant-derived (zein, soya) and animal-derived (gelatin, collagen) biopolymers [3, 4].

CHEMISTRY AND EXTRACTION OF BIOPOLYMERS

Starch

Starch is a storage polysaccharide found in corn, potatoes, rice, wheat, and barley, and is the second most common biomass material. It is natural, biodegradable, renewable, abundant, low-cost, and easily available. Starch has been used since ancient Egypt for adhesives and is suitable for manufacturing films and coatings [5, 6]. Native starches are mainly extracted from plants and can be modified (modified starch) *via* synthetic routes to enhance their properties. Food, plastic, and pharmaceutical industries often utilize starch for its gelling, film-forming, and biodegradable properties. Hence, the interest in using starch and other organic macromolecules is more widespread due to the different qualities offered by different types of starch for advanced applications [7, 8].

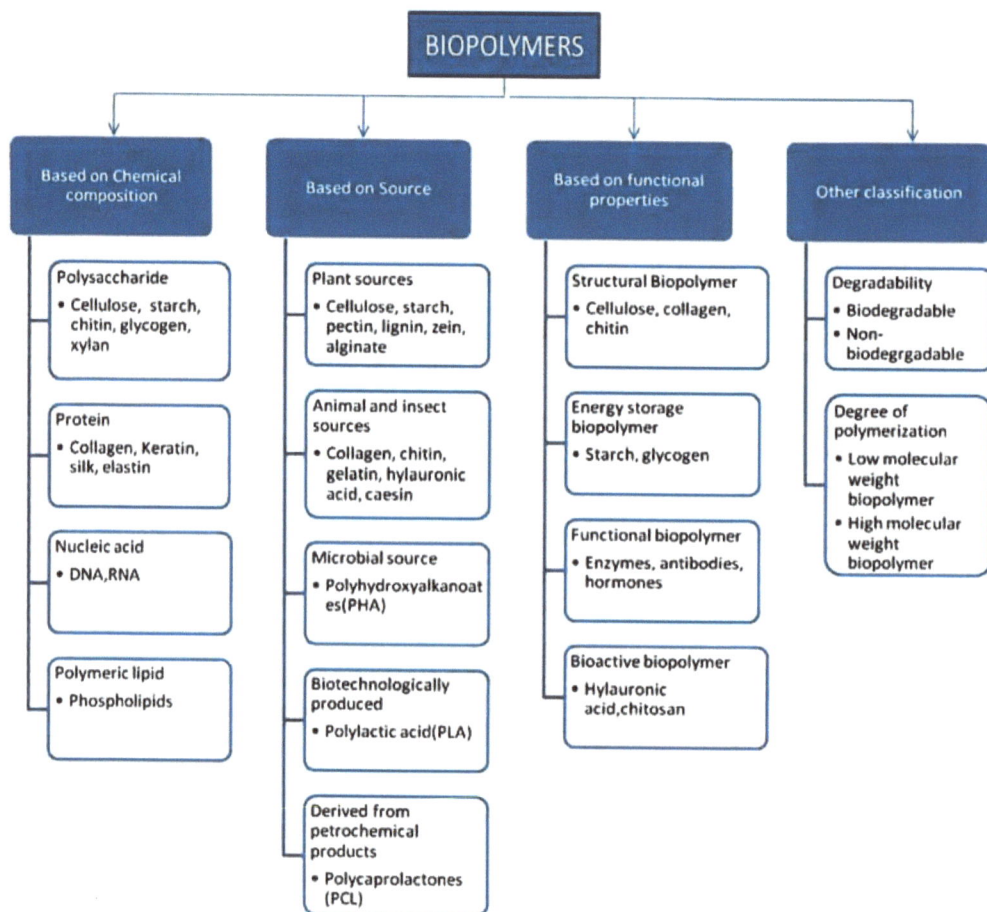

BIOPOLYMERS

Based on Chemical composition

Polysaccharide
• Cellulose, starch, chitin, glycogen, xylan

Protein
• Collagen, Keratin, silk, elastin

Nucleic acid
• DNA,RNA

Polymeric lipid
• Phospholipids

Based on Source

Plant sources
• Cellulose, starch, pectin, lignin, zein, alginate

Animal and insect sources
• Collagen, chitin, gelatin, hylauronic acid, caesin

Microbial source
• Polyhydroxyalkanoat es(PHA)

Biotechnologically produced
• Polylactic acid(PLA)

Derived from petrochemical products
• Polycaprolactones (PCL)

Based on functional properties

Structural Biopolymer
• Cellulose, collagen, chitin

Energy storage biopolymer
• Starch, glycogen

Functional biopolymer
• Enzymes, antibodies, hormones

Bioactive biopolymer
• Hylauronic acid,chitosan

Other classification

Degradability
• Biodegradable
• Non-biodegrgadable

Degree of polymerization
• Low molecular weight biopolymer
• High molecular weight biopolymer

Fig. (1). Classification of biopolymers.

Structure of Starch

Starch consists of two components: amylose, a long and linear chain polysaccharide (has a greater tendency to entangle), and amylopectin, a short and highly branched-chain polysaccharide (Table **1**). Amylopectin has 5% branches, and the degree of polymerization (DP) ranges from 5,000 to 1 million, whereas for amylose, the DP is below 5,000. Native granules of amylopectin exist as double helices and crystallites, forming a semi-crystalline structure that exhibits alternating layers of crystalline and amorphous regions, and also displays birefringence under polarized light. Amylose constitutes 20–30% of starch and is made up of glucose molecules linked by α-1,4-glycosidic linkages. This component thickens upon heating with water and forms a gel. Amylopectin constitutes 70–80% of starch and is a branched polymer with both α-1,4- and α-1,6-glycosidic linkages [9, 10].

Properties of Starch

The amylose-to-amylopectin ratio varies among different species and varieties. In addition, starch contains trace amounts of protein, fiber, and lipids. When heated above 52°C, gelatinized starch paste is formed as granules swell due to the breakdown of hydrogen bonds, allowing water to penetrate the amorphous regions of the granule. This process reduces crystallinity, increases water absorption, causes granule expansion, loss of birefringence, and results in a starch paste with increased viscosity. Light transmittance at 640 nm, which indicates the clarity of the starch paste, is influenced by phospholipids bound to amylose, resulting in an opaque paste. In contrast, the monoesters of phosphate attached to amylopectin enhance clarity. Granule size also affects the light transmittance of starch [11, 12].

Extraction of Starch

Modern starch extraction processes incorporate traditional, mechanical, and advanced technologies, such as ultrasound and pulsed electric fields, utilizing wet, dry, chemical, and enzymatic approaches. These methods vary in efficiency, benefits, drawbacks, and the sources of common starches. Pulses contain 35–60% starch and 14.9–39.4% protein, with wet and dry milling being common extraction methods. Dry milling causes greater starch fragmentation and higher starch damage, which affects its physicochemical properties [13].

Structure-property Relationship in Biopolymers

The chemical structure of biopolymers, such as starch and polyhydroxyalkanoates (PHAs), directly influences their performance in various applications. Starch is composed of amylose and amylopectin: amylose's linear structure enhances tensile strength and barrier properties but reduces flexibility, whereas amylopectin's branched structure improves elasticity and water absorption, making it suitable for food and thickening applications. Crystallinity in starch affects solubility, water uptake, and film transparency. Similarly, PHAs vary in side-chain length and crystallinity, which influence their mechanical properties and biodegradation rates. Short-chain PHAs (*e.g.*, P(3HB)) exhibit high crystallinity, making them stiff and brittle, with slow degradation, which limits their use in flexible packaging. In contrast, medium-chain PHAs (*e.g.*, P(3HHx), P(3HO)) exhibit lower crystallinity, which results in greater flexibility and faster biodegradation, making them ideal for use in medical implants and drug delivery applications. Molecular weight also plays a crucial role: high-molecular-weight PHAs offer greater strength but are more difficult to process, while lower-molecular-weight PHAs degrade more quickly but have reduced mechanical strength. Understanding these structure–property relationships is essential for optimizing the use of biopolymers in packaging, biomedical, and sustainable material applications.

Traditional Methods

Primary methods for starch extraction include traditional washing and grinding, followed by filtration and sedimentation, typically *via* wet and dry pathways.

- **Wet Pathway:** For starch extraction from taro, the tubers are cleaned, peeled, and chopped. The resulting puree is filtered, and the starch slurry is allowed to settle. The wet starch is then sun-dried, ground into a powder, and stored for testing. An alternative method involves soaking taro flour in water at 35°C for 12 hours, followed by homogenization, sieving, and settling for 24 hours. The starch is then washed, dried at 50°C for 48 hours, and stored in sealed bags for analysis. This method yields pure white starch with very low chemical contamination. In contrast, starch extraction from other tubers containing mucilage and latex can interfere with sedimentation and compromise starch quality [14].
- **Dry Pathway:** Starch extraction (Yucca) using a dry pathway involves washing, peeling, grating, pre-dehydration, pre-milling, dehydrating, milling, and sifting. Grating releases the starch, pre-dehydration reduces moisture, and pre-milling separates the fibers from the starch. Final dehydration and sifting refine the starch to the desired consistency.

Mechanical Methods

Mechanical methods encompass traditional techniques, such as wet and dry pathways, along with crushing, pressing, and centrifugation to enhance starch yield.

Enzymatic and Chemical Methods

The use of solvents and specific enzymes can enhance starch yield from tubers and rhizomes. When combined with traditional and mechanical methods, these approaches create more favorable conditions for efficient starch extraction [15].

New Methods for Starch Extraction

Ultrasound and Pulsed Electric Field (PEF) technologies have improved extraction efficiency and yields, particularly for compounds such as polyphenols. These green, non-thermal extraction methods significantly reduce extraction time and energy consumption compared to traditional techniques. Applying pulsed electric fields of up to 80 kV/cm for starch extraction or modification qualifies as a non-thermal process. This technique, when combined with PEF technology, enhances the efficiency of starch extraction and modification, offering a sustainable, chemical-free alternative that operates at lower temperatures and requires shorter processing times. These advanced technologies present promising options for starch processing. Ultrasound improves extraction yield, while PEF excels in starch modification, making the process more cost-effective, efficient, and environmentally friendly [16].

Gelatin

Gelatin, a hydrolyzed collagen polymer, is known for its health benefits and is commonly administered in the form of pills, powders, and as a scaffolding material for tissue engineering and 3D cell growth. It is rich in protein and is extracted from the skin, bones, tendons, and ligaments of animals [17].

Structure of Gelatin

It is an amorphous substance composed of soluble, free chains (α, β, γ chains) stabilized by hydrogen bonds. During gelatin production, these hydrogen bonds are broken, unraveling the triple helices of collagen, including both intramolecular and intermolecular bonds, as well as peptidebonds. As a result, collagen fibrils are degraded into gelatin, forming a viscous colloidal solution. Depending on the processing parameters, two types of gelatin exist: Types A and B. Type A gelatin, derived from pig and fish skins, contains fewer crosslinked structures and typically has an isoelectric pH at 8–9. Type B gelatin, obtained from alkali-treated products like cow hides, has an isoelectric pH of 4.8–5.5. Type A gelatin has a smaller peptide size and a broader molecular weight range compared to Type B. The molecular weight fractions of Type B gelatin are quite large, about 100 kDa, and exhibit higher gelling strength. The methods of extraction and hydrolysis influence the molecular spectrum and functional properties of gelatin [18].

Properties of Gelatin

Gelatin contains 18 amino acids, of which glycine, proline, and hydroxyproline together make up 57% (Table 1). The others include aspartic acid, glutamic acid, arginine, and alanine; its elemental composition is 50.5% carbon, 17% nitrogen,

25.2% oxygen, and 6.8% hydrogen. Gelatin also contains hydrophilic random coils. The chemical structure of gelatin consists of polypeptide chains, which include α-chains (90 kDa), β-chains (180 kDa), and γ-chains (300 kDa). When heated, it disintegrates into colloids but becomes gel-like below 35–40°C. Prolonged boiling irreversibly breaks it down. Temperature, pH, electrolyte conditions, and molecular mass distribution influence the viscosity and gel strength of gelatin [19, 20].

Extraction of Gelatin

Collagen is a fibrous protein found in animal skin, bones, cartilage, and tendons, which forms the structure of tissues and organs. Gelatin is obtained from collagen through thermal denaturation, a process that hydrolyzes it under acidic or alkaline conditions. Acid or base treatment converts insoluble collagen into a soluble form, swells its native structure, and then, upon heating, breaks hydrogen and covalent bonds, resulting in a helix-to-coil transition that creates gelatin.

The three major processes of gelatin manufacture are raw material pre-treatment, extraction, purification, and drying. Denaturation, whereby collagen becomes soluble in water, requires enzymatic as well as acidic and alkaline treatment. Controlled heat denaturation then breaks it down into collagen, thus converting it into gelatin. Gelatin from colloids is purified using ion exchangers, which eliminate excess salts, allowing for ultrafiltration and sterilization to remove impurities. The solution, once cooled, forms a gel which, in the final drying stage, is subjected to evaporation and pulled apart into thin threads. The final properties of gelatin depend on whether it is pretreated with acid (Type A) or alkali (Type B). Type A gelatin falls in the pH range of 7.0–9.4, whereas Type B is between 4.8 and 5.5. Both have a wide molecular weight range of 10 to 400 kDa; the larger the molecular weight, the stronger the gel [21, 22].

Pectin

Plant cell walls contain pectin, a polymer mostly composed of galacturonic acid that is attached to cellulose, hemicellulose, and lignin. Its major concentrations are found in the middle lamella and main cell wall, which support vital processes, such as tissue stiffness, cell adhesion, and stress tolerance during cold and drought conditions. Fluid management in quickly expanding plant parts is another way that pectin affects the texture of fruits and vegetables throughout the development and storage stages [23, 24].

One of the reasons pectin is so essential in food processing is that it is widely utilized in industry as an emulsifier, thickener, stabilizer, and gelling agent. Along with its potential health benefits, including improved digestion and other medical effects, it is a rich source of dietary fiber. Pectin is utilized in medicine and biology for wound healing, controlled drug delivery systems, and specifically for oral, ocular, and nasal drug administration. Edible coatings based on pectin are emerging as a new sustainable alternative to packaging materials derived from petroleum [23, 24].

Chemistry of Pectin

Higher plants' cell walls contain three different forms of pectin: rhamnogalacturonan-I (RG-I), rhamnogalacturonan-II (RG-II), and homogalacturonan (HG). The "smooth region," or HG, is a linear homopolymer made up of D-galacturonic acid units connected by α-(1,4)-glycosidic linkages. The linear chain of L-rhamnopyranosyl units and D-galacturonic acid that comprises RG-I is known as the "hairy region." With an average molecular weight ranging from 300,000 to 350,000 g/mol, it contains 12 distinct monosaccharides and roughly 20 linkage types. Its degree of esterification (DE), or the proportion of methoxylation of carboxylic acid units that are present in pectin, has a significant influence on the structure and functionality of pectin. Pectin structure analysis is complicated because the structure is variable, as it is composed of α-(1,4)-linked D-galacturonic acid units [23 - 25].

Properties of Pectin

While di- and trivalent cation ions of pectin are extremely weak or insoluble, monovalent cation salts of pectin are soluble in distilled water. The dry powdered pectin tends to lump when hydrated. Lumping can be avoided by adding a water-soluble carrier or by producing pectin specifically to enhance its dispersibility during production. At higher concentrations, the Newtonian flow of a highly concentrated pectin solution shifts to a pseudoplastic one. Pectin's viscosity, solubility, and gelation are affected by variations in gelling. Pectin's monovalent cation salts are present in solution in high ionization states, which keeps the molecule stretched by coulombic repulsion [26]. Pectin is mostly used as a gelling agent in food. It has been observed that high-molecular-weight pectin gels are affected when acids and sugars are present. However, in contrast to low molecular weight pectin, it has too few acidic groups to gel with calcium ions; yet, precipitation may occasionally result from ions like copper or aluminum. It is reported that hydrogen bonding and hydrophobic interactions drive the aggregation of pectin molecules. Gelation occurs due to the formation of hydrogen bonds between the carboxyl and hydroxyl groups of adjacent pectin molecules. Adding acid to a neutral pectin dispersion causes carboxyl ions to become unionized, reducing pectin's water attraction and negative charges. The rate of gelation is inversely proportional to DE, meaning HM pectins with a higher DE set more rapidly [26].

Extraction of Pectin

Pectin, a polysaccharide found in nearly all plants, functions as a binding agent in cell walls, where protopectin is an insoluble component. Pectin is extracted by hydrolyzing protopectin at high temperatures to break the sugar-cell wall bonds. Traditionally, pectin was extracted by boiling apple pomace and citrus peels in an acidic solution. Now, the extraction of pectin from agro-industrial waste involves methodologies such as deep eutectic solvents, enzyme-assisted extraction, subcritical fluid extraction, ultrasound, microwave-based extraction, and hybrids [27].

Conventional Extraction

Since it is "green" chemistry, citric acid is taken into consideration for pectin extraction. Historically, sulfuric, hydrochloric, and nitric acids have been used as mineral acids. These acids increase industrial costs and pose environmental hazards. Organic acids are more "food-friendly" and suitable for clean labeling. Despite having lower hydrolysis efficiency, organic acids, such as citric, tartaric, malic, and acetic acids, can still produce pectin yields comparable to those of mineral acids. For example, the pectin yields following treatment with tartaric, malic, and citric acids were 6.2%, 5.4%, and 5.3%, respectively. The molecular weight of pectin extracted with citric acid was 4.8×10^{-5} g/mol, which was much greater than that of pectin extracted with hydrochloric acid. However, hydrochloric acid has a higher dissociation constant than citric acid, making it a weaker degrading agent. Consequently, pectin may be hydrolyzed by higher hydrogen ion concentrations, producing lower molecular weight polymers [28, 29].

Ultrasound Extraction of Pectin

Mechanical waves that occur above the human hearing frequency range of 20 Hz to 100 kHz are known as ultrasonic waves. The frequencies of ultrasonic-assisted extraction typically range from 20 to 100 kHz. The frequency of ultrasonic waves influences extraction by modulating mass transfer resistance and the size of the microbubbles. As the frequency increases, cavitation strength in the liquid reduces. In comparison with heat extraction, ultrasonic-assisted extraction enhances yield and decreases the time required for extraction. Powerful ultrasonics break plant cells more easily due to extreme cavitation and bubble collapse, which increases the extractability of pectin [30, 31].

Microwave-assisted Extraction

Shorter processing times, less solvent use, and often higher yields and higher-quality material are among the benefits of microwave-assisted extraction. This procedure uses a microwave field to heat a dielectric substance by ionic conduction and dipole rotation. While the shifting polarity of polar molecules causes dipole rotation, microwave radiation induces the electrophoretic transfer of ions and electrons, creating an electric field that facilitates the movement of particles. Because microwaves induce water dipoles to rotate, this enables appropriate heating. Nonpolar molecules, when polarized by electric fields, absorb less microwave energy than water. When the positive and negative ions travel to their opposing charge sites during extraction, ionic conduction generates heat. Faster temperatures encourage faster diffusion rates, which directly translates into higher extraction efficiency, necessitating this heat generation. The solubility of the target compounds in the solvent and the dielectric characteristics of both the sample and solvent are additional factors that influence extraction rates and the quality of the extracted chemicals. Watts of microwave power is a crucial factor in pectin extraction effectiveness. The efficiency of the extraction process increases with the increase in microwave power [32].

Alginate

Alginate, obtained from *Ascophyllum* and *Laminaria,*is a type of brown algae. Alginate consists of repeating units of monosaccharides that can combine to form polysaccharides, thereby aiding in the formation of cell structure. When mixed with water, it is capable of absorbing 200–300 times its weight of water, making it more useful in pharmaceuticals, thickening, and medical applications. The adjustable thickness, together with its biodegradable nature, makes it ideal for therapeutic use. Two main bacteria, *Pseudomonas aeruginosa* and *Azotobacter vinelandii*, are primarily responsible for producing bacterial alginates. The biosynthesis of alginates by bacteria occurs in four steps: synthesis of a precursor, polymerization and membrane transfer, modification in the periplasm, and export. Alginates produced by the two bacteria exhibit material properties and regulatory differences despite sharing identical biosynthetic genes [33].

Structure of Alginates

The primary source of alginates (a carbohydrate polymer) is brown algae. It contains salts of alginic acid and Mg^{+2}, Na^+, K^+, and Ca^{+2} salts, which collectively make the cell walls of brown algae, forming unbranched binary polymers of α-1-glucuronic (G) and β-d-mannuronic (M) acids bonded *via* 1,4 bonds (Table 1). The alginate composition varies depending on the source, primarily due to the

G:M ratio, which also alters morphology, volume fraction, molecular mass, and the availability of cations.

Properties of Alginates

Alginates occur in different forms and quantities within the cell walls of several species of seaweed. They have molecular weights ranging from 50,000 to 500,000. Seaweeds are flexible and resistant to tides and currents due to the presence of alginates. Since the protonation of guluronic acid creates hydrogen bonding, the viscosity of alginate is pH-responsive. Its viscosity increases with a decrease in pH, peaking at around 3.5, which depends on the pre-gel viscosity or post-gel dispersion strength. A combination of high- and low-molecular-weight alginates helps retain solution viscosity. They can be either white or golden brown, depending on the metals present with which they are associated. Two of the most commonly described forms are sodium alginate and sodium alginic acid. Alginic acid is also an important component of the biofilms produced by *Pseudomonas aeruginosa* [34, 35]. Alginates are among the most commonly utilized biodegradable biopolymers in pharmaceutical and biomedical applications due to their acidic properties. Alginates gel faster when guluronic acid (G) monomers are present, especially in the presence of Ca^{2+} ions [36]. One great application for the alginate gelling property is encapsulating cells or fragments with few drawbacks. Alginate can be chemically modified with carboxylic groups, considering various applications [37].

Extraction of Alginates

The extraction of alginates from brown algae is a multi-step process. Algae are harvested, dried, or processed wet for *M. pyrifera*, and treated with mineral acids to remove counter ions. The insoluble fractions are neutralized with alkalis, such as NaOH or Na_2CO_3, to extract alginic acid. Precipitates obtained through centrifugation or flotation are purified using alcohols or acids to get sodium alginate. Alternatively, Ba or Ca ions form stable gels. Lastly, sodium alginate, when treated with alkaline solutions, gives pure sodium alginate [38].

Chitin

Chitin is the second most common biopolymer after cellulose. In addition to the cell walls of fungi and algae, it also forms the exoskeletons of insects and crustaceans. By providing these creatures with structural stability and protection, this naturally occurring polysaccharide serves a crucial purpose. The amount of chitin found in arthropods alone is sufficient for complete sustainability. Chitin is deacetylated to produce chitosan, a biodegradable, biocompatible derivative with antibacterial properties that finds use in food science, medicine, and agriculture due to its capacity to form gels [39].

Structure of Chitin

Chitin contains N-acetyl-D-glucosamine units, connected by glycosidic linkages, a $\beta(1\rightarrow4)$ bond. The presence of the acetamido group at C-2 makes it resemble cellulose quite a lot. It is an effective structural biopolymer in fungal walls and exoskeletons of crustaceans (Table **1**). There are three crystalline forms of the polymer chains: α-chitin, which has antiparallel and very hard chains; β-chitin, which has parallel chains; and γ-chitin, which has helical chains supported by hydrogen bonds. Chitin, a naturally occurring cationic polymer, undergoes partial deacetylation to produce chitosan. The degree of deacetylation is often greater than 50%. This polymer finds applications in drug delivery, wound healing, and as a biopesticide in agriculture [40, 41].

Properties of Chitin

Chitosan, a derivative of chitin, is a highly versatile biopolymer. Chitin has applications in biomedical fields, including medication administration and wound healing, due to its non-toxic, biocompatible, and strong structural makeup. It is insoluble in water and decomposes naturally. As chitin is partially deacetylated, chitosan has antibacterial qualities and dissolves in acidic solutions. Due to their positively charged nature, chitin and chitosan help improve drug delivery, preserve food, and aid in wound healing by forming flexible films that promote cell growth. These films are utilized in various fields, including environmental science, agriculture, and medicine, as well as in the development of sustainable, innovative materials [42].

Biosynthesis of Chitin

The first step in Chitin production in fungi and insects involves Chitin synthase, a glycosyltransferase that polymerizes N-acetylglucosamine units into long-chain chitin. Under certain cellular circumstances, each of the seven different classes of chitin synthases (CHS) makes a distinctive contribution to the synthesis of chitin. Chitin and cell wall production in fungi are regulated during the cell cycle and occur in regions of polarized growth. It begins by generating the substrate for chitin production, UDP-N-acetylglucosamine. Microfibrils are formed when chitin chains polymerize and the resulting product moves across the plasma membrane through hydrogen bonding. The latter provides structural support by integrating with fungal cell walls or arthropod cuticles [43 - 45].

Extraction of Chitin

Chitin can be extracted from shell waste through chemical, enzymatic, or microbiological processes (Fig. **2**). The chemical process of demineralization and deproteination is arguably the most widely used commercial approach. In demineralization, inorganic residues are removed by using diluted acids, mostly hydrochloric acid (HCl); in deproteination, proteins are dissolved by strong alkaline solutions, primarily sodium hydroxide (NaOH), which preserve the pure chitin. The integrity of the chitin is maintained through enzymatic hydrolysis, which is accomplished by proteolytic enzymes, such as papain or trypsin, resulting in a gentler extraction. Since the microbial extraction method utilizes organic acids produced by bacterial activity, it is a sustainable technology with great potential for chitin extraction [44, 45].

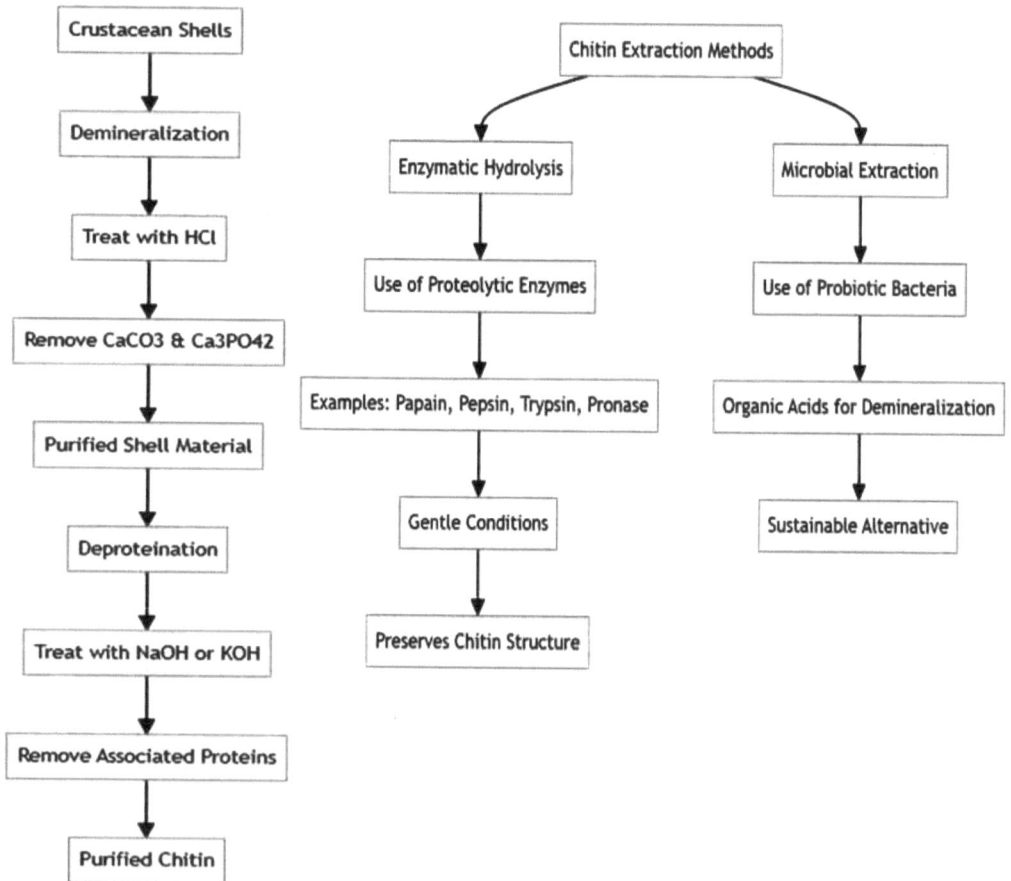

Fig. (2). Different methods of chitin extraction.

Polyhydroxyalkanoates

Polyhydroxyalkanoates (PHAs) are a type of biodegradable polyester that mainly consists of hydroxy fatty acids. In the presence of abundant carbon sources or nutritional imbalance, a wide variety of bacterial species produce and sequester these biopolymers in the cytoplasm as intracellular hydrophobic granules. These granules function as a form of energy and carbon storage, providing bacteria with a buffer they can draw upon in times of low external food supply. Therefore, by stabilizing and optimizing energy supplies, PHAs play a critical role in the bacterial cellular economy [46 - 48].

Chemistry of PHAs

PHAs are a type of biopolyester made up of linear head-to-tail, aliphatic polymers of the monomer 3-hydroxy fatty acids. The pendant groups attached to the (R)-3-hydroxy fatty acid monomers define the PHA structure. These pendant groups can carry unsaturated groups or hydroxyl groups at various locations, or they can be as simple as a methyl (C1) group or as complex as a tridecyl (C13) group. The number of carbon atoms in the monomer chains determines the classification of PHAs into three groups. PHB, PHV, and PHBV are examples of SCL-PHAs, which have a short chain length with three to five carbon atoms; MCL-PHAs, such as PHO and PHN, have six to fourteen carbon atoms; and LCL-PHAs have fifteen or more carbon atoms. This material may exhibit varying functionality due to its different qualities, including flexibility and crystallinity [49 - 51].

Characteristics of PHAs

Depending on the microbial producer, PHAs can vary in molecular mass from 50,000 to 1,000,000 Da. Due to its high molecular weight, bacterial-based poly(3-hydroxybutyrate), or P(3HB), exhibits characteristics similar to those of commercial polymers, such as polypropylene. P(3HB) becomes hard, brittle, stiff, and crystalline when it is removed from the bacterial cells and has little resilience to stress. Processing of PHA becomes difficult due to its high melting point of 170 °C, which is close to the temperature at which it undergoes thermal degradation. Compared to others, the MCL-PHAs are more difficult to process because they are more viscous and sticky. Compared to P(3HB), MCL-PHAs are softer, less crystalline, and have a wide range of applications. Another significant characteristic of PHAs is their ability to be biodegraded by particular strains of microbes, namely, PHA hydrolases and depolymerases [49 - 51].

Biosynthesis of PHAs

Polyhydroxyalkanoates (PHAs) are linear polyesters derived from renewable carbon sources, such as fermentation feedstocks, industrial waste, and agricultural residues. Both gram-positive and gram-negative bacteria naturally produce these biopolymers in response to environmental stressors, including physical or dietary limitations [47 - 49].

In the absence of thiolase and reductase (alternative pathways for PHA production), some bacteria hydrate crotonyl-CoA or hexenoyl-CoA to (R)-3-hydroxy monomers *via* enoyl-CoA hydratase. Using 3-hydroxyvalerate as an intermediate, several bacterial species can biosynthesize P (3HB-3HV) from carbohydrates. Fatty acid metabolism intermediates are hydrated to (R)-3-hydroxyacyl-CoA for mcl-PHAs. When acetyl-CoA is produced, transacylase PhaG redirects the intermediates. Enoyl-CoA hydratase (PhaJ) transforms the product into (R)-3-hydroxyacyl-CoA during β-oxidation. These monomers are then polymerized into PHAs by PHA synthase (PhaC) [51].

Extraction of PHAs

Industrial waste provides a raw material for the production of Polyhydroxyalkanoates (PHAs). Bacterial cells are first concentrated by flocculation, filtration, or centrifugation, and then they are isolated.

Solvent Extraction: In this method, the biomass (previously concentrated) is freeze-dried or spray-dried, and then crushed. It is then dissolved in dichloromethane, chloroform, or other suitable solvents, and PHAs are subsequently precipitated using non-solvents, such as cold ethanol. The resulting solution is filtered to remove impurities. Critical steps include heating the mixture, removing residues, and recovering PHAs through evaporation or precipitation. High-boiling solvents condense, and solvent recovery depends on the boiling point. Pervaporation can be used to recover low-boiling solvents.

Digestion Method: Digestion methods can be used to extract polyhydroxyalkanoates (PHAs) from cell biomass without causing environmental harm. This may involve enzymatic digestion, which utilizes proteolytic enzymes such as lysozyme, or chemical digestion, which employs oxidants and surfactants to break down non-PHA cellular material (NPCM). The process usually entails boiling the biomass slurry and adding chemicals. Afterward, the mixture is separated into liquid and solid PHA phases, which are then dried and purified. Wastewater containing additives and solubilized biomass is treated in the liquid phase. Other extraction techniques include mechanical extraction and supercritical fluid extraction. The most effective method is supercritical fluid extraction, which often utilizes carbon dioxide due to its ability to effectively break down bacterial cells and its environmentally friendly benefits. Bead milling and homogenization are typically employed in mechanical extraction to generate solvent-free forces that facilitate PHA recovery [52, 53].

Polylactic Acid

Polylactic acid (PLA) is a basic hydrocarboxylic acid used in dairy and fermentation for flavoring and food preservation. It is essential to the textile, cosmetic, and pharmaceutical sectors. It is a sustainable substitute for conventional plastics, solving the issue of plastic waste since it is ecologically benign and possesses the required qualities, such as mechanical and thermal plasticity [54 - 57].

Chemistry of PLA

Lactic acid is the only monomer found in polylactic acid or PLA. It is the most common carboxylic acid found in nature and can be produced through chemical synthesis or fermentation. Since it has a hydroxyl group next to its carboxyl, it is known as an alpha-hydroxy acid, or AHA. Lactic acid exists in two optical isomers: the physiologically significant L-(+)-lactic acid, also called (S)-lactic acid, and the D-(−)-lactic acid, sometimes called (R)-lactic acid, which is produced by racemization or certain microbes. *Bacillus acidilacti, Lactobacillus delbrueckii,* and *Lactobacillus bulgaricus* are the common organisms used for fermentation; the choice of organism depends on the carbohydrate substrate [54, 56, 58].

Three distinct types of PLA are produced by polymerization of the two enantiomers, L-, D-, and meso-forms of lactic acid. These types differ from one another in their chemical and physical characteristics and, therefore, in their range of applications [59].

Characteristics of PLA

PLA has a melting point between 159 °C and 178°C and a glass transition temperature between 50 °C and 59 °C. It is a potent substitute for biopolymers like PHAs, PEG, and PCL as it is recyclable, environmentally friendly, and processed with minimal energy use. It has some drawbacks, including hydrolysis and humidity sensitivity, brittleness with less than 10% elongation at break, and gradual deterioration due to its high hydrophobicity. Nevertheless, it possesses several intriguing qualities, including a high tensile modulus, appealing transparency, low-temperature thermal stability, and resistance to oil and grease [59].

Biosynthesis and Extraction of PLA

The fermentation process is the most commonly used technique (Fig. **3**). First, fermentable sugars, such as starch, glucose, lactose, and maltose, are extracted from renewable resources like sugarcane and maize starch. Certain bacteria, such as *Lactobacillus delbrueckii*, *Lactobacillus bulgaricus*, and *Bacillus acidilacti*, are used to ferment these sugars for three to five days at a regulated temperature of approximately 40°C and a pH of 5.0. The presence of a large amount of lactic acid makes the fermentation process hazardous. Following fermentation, calcium hydroxide or calcium carbonate is typically added to purify crude lactic acid, producing calcium lactate, which is then crystallized and acidified to produce lactic acid. There are two ways to polymerize lactic acid: (1) Ring-opening Polymerization (ROP), which converts lactic acid into lactide to create PLA with higher molecular weights, or (2) Condensation Polymerization, which directly polymerizes lactic acid monomers to create typically lower molecular weight PLA [58 - 60].

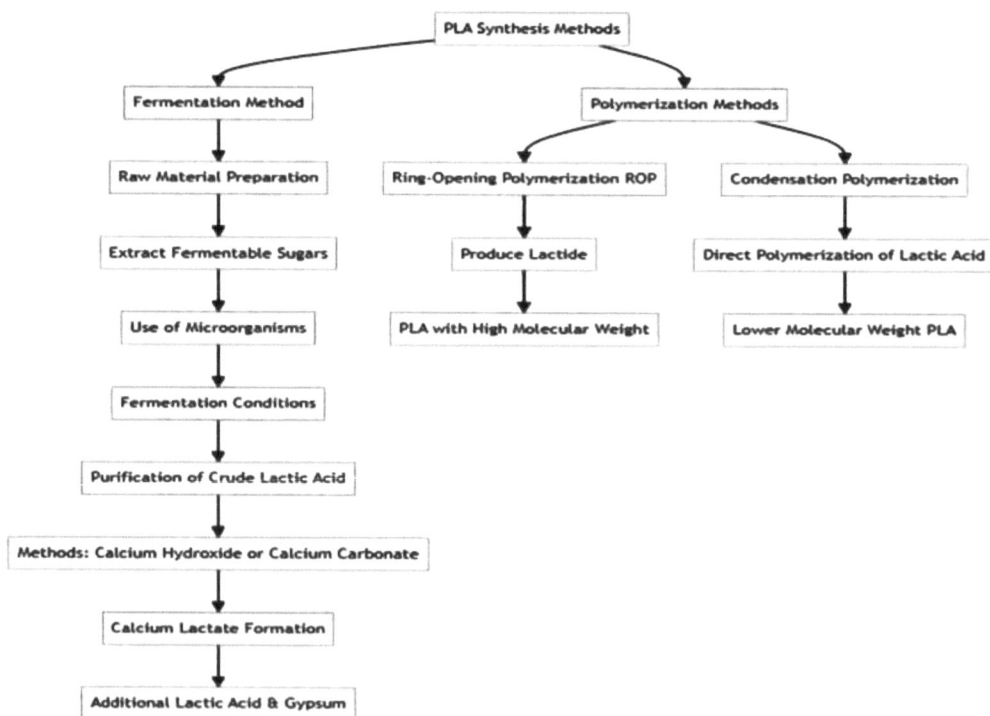

Fig. (3). Methods of PLA synthesis.

Although the chemical production method of lactic acid is less frequently used due to sustainability concerns, it involves three main steps that rely on petrochemical sources (Fig. **4**). In the lactonitrile method, hydrolysis of lactonitrile with sulfuric acid yields crude lactic acid, which is then esterified with ethanol and purified by distillation. This process also uses acrylonitrile as a feedstock (in the acrylonitrile method). The steps involved in refining include sulfuric acid hydrolysis, reaction with methanol, and distillation. In the propionic acid method, crude lactic acid is hydrolyzed with propionic acid, and the final product is produced through a series of steps, including hydrolysis, chlorination of the crude lactic acid, esterification, and rectification.

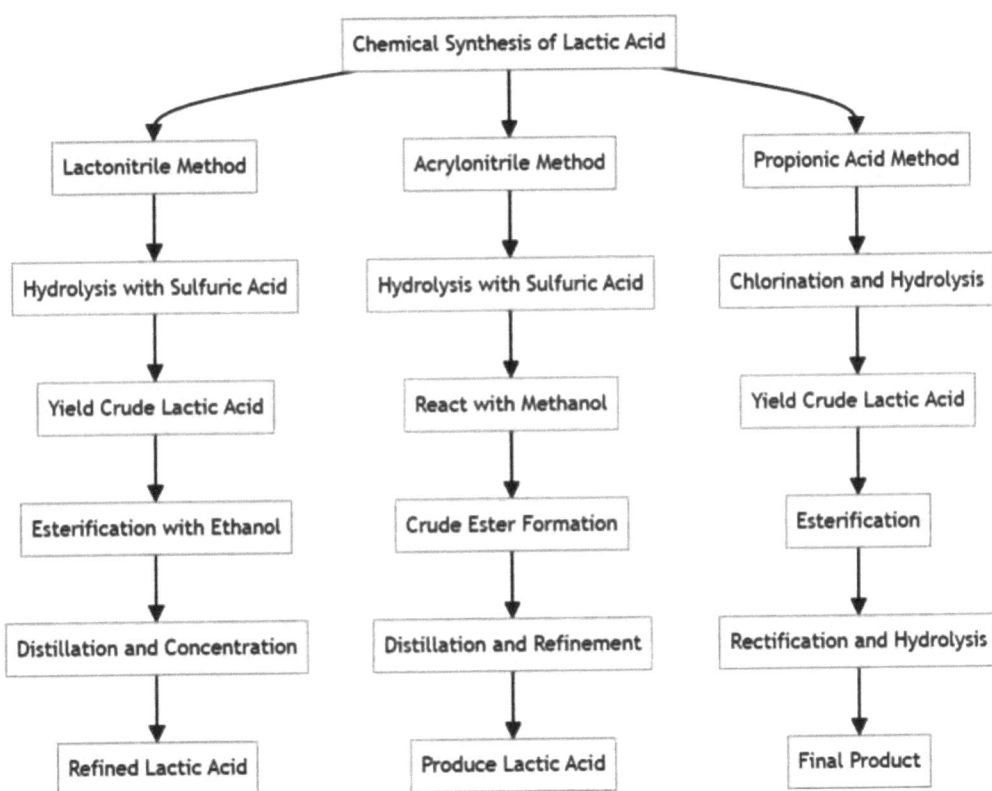

Fig. (4). Chemical synthesis of lactic acid.

Instead of using conventional techniques, PLA extraction essentially entails purifying the polymer during its manufacturing. The liquid is chilled and then pelleted or granulated following fermentation or chemical synthesis. Following a solvent wash to eliminate any unreacted monomers and byproducts, the solidified PLA is filtered and dried to eliminate any remaining moisture or solvent residue.

For storage, the refined PLA is further crushed and formed into pellets. The final PLA product has been subjected to quality control procedures, including GPC and NMR spectroscopy, to determine and ascertain its molecular weight, purity, and other characteristics for various applications, such as medical devices and biodegradable packaging [58].

Poly(Esteramide)/Pea

Polyesters are ester-bonded polymers with widespread utility as plastics and biomaterials due to their strength, hydrolyzability, and biocompatibility. PEAs are analogous to properties found in polyesters but possess the strength and thermal properties of PAs, making them valuable, biodegradable polymers. Within the last decade, PEAs have come into the center of attention of numerous scientists as a material for medical applications, like hydrogels, gene carriers, drug delivery, innovative materials, adhesives, and scaffolds for tissue engineering (TE). Synthetic polymers are primarily used in tissue replacement, restoration, and regeneration, whereas TE scaffolds use metals, ceramics, and glasses. A myriad of medical applications, such as drug delivery, implants, and TE, involve the use of many synthetic polymers [61 - 64].

Chemistry of PEA

The production of poly(ester amide)s involves the use of dicarboxylic acids, α-amino acids, and dianhydrohexitols. The established class of biodegradable polymers features amide and ester groups in the polymer chain, offering both good mechanical and thermal properties suitable for both specialty and commodity applications [62 - 64].

Properties of PEA

According to the reports, dynamic DSC measurements revealed that the addition of low aliphatic substituents increased the Tg. Methyl and ethyl groups affected secondary bonding between polymer chains, specifically by breaking hydrogen bonds and influencing the glass transition temperature (Tg). The Tg of the Me-Me-substituted polymer was decreased because its molecular weight was lower and its backbone was more symmetrical. Except for the crystalline PEA, the aliphatic poly(ester amide)s are elastomers with Tgs below room temperature.

Synthesis of PEA

PEAs can be prepared by polycondensation of linear monomers or by ring-opening polymerization (ROP) of cyclic monomers. ROP can also produce various PEAs from lactams and lactones. Regular, segmented, or random PEAs can be synthesized using different polycondensation techniques, including melt, solution, and interfacial polymerization [65, 66]. Segmented PEAs are best prepared by reacting diesters with diamide diol units, whereas conventional PEAs are prepared by thermal polycondensation of diols and diamide-diesters. Random PEAs can be obtained through the polycondensation of diols, dicarboxylic acids, and amino acids. Other methods include reacting chloroacetate with amino acids or polycondensing diamide-diols with dicarboxylic acids. PEAs can also be synthesized *via* solution polymerization by reacting ester or diester diamine salts with activated dicarboxylic acids [67, 68].

Polydepsipeptides (PDPs) are afforded by the reaction of nitrophenyl esters from α-hydroxy acids with the p-toluene sulfonic acid salt of bis-α-(l-amino acid)-α,ω-alkylene diesters (BAAD). This process is carried out in the following steps: an α-hydroxy acid is reacted with diacyl chloride in the presence of pyridine and then thionyl chloride to yield a dicarboxylic acid [69].

Polyglycolic Acid (PGA)

PGA is a highly biocompatible and biodegradable material that is both strong and flexible, with the potential to significantly transform medical science. Over half a century ago, more than 50 million procedures involved PGA in orthopedic, gynecological, cardiovascular, and general surgeries [70]. PGA is a crystalline polymer with a glass transition temperature (Tg) of 35–40°C and a melting point between 220–225°C. The polymer exhibits strong mechanical properties and undergoes rapid degradation. Due to its high degree of crystallinity, PGA is poorly soluble in highly fluorinated solvents, such as hexafluoroisopropanol (HFIP), for molecular weights up to 45,000 g/mol [71]. Industrial synthesis of PGA is quite challenging because the material is prone to degradation and is sensitive by nature, making it difficult to obtain high molecular weights. The most common monomers used in PGA synthesis are glycolide and glycolic acid. Ring-opening polymerization of glycolide can yield high-molecular-weight PGA, but this process is expensive and limited in scale. An alternative, less expensive method employs the dehydrated condensation of glycolic acid, but it suffers from difficulties in chain extension and achieving a high molecular weight PGA [72, 73]. Poly(glycolic acid) (PGA) also possesses good gas barrier properties, high tensile strength, resistance to organic solvents, a high heat distortion temperature, and rapid biodegradability. Blending PGA with other bioplastics could help reduce costs and enhance material properties while retaining environmentally friendly characteristics [74].

Structure of PGA

The melting point of PGA is 224–227°C, and the glass transition temperature is 35–40°C. The degree of crystallinity ranges from 0% to 52%. X-ray diffraction studies revealed that the crystal structure is orthorhombic with tightly packed macromolecular chains forming a sheet-like structure. The high crystalline density is responsible for a value of 1.69 g/cm^3, along with closely packed ester groups that result in a high melting point and stability for PGA [75, 76].

Properties of PGA

PGA is a strong, stiff polymer with high thermal stability, having a melting point of 220–230°C and a crystallinity degree of 45–55%. It is 100% compostable and biodegradable and has a biodegradation profile similar to that of cellulose. It is predominantly used as a copolymer of PLGA. It is insoluble in most organic solvents due to its high molecular weight, but it can be dissolved in highly fluorinated solvents for processing. PGA fibers are strong, stiff, and have a high modulus of 7 GPa [77].

Synthesis of PGA

Glycolate is mixed with various catalysts in a flask and kept in an argon environment. The water produced during the condensation reaction should be removed using a Dean-Stark trap. For azeotropic condensation polymerization, toluene and anisole can be employed as solvents. The corresponding polymerization temperatures for each solvent should be maintained. During the reaction, the mixture should be magnetically agitated and heated to the boiling point of the solvent. The resultant solid is precipitated and filtered once the reaction is complete. After the ethyl acetate is removed, the resultant polymer is filtered and vacuum-dried.

APPLICATIONS OF BIOPOLYMERS

A comparative analysis of various biopolymers used in drug delivery applications aids in selecting the most suitable material based on factors, such as drug release profile, biocompatibility, and biodegradation rate. Table **2** summarizes the key properties of different biopolymers used in drug delivery applications.

Tissue Engineering Applications

The biocompatible and biodegradable scaffolds known as polyhydroxyalkanoates (PHAs) promote cell adhesion, proliferation, and differentiation while reducing immune responses. They are effective for heart valves, neural tissues, and vascular grafts, particularly when combined with growth factors or stem cells to promote tissue regeneration [78 - 80]. Zein is also an auspicious scaffold material because of its antibacterial properties, structural flexibility, and biodegradability. Electrospun zein/poly(ε-caprolactone) scaffolds have been reported to exhibit better hydrophilicity and customizable degradation rates, thereby enhancing cellular compatibility, and have been employed for liver cell culture and bone defect repair [81, 82]. Sodium alginate is a naturally occurring polymer that is used in tissue engineering and wound healing. Alginate hydrogels speed up and accelerate healing, lower infection, and retain moisture. To treat chronic wounds, dressings containing ZnO-alginate increase the mechanical strength and antibacterial activity [83 - 85]. To create 3D scaffolds and nanofibers, gelatin is typically combined with polymers or ceramics. Its mechanical and electrical qualities make it appropriate for tissue engineering of the heart and nerves [86].

Table 2. Comparative analysis and applications of biopolymers.

Biopolymer	Drug Release Profile	Biocompatibility & Biodegradation Rate	Advantages	Disadvantages
Alginate	Sustained release, pH-responsive	High & Moderate	Excellent mucoadhesive properties and is widely used in encapsulation.	Sensitive to pH variations and can form brittle gels.
Chitosan	Controlled release, enzyme-responsive	High & Moderate to high	It has antimicrobial properties and enhances permeation for oral drugs.	Limited solubility at neutral pH and requires chemical modifications.
Gelatin	Fast release, enzymatic degradation	High & High	Biocompatible, used in injectable formulations.	Low mechanical strength and rapid degradation.
Polylactic Acid (PLA)	Sustained release	High & Slow	Good mechanical strength and FDA-approved for implants.	Hydrophobic, slow degradation may delay drug release.
Polyhydroxyalkanoates (PHAs)	Variable (depends on composition)	High & Moderate to slow	Versatile degradation rates and tunable mechanical properties.	Expensive production and limited commercial availability.

Biopolymer	Drug Release Profile	Biocompatibility & Biodegradation Rate	Advantages	Disadvantages
Pectin	Controlled release, gel-forming	High & Moderate	Non-toxic, bioadhesive, suitable for colon-targeted delivery.	Sensitive to ionic strength and pH.
Polyglycolic Acid (PGA)	Rapid degradation, burst release	High & High	Biodegrades quickly, ideal for short-term implants.	Poor mechanical strength and requires blending with other polymers.

Applications in Bone and Cartilage Repair

The osteogenic and angiogenic qualities of polylactic acid (PLA) promote bone repair. When hydroxyapatite (HA) integrates with PLA scaffolds, osteogenesis is further enhanced, and cell functions are optimized, thereby enabling bone repair [87, 88]. A fresh perspective on cartilage regeneration applications includes 3D PHA nanofiber scaffolds [79]. Alginate-based hydrogels facilitate chondrogenic differentiation, promoting the effective distribution of cells and therapeutic substances for the treatment of osteoarthritis and cartilage regeneration [83, 85]. The most effective PEA concentration is 15%, while HA-MA and Arg-UPEA hybrid hydrogels encourage calcium depositions and osteogenesis in bone regeneration.

Additionally, gelatin methacryloyl hydrogels, particularly when combined with bioactive nanomaterials, can effectively mimic extracellular matrices to support osteogenic differentiation and bone regeneration. The most popular type of scaffolds, fibers, and microspheres for bone regeneration utilize PLGA, an FDA-approved polymer that breaks down gradually. Together, these biomaterials support cutting-edge tissue engineering techniques that improve bone and cartilage repair.

Application in Drug Delivery

PHAs' biodegradability and biocompatibility make them suitable materials for drug delivery systems. Because polyhydroxybutyrate (PHB) has a low melting point, it can be fabricated as nanoparticles, microspheres, and films for controlled release in transdermal delivery. Polylactic acid (PLA) is another useful polymer in drug administration due to its adjustable degradation rate, which enables prolonged drug release and improved patient safety, particularly in cancer and chronic illnesses [87, 88]. Due to its stability and harmless degradation products, poly(lactic-co-glycolic acid) (PLGA) is widely used in pulmonary drug delivery systems and can be employed to treat respiratory diseases, such as lung cancer and asthma. Chitosan is gaining increasing popularity for gene transfer and anticancer treatments. Zein utilizes its mucoadhesive properties in pharmaceutical films and microspheres to target medications and control their release, which is beneficial for sparingly soluble drugs [82]. Another popular polymer for drug delivery is alginate, which can be conjugated with RGD peptides to improve cell adhesion and proliferation, and used to create crosslinked gels for controlled release. Newer hydrogels, such as gelatin-PANI, are utilized to treat Parkinson's disease by specifically delivering BMSCs, while more recent excipients, like PADAS, a polyesteramide matrix, delay drug release. Biodegradable polymers appear as catalysts for transforming drug delivery methods used in the medical field.

Application in Packaging

Chitosan possesses potent antibacterial properties that prolong shelf life and enhance the barrier qualities of food packaging films, making it a valuable natural preservative in the food industry. In addition to being biodegradable, which helps create sustainable packaging options, it can also be used as a food additive to enhance dietary fiber and as a flavoring and coloring agent [42, 43]. Polylactic acid (PLA), a bioplastic, is a sustainable material for food packaging. PLA absorbs carbon dioxide during production and is recyclable and compostable. Although it is more expensive, the FDA approves its use for food contact packaging [87, 88]. Zein has good gas permeability and works well with natural antioxidants and antimicrobials, making it an excellent material for food packaging. It can also enhance the quality of gluten-free products such as bread and noodles when used as a coating or additive. Due to its physical characteristics, gelatin is increasingly used as a biodegradable substitute for food packaging. The mechanical properties of gelatin-based films are enhanced by the addition of food components, expanding their uses in food packaging.

Application in Medical Devices

Due to their exceptional mechanical properties, biocompatibility, and biodegradability, polyhydroxyalkanoates (PHAs) are commonly used in medical devices. Some applications of PHAs include adhesion barriers, cartilage regeneration, cardiovascular patches, and surgical implants such as sutures and fixation devices. These materials are particularly useful for biodegradable implants, which can reduce the duration of follow-up procedures and lower the risk of infection. PHAs are also used in novel wound dressings designed to treat chronic wounds, as they are believed to promote healing, enhance cell migration, and deliver growth factors or antibacterial agents for faster recovery [78 - 80]. In biomedical implants, bacterial cellulose (BC) plays a crucial role, especially in the development of artificial blood vessels. Compared to conventional materials, BC shows reduced occlusion and thrombosis in vascular grafts. Due to their anticoagulant properties, heparin-hybridized BC scaffolds are well-suited for valve replacement and vascular tissue engineering.

Application in Nanoparticles

In drug delivery, zein-based nanoparticles (NPs), which are non-toxic, biocompatible, and biodegradable, increase oral bioavailability and stability, regulate drug release, and improve targeting. Zein NPs have been effectively used to target the liver with 5-fluorouracil, achieving a targeting efficiency of more than 31% [89, 90]. Starch nanoparticles (SNPs) have an increased surface area for bacterial adhesion and thus exhibit enhanced bactericidal potential. SNPs improve curcumin's effectiveness against *S. aureus* and *E. coli* bacteria. They are also employed as delivery systems for bioactive molecules in food and medicinal products. The FDA has approved PLGA biodegradable nanoparticles for drug delivery. It has been demonstrated that this controlled-release system targets tumors more effectively and produces fewer adverse effects [91].

Application in Disease Treatment

PLA-based drug delivery systems are utilized as effective strategies for cancer treatment due to their adjustable degradation rates, which result in prolonged drug release. Electrospun poly-L-lactic acid scaffolds containing doxorubicin, bicarbonate, and ibuprofen have shown improved treatment outcomes in liver malignancies by reducing inflammation and enabling controlled medication release. PLA's excellent use in bone regeneration and its mechanical properties in dental resins are attributed to its biocompatibility and flexibility. Dental restorations may last longer with nano-PLA/Al_2O_3 scaffolds, which are stronger than traditional resins.

Pectin has demonstrated great potential in treating chronic illnesses such as diabetes, cancer, high blood pressure, Alzheimer's disease, and liver disorders. It induces apoptosis and inhibits the proliferation of malignant cells; modified pectins, especially Pec-MA, are more effective than native pectin. Citrus pectin improves insulin sensitivity, lowers fasting blood glucose, and enhances glucose tolerance, possibly by influencing the PI3K/Akt signaling pathway. Some studies also indicate that pectin lowers blood pressure, particularly in hypertensive rats, where pear pectin appears more effective than apple pectin. Amyloid-beta interacts with pectin [92], which could be used to treat Alzheimer's disease by reducing plaque formation. Pectin also supports gut microbial flora, helping to lessen alcoholic liver damage and hepatic steatosis in liver disorders.

Gelatin methacryloyl (GelMA) hydrogels show promise for treating musculoskeletal disorders and support osteogenic differentiation and cell monitoring in bone research. After esophageal cancer resection, polyglycolic acid (PGA) sheets combined with fibrin glue have been effectively used to prevent anastomotic leaks. Notably, nearly 90% of fistulas have been successfully closed using the pre-soak technique, compared to 25% closure with the traditional method.

Application in Tissue Regeneration

PLA is an essential component in tissue regenerative medicine, particularly in wound healing. It can encapsulate bioactive medications into scaffolds that promote healing. To enhance cell adhesion and proliferation, Cheng *et al.* co-delivered ginsenoside-Rg3 and bFGF, benefiting from their synergistic effect. In other words, PLA is utilized in composite scaffolds for potential regenerative medicine applications [92].

In agriculture, chitin and chitosan serve as biopesticides, enhancing soil health, reducing pesticide use, and promoting plant growth and disease resistance. They also contribute to sustainable agriculture and aid in removing contaminants from wastewater. Cellulose, a plant-based biopolymer, is used as a filler in medicinal formulations and in biofuel production, particularly for renewable energy generation. Another biopolymer, zein, is increasingly employed in tissue engineering, drug delivery, and the food industry. It improves the solubility and bioavailability of medications and is renewable and biodegradable. Lastly, the biodegradable polymer starch provides stability and resistance to degradation in the digestive tract, making it a viable delivery system for nutraceuticals [93].

CHALLENGES AND ENVIRONMENTAL IMPACTS OF BIOPOLYMER EXTRACTION

Existing biopolymer extraction techniques face challenges related to sustainability, efficiency, and environmental impact. Although chemical extraction is effective, it consumes significant energy and generates hazardous waste; therefore, enzymatic hydrolysis and green solvents are needed. Utilizing ionic liquids or supercritical fluid extraction can help reduce air and water pollution risks associated with traditional solvent-based methods [94]. While mechanical methods are generally safer for the environment, they often yield lower extraction efficiency and require optimization to reduce energy consumption. Enzymatic procedures, although slow and costly, are environmentally friendly, highlighting the need for genetically engineered enzymes. Supercritical fluid extraction decreases solvent waste but demands high energy input, which hybrid methods and renewable CO_2 sources can help mitigate [95]. By addressing these challenges with sustainable technologies, biopolymer production can be enhanced while minimizing environmental hazards.

FUTURE RESEARCH DIRECTIONS IN BIOPOLYMER SCIENCE

Future research in biopolymer science should focus on enhancing synthesis, extraction, and functional properties to improve efficiency and sustainability. Genetic engineering can tailor biopolymers for improved mechanical strength and biodegradability, while bio-based nanocomposites enhance their thermal and barrier properties. Microbial and algal biotechnology offer scalable and eco-friendly production methods. Novel extraction technologies, including nanotechnology-based and enzyme-assisted techniques, can increase yield while minimizing environmental impact [96]. Artificial intelligence can optimize processing conditions for greater efficiency. Molecular modifications and hybrid biopolymer composites can introduce new functionalities for medical and packaging applications, while advancements in 3D printing enable precise fabrication of biopolymer-based materials for biomedical and industrial uses. These innovations will drive the future of sustainable and high-performance biomaterials [97].

CONCLUSION

Biopolymers are polymers derived from biological sources, such as plants, animals, or minerals, or synthesized in laboratories using these raw materials. They have recently garnered significant research interest because synthetic polymers, which have been widely used for a long time, pose risks to both human health and the environment. This chapter describes the chemistry, properties, and extraction methods of various biopolymers, including starch, gelatin, alginates, polyhydroxyalkanoates, polylactic acid, and polyesteramides. These biopolymers are known for their versatile applications in drug delivery, formulation and development, tissue engineering, regenerative medicine, and the food industry. Moreover, they are safe, eco-friendly, non-toxic, cost-effective, and contribute to environmental sustainability.

LIST OF ABBREVIATIONS

3D	Three-dimensional
AHA	Alpha hydroxy acid
Ba	Barium
BAAD	bis-α-(L-amino acid)-α,ω-alkylene diesters

BP	Biopolymers
Ca	Calcium
CHS	Chitin synthases
Da	Dalton
DE	Degree of esterification
DNA	Deoxyribonucleic acid
DP	Degree of polymerization
DSC	Differential Scanning Calorimetry
HG	Homogalacturonan
HM	High molecular weight
Hz	Hertz
K	Potassium
kDa	Kilodalton
MCL-PHAs	Medium Chain Length-Polyhydroxyalkanoates
Mg	Magnesium
Na	Sodium
Na$_2$CO$_3$	Sodium carbonate
NaOH	Sodium hydroxide
NCPM	Non-PHA cellular material
PA	Polyamide
PC	Polycarbonate
PCL	Polycaprolactone
PDS	Polydioxanone
PE	Polyethylene
PEA	Polyesteramide
PEF	Pulsed electric field
PGA	Polyglycolic acid
PHAs	Polyhydroxyalkanoates
PHB	Polyhydroxybutyrate
PHBV	Poly-(3-hydroxybutyrate-co-hydroxyvalerate)
PHV	Poly-(3-hydroxyvalerate)
PLA	Polylactic acid
PMMA	Polymethyl methacrylate
PDP	Polydepsipeptide
PP	Polypropylene

PPF	Polypropylene fumarate
PU	Polyurethane
PVC	Polyvinyl chloride
RG	Rhamnogalacturonan
ROP	Ring-Opening Polymerization
RNA	Ribonucleic acid
SCL-PHAs	Short Chain Length-Polyhydroxyalkanoates
TE	Tissue engineering

REFERENCES

[1]   Azadi E, Dinari M, Derakhshani M, Reid KR, Karimi B. Sources and extraction of biopolymers and manufacturing of bio-based nanocomposites for different applications. Molecules 2024; 29(18): 4406.
[http://dx.doi.org/10.3390/molecules29184406] [PMID: 39339400]

[2]   Baranwal J, Barse B, Fais A, Delogu GL, Kumar A. Biopolymer: A sustainable material for food and medical applications. Polymers (Basel) 2022; 14(5): 983.
[http://dx.doi.org/10.3390/polym14050983] [PMID: 35267803]

[3]   Das A, Ringu T, Ghosh S, Pramanik N. A comprehensive review on recent advances in preparation, physicochemical characterization, and bioengineering applications of biopolymers. Polym Bull 2023; 80(7): 7247-312.
[http://dx.doi.org/10.1007/s00289-022-04443-4] [PMID: 36043186]

[4]   Attanayake NAB, Chandrasiri MTMS, Asela AU, Pitawala HMJC, Senevirathna MASR. Biopolymers: Structure, properties, extraction methods and applications Sri Lankan. J Appl Sci 2022; 1(1): 18-30.

[5]   Kaur P, Kaur K, Basha SJ, Kennedy JF. Current trends in the preparation, characterization and applications of oat starch — A review. Int J Biol Macromol 2022; 212: 172-81.
[http://dx.doi.org/10.1016/j.ijbiomac.2022.05.117] [PMID: 35598726]

[6]   Salimi M, Channab B, El Idrissi A, Zahouily M, Motamedi E. A comprehensive review on starch: Structure, modification, and applications in slow/controlled-release fertilizers in agriculture. Carbohydr Polym 2023; 322: 121326.
[http://dx.doi.org/10.1016/j.carbpol.2023.121326] [PMID: 37839830]

[7]   Hassan NA, Darwesh OM, Smuda SS, *et al.* Recent trends in the preparation of nano-starch particles. Molecules 2022; 27(17): 5497.
[http://dx.doi.org/10.3390/molecules27175497] [PMID: 36080267]

[8]   Palanisamy CP, Cui B, Zhang H, Jayaraman S, Kodiveri Muthukaliannan G. A comprehensive review on corn starch-based nanomaterials: properties, simulations, and applications. Polymers (Basel) 2020; 12(9): 2161.
[http://dx.doi.org/10.3390/polym12092161] [PMID: 32971849]

[9]   Obadi M, Qi Y, Xu B. High-amylose maize starch: Structure, properties, modifications and industrial applications. Carbohydr Polym 2023; 299: 120185.
[http://dx.doi.org/10.1016/j.carbpol.2022.120185] [PMID: 36876800]

[10]   Li C, Dhital S, Gilbert RG, Gidley MJ. High-amylose wheat starch: Structural basis for water absorption and pasting properties. Carbohydr Polym 2020; 245: 116557.
[http://dx.doi.org/10.1016/j.carbpol.2020.116557] [PMID: 32718645]

[11]   Bashir K, Aggarwal M. Physicochemical, structural and functional properties of native and irradiated starch: a review. J Food Sci Technol 2019; 56(2): 513-23.
[http://dx.doi.org/10.1007/s13197-018-3530-2] [PMID: 30906009]

[12]   Wang L, Tong L. Production and properties of starch: current research. Molecules 2024; 29(3): 646.
        [http://dx.doi.org/10.3390/molecules29030646] [PMID: 38338392]

[13]   Dorantes-Fuertes MG, López-Méndez MC, Martínez-Castellanos G, Meléndez-Armenta RÁ, Jiménez-
        Martínez HE. Starch extraction methods in tubers and roots: a systematic review. Agronomy (Basel)
        2024; 14(4): 865.
        [http://dx.doi.org/10.3390/agronomy14040865]

[14]   Sit N, Deka SC, Misra S. Optimization of starch isolation from taro using combination of enzymes and
        comparison of properties of starches isolated by enzymatic and conventional methods. J Food Sci
        Technol 2015; 52(7): 4324-32.
        [http://dx.doi.org/10.1007/s13197-014-1462-z] [PMID: 26139897]

[15]   Punia Bangar S, Ashogbon AO, Singh A, Chaudhary V, Whiteside WS. Enzymatic modification of
        starch: A green approach for starch applications. Carbohydr Polym 2022; 287: 119265.
        [http://dx.doi.org/10.1016/j.carbpol.2022.119265] [PMID: 35422280]

[16]   Wang J, Lan T, Lei Y, *et al.* Optimization of ultrasonic-assisted enzymatic extraction of kiwi starch
        and evaluation of its structural, physicochemical, and functional characteristics. Ultrason Sonochem
        2021; 81: 105866.
        [http://dx.doi.org/10.1016/j.ultsonch.2021.105866] [PMID: 34896805]

[17]   Alipal J, Mohd Pu'ad NAS, Lee TC, *et al.* A review of gelatin: Properties, sources, process,
        applications, and commercialisation. Mater Today Proc 2021; 42: 240-50.
        [http://dx.doi.org/10.1016/j.matpr.2020.12.922]

[18]   Lukin I, Erezuma I, Maeso L, *et al.* Progress in gelatin as biomaterial for tissue engineering.
        Pharmaceutics 2022; 14(6): 1177.
        [http://dx.doi.org/10.3390/pharmaceutics14061177] [PMID: 35745750]

[19]   Samatra MY, Noor NQIM, Razali UHM, Bakar J, Shaarani SM. Bovidae-based gelatin: Extractions
        method, physicochemical and functional properties, applications, and future trends. Compr Rev Food
        Sci Food Saf 2022; 21(4): 3153-76.
        [http://dx.doi.org/10.1111/1541-4337.12967] [PMID: 35638329]

[20]   Li F, Jia D, Yao K. Amino acid composition and functional properties of collagen polypeptide from
        Yak (*Bos grunniens*) bone. Lebensm Wiss Technol 2009; 42(5): 945-9.
        [http://dx.doi.org/10.1016/j.lwt.2008.12.005]

[21]   Sultana S, Ali ME, Ahamad MNU. Gelatine, collagen, and single cell proteins as natural and newly
        emerging food ingredients. In: Ali ME, Nizar NNA, eds, Preparation and Processing of Religious and
        Cultural Foods. Woodhead Publishing 2018; pp. 215-39.
        [http://dx.doi.org/10.1016/B978-0-08-101892-7.00011-0]

[22]   Ahmad T, Ismail A, Ahmad SA, *et al.* Recent advances on the role of process variables affecting
        gelatin yield and characteristics with special reference to enzymatic extraction: A review. Food
        Hydrocoll 2017; 63: 85-96.
        [http://dx.doi.org/10.1016/j.foodhyd.2016.08.007]

[23]   Willats WGT, McCartney L, Mackie W, Knox JP. Pectin: cell biology and prospects for functional
        analysis. Plant Mol Biol 2001; 47(1/2): 9-27.
        [http://dx.doi.org/10.1023/A:1010662911148] [PMID: 11554482]

[24]   Roy S, Priyadarshi R, Łopusiewicz Ł, Biswas D, Chandel V, Rhim JW. Recent progress in pectin
        extraction, characterization, and pectin-based films for active food packaging applications: A review.
        Int J Biol Macromol 2023; 239: 124248.
        [http://dx.doi.org/10.1016/j.ijbiomac.2023.124248] [PMID: 37003387]

[25]   Parre E, Geitmann A. Pectin and the role of the physical properties of the cell wall in pollen tube
        growth of Solanum chacoense. Planta 2005; 220(4): 582-92.
        [http://dx.doi.org/10.1007/s00425-004-1368-5] [PMID: 15449057]

[26]   Harholt J, Suttangkakul A, Vibe Scheller H. Biosynthesis of Pectin. Plant Physiol 2010; 153(2): 384-95.
        [http://dx.doi.org/10.1104/pp.110.156588] [PMID: 20427466]

[27]   Adetunji LR, Adekunle A, Orsat V, Raghavan V. Advances in the pectin production process using novel extraction techniques: A review. Food Hydrocoll 2017; 62: 239-50.
        [http://dx.doi.org/10.1016/j.foodhyd.2016.08.015]

[28]   Jacob EM, Borah A, Jindal A, *et al.* Synthesis and characterization of citrus-derived pectin nanoparticles based on their degree of esterification. J Mater Res 2020; 35(12): 1514-22.
        [http://dx.doi.org/10.1557/jmr.2020.108]

[29]   Cho EH, Jung HT, Lee BH, Kim HS, Rhee JK, Yoo SH. Green process development for apple-peel pectin production by organic acid extraction. Carbohydr Polym 2019; 204: 97-103.
        [http://dx.doi.org/10.1016/j.carbpol.2018.09.086] [PMID: 30366548]

[30]   Rutkowska M, Namieśnik J, Konieczka P. Ultrasound-assisted extraction. In: Morrissey K, Ed. The Application of Green Solvents in Separation Processes. Elsevier 2017; pp. 301-24.
        [http://dx.doi.org/10.1016/B978-0-12-805297-6.00010-3]

[31]   Hosseini SS, Khodaiyan F, Yarmand MS. Optimization of microwave assisted extraction of pectin from sour orange peel and its physicochemical properties. Carbohydr Polym 2016; 140: 59-65.
        [http://dx.doi.org/10.1016/j.carbpol.2015.12.051] [PMID: 26876828]

[32]   Hu W, Zhao Y, Yang Y, *et al.* Microwave-assisted extraction, physicochemical characterization and bioactivity of polysaccharides from *Camptotheca acuminata* fruits. Int J Biol Macromol 2019; 133: 127-36.
        [http://dx.doi.org/10.1016/j.ijbiomac.2019.04.086] [PMID: 30986453]

[33]   Zhang H, Cheng J, Ao Q. Preparation of alginate-based biomaterials and their applications in biomedicine. Mar Drugs 2021; 19(5): 264.
        [http://dx.doi.org/10.3390/md19050264] [PMID: 34068547]

[34]   Liew CV, Chan LW, Ching AL, Heng PWS. Evaluation of sodium alginate as drug release modifier in matrix tablets. Int J Pharm 2006; 309(1-2): 25-37.
        [http://dx.doi.org/10.1016/j.ijpharm.2005.10.040] [PMID: 16364576]

[35]   Giovagnoli S, Luca G, Blasi P, *et al.* Alginates in pharmaceutics and biomedicine: Is the future so bright? Curr Pharm Des 2015; 21(33): 4917-35.
        [http://dx.doi.org/10.2174/1381612821666150820105639] [PMID: 26290204]

[36]   Shoichet MS, Li RH, White ML, Winn SR. Stability of hydrogels used in cell encapsulation: An *in vitro* comparison of alginate and agarose. Biotechnol Bioeng 1996; 50(4): 374-81.
        [http://dx.doi.org/10.1002/(SICI)1097-0290(19960520)50:4<374::AID-BIT4>3.0.CO;2-I]   [PMID: 18626986]

[37]   Kruk K, Winnicka K. Alginates combined with natural polymers as valuable drug delivery platforms. Mar Drugs 2022; 21(1): 11.
        [http://dx.doi.org/10.3390/md21010011] [PMID: 36662184]

[38]   Jha A, Kumar A. Biobased technologies for the efficient extraction of biopolymers from waste biomass. Bioprocess Biosyst Eng 2019; 42(12): 1893-901.
        [http://dx.doi.org/10.1007/s00449-019-02199-2] [PMID: 31542821]

[39]   Chen DD, Wang ZB, Wang LX, Zhao P, Yun CH, Bai L. Structure, catalysis, chitin transport, and selective inhibition of chitin synthase. Nat Commun 2023; 14(1): 4776.
        [http://dx.doi.org/10.1038/s41467-023-40479-4] [PMID: 37553334]

[40]   Pakizeh M, Moradi A, Ghassemi T. Chemical extraction and modification of chitin and chitosan from shrimp shells. Eur Polym J 2021; 159: 110709.
        [http://dx.doi.org/10.1016/j.eurpolymj.2021.110709]

[41] Abdou ES, Nagy KSA, Elsabee MZ. Extraction and characterization of chitin and chitosan from local sources. Bioresour Technol 2008; 99(5): 1359-67.
[http://dx.doi.org/10.1016/j.biortech.2007.01.051] [PMID: 17383869]

[42] Kozma M, Acharya B, Bissessur R. Chitin, chitosan, and nanochitin: Extraction, synthesis, and applications. Polymers (Basel) 2022; 14(19): 3989.
[http://dx.doi.org/10.3390/polym14193989] [PMID: 36235937]

[43] Elieh-Ali-Komi D, Hamblin MR. Chitin and chitosan: Production and application of versatile biomedical nanomaterials. Int J Adv Res (Indore) 2016; 4(3): 411-27.
[PMID: 27819009]

[44] Lenardon MD, Munro CA, Gow NAR. Chitin synthesis and fungal pathogenesis. Curr Opin Microbiol 2010; 13(4): 416-23.
[http://dx.doi.org/10.1016/j.mib.2010.05.002] [PMID: 20561815]

[45] Merzendorfer H. The cellular basis of chitin synthesis in fungi and insects: Common principles and differences. Eur J Cell Biol 2011; 90(9): 759-69.
[http://dx.doi.org/10.1016/j.ejcb.2011.04.014] [PMID: 21700357]

[46] Samrot AV, Samanvitha SK, Shobana N, *et al.* The synthesis, characterization and applications of polyhydroxyalkanoates (PHAs) and PHA-based nanoparticles. Polymers (Basel) 2021; 13(19): 3302.
[http://dx.doi.org/10.3390/polym13193302] [PMID: 34641118]

[47] Muhammadi S, Shabina , Afzal M, Hameed S. Bacterial polyhydroxyalkanoates-eco-friendly next generation plastic: Production, biocompatibility, biodegradation, physical properties and applications. Green Chem Lett Rev 2015; 8(3-4): 56-77.
[http://dx.doi.org/10.1080/17518253.2015.1109715]

[48] Ray S, Kalia VC. Biomedical applications of polyhydroxyalkanoates. Indian J Microbiol 2017; 57(3): 261-9.
[http://dx.doi.org/10.1007/s12088-017-0651-7] [PMID: 28904409]

[49] Madison LL, Huisman GW. Metabolic engineering of poly(3-hydroxyalkanoates): from DNA to plastic. Microbiol Mol Biol Rev 1999; 63(1): 21-53.
[http://dx.doi.org/10.1128/MMBR.63.1.21-53.1999] [PMID: 10066830]

[50] Li Z, Yang J, Loh XJ. Polyhydroxyalkanoates: opening doors for a sustainable future. NPG Asia Mater 2016; 8(4): e265.
[http://dx.doi.org/10.1038/am.2016.48]

[51] Thorat Gadgil BS, Killi N, Rathna GVN. Polyhydroxyalkanoates as biomaterials. MedChemComm 2017; 8(9): 1774-87.
[http://dx.doi.org/10.1039/C7MD00252A] [PMID: 30108887]

[52] Mozejko-Ciesielska J, Moraczewski K, Czaplicki S, Singh V. Production and characterization of polyhydroxyalkanoates by *Halomonas alkaliantarctica* utilizing dairy waste as feedstock. Sci Rep 2023; 13(1): 22289.
[http://dx.doi.org/10.1038/s41598-023-47489-8] [PMID: 38097607]

[53] Pagliano G, Galletti P, Samorì C, Zaghini A, Torri C. Recovery of polyhydroxyalkanoates from single and mixed microbial cultures: A review. Front Bioeng Biotechnol 2021; 9: 624021.
[http://dx.doi.org/10.3389/fbioe.2021.624021] [PMID: 33644018]

[54] Ulery BD, Nair LS, Laurencin CT. Biomedical applications of biodegradable polymers. J Polym Sci, B, Polym Phys 2011; 49(12): 832-64.
[http://dx.doi.org/10.1002/polb.22259] [PMID: 21769165]

[55] Ahmed J, Varshney SK. Polylactides—Chemistry, properties, and green packaging technology: A review. Int J Food Prop 2011; 14(1): 37-58.
[http://dx.doi.org/10.1080/10942910903125284]

[56] Hu Y, Daoud WA, Cheuk KKL, Lin CSK. Newly developed techniques on polycondensation, ring-opening polymerization and polymer modification: focus on poly(lactic acid). Materials. 2016; 9(3): 133.
[http://dx.doi.org/10.3390/ma9030133]

[57] Khouri NG, Bahú JO, Blanco-Llamero C, Severino P, Concha VOC, Souto EB. Polylactic acid (PLA): Properties, synthesis, and biomedical applications – A review of the literature. J Mol Struct 2024; 1309: 138243.
[http://dx.doi.org/10.1016/j.molstruc.2024.138243]

[58] Ramezani Dana H, Ebrahimi F. Synthesis, properties, and applications of polylactic acid-based polymers. Polym Eng Sci 2022; 63, 1, 22-43.

[59] Li G, Zhao M, Xu F, *et al.* Synthesis and biological application of polylactic acid. Molecules 2020; 25(21): 5023.
[http://dx.doi.org/10.3390/molecules25215023] [PMID: 33138232]

[60] Ranakoti L, Gangil B, Mishra SK, *et al.* Critical review on polylactic acid: Properties, structure, processing, biocomposites, and nanocomposites. Materials (Basel) 2022; 15(12): 4312.
[http://dx.doi.org/10.3390/ma15124312] [PMID: 35744371]

[61] Winnacker M, Rieger B. Biobased polyamides: Recent advances in basic and applied research. Macromol Rapid Commun 2016; 37(17): 1391-413.
[http://dx.doi.org/10.1002/marc.201600181] [PMID: 27457825]

[62] Fonseca AC, Gil MH, Simões PN. Biodegradable poly(ester amide)s – A remarkable opportunity for the biomedical area: Review on the synthesis, characterization and applications. Prog Polym Sci 2014; 39(7): 1291-311.
[http://dx.doi.org/10.1016/j.progpolymsci.2013.11.007]

[63] Ghosal K, Latha MS, Thomas S. Poly(ester amides) (PEAs)—scaffold for tissue engineering applications. Eur Polym 2014; 60: 58-68.
[http://dx.doi.org/10.1016/j.eurpolymj.2014.08.006]

[64] Gilmore KA, Lampley MW, Boyer C, Harth E. Matrices for combined delivery of proteins and synthetic molecules. Adv Drug Deliv Rev 2016; 98: 77-85.
[http://dx.doi.org/10.1016/j.addr.2015.11.018] [PMID: 26656604]

[65] Zhou QH, Li M, Yang P, Gu Y. Effect of hydrogen bonds on structures and glass transition temperatures of maleimide–isobutene alternating copolymers: Molecular dynamics simulation study. Macromol Theory Simul 2013; 22(2): 107-14.
[http://dx.doi.org/10.1002/mats.201200057]

[66] Killi N, Pawar AT, Gundloori RVN. Polyesteramide of neem oil and its blends as an active nanomaterial for tissue regeneration. ACS Appl Bio Mater 2019; 2(8): 3341-51.
[http://dx.doi.org/10.1021/acsabm.9b00354] [PMID: 35030776]

[67] Rodriguez-Galan A, Franco L, Puiggali J. Degradable Poly(ester amide)s for Biomedical Applications. Polymers (Basel) 2010; 3(1): 65-99.
[http://dx.doi.org/10.3390/polym3010065]

[68] Katsarava R. Active polycondensation: from pep tide chemistry to amino acid based biodegradable polymers. InMacromolecular Symposia 2003; 199(1): 419-430.
[http://dx.doi.org/10.1002/masy.200350935]

[69] Katsarava R, Beridze V, Arabuli N, Kharadze D, Chu CC, Won CY. Amino acid-based bioanalogous polymers. Synthesis, and study of regular poly(ester amide)s based on bis(?-amino acid)?? -alkylene diesters, and aliphatic dicarboxylic acids. J Polym Sci A Polym Chem 1999; 37(4): 391-407.
https://doi.org/10.1002/(SICI)1099-0518(19990215)37:4<391::AID-POLA3>3.0.CO;2-E

[70] Soni S, Gupta H, Kumar N, *et al.* Biodegradable biomaterials. Recent Pat Biomed Eng 2010; 3(1): 30-40.

[http://dx.doi.org/10.2174/1874764711003010030]

[71] Lu Y, Schmidt C, Beuermann S. Fast synthesis of high-molecular-weight polyglycolide using diphenyl bismuth bromide as catalyst. Macromol Chem Phys 2015; 216(4): 395-9.
[http://dx.doi.org/10.1002/macp.201400474]

[72] Singh V, Tiwari M. Structure-processing-property relationship of poly(glycolic acid) for drug delivery systems 1: Synthesis and catalysis. Int J Polym Sci 2010; 2010: 1-23.
[http://dx.doi.org/10.1155/2010/652719]

[73] Reyhanoglu Y, Gokturk E. Synthesis of polyglycolic acid copolymers from cationic copolymerization of C1 feedstocks and long chain epoxides. J Saudi Chem Soc 2019; 23(7): 879-86.
[http://dx.doi.org/10.1016/j.jscs.2019.01.008]

[74] Sanko V, Sahin I, Aydemir Sezer U, Sezer S. A versatile method for the synthesis of poly(glycolic acid): high solubility and tunable molecular weights. Polym J 2019; 51(7): 637-47.
[http://dx.doi.org/10.1038/s41428-019-0182-7]

[75] Mahar R, Chakraborty A, Nainwal N, Bahuguna R, Sajwan M, Jakhmola V. Application of PLGA as a biodegradable and biocompatible polymer for pulmonary delivery of drugs. AAPS PharmSciTech 2023; 24(1): 39.
[http://dx.doi.org/10.1208/s12249-023-02502-1] [PMID: 36653547]

[76] Hurrell S, Milroy GE, Cameron RE. The distribution of water in degrading polyglycolide. Part I: Sample size and drug release. J Mater Sci Mater Med 2003; 14(5): 457-64.
[http://dx.doi.org/10.1023/A:1023271003571] [PMID: 15348450]

[77] Pillai CKS, Sharma CP. Review paper: absorbable polymeric surgical sutures: chemistry, production, properties, biodegradability, and performance. J Biomater Appl 2010; 25(4): 291-366.
[http://dx.doi.org/10.1177/0885328210384890] [PMID: 20971780]

[78] Ching KY, Andriotis OG, Li S, *et al.* Nanofibrous poly(3-hydroxybutyrate)/poly(3-hydroxyoctanoate) scaffolds provide a functional microenvironment for cartilage repair. J Biomater Appl 2016; 31(1): 77-91.
[http://dx.doi.org/10.1177/0885328216639749] [PMID: 27013217]

[79] Riaz S, Rhee KY, Park SJ. Polyhydroxyalkanoates (PHAs): Biopolymers for biofuel and biorefineries. Polymers (Basel) 2021; 13(2): 253.
[http://dx.doi.org/10.3390/polym13020253] [PMID: 33451137]

[80] Nigmatullin R, Thomas P, Lukasiewicz B, Puthussery H, Roy I. Polyhydroxyalkanoates, a family of natural polymers, and their applications in drug delivery. J Chem Technol Biotechnol 2015; 90(7): 1209-21.
[http://dx.doi.org/10.1002/jctb.4685]

[81] Shi W, Dumont MJ. Review: bio-based films from zein, keratin, pea, and rapeseed protein feedstocks. J Mater Sci 2014; 49(5): 1915-30.
[http://dx.doi.org/10.1007/s10853-013-7933-1]

[82] Gupta J, Wilson BW, Vadlani PV. Evaluation of green solvents for a sustainable zein extraction from ethanol industry DDGS. Biomass Bioenergy 2016; 85: 313-9.
[http://dx.doi.org/10.1016/j.biombioe.2015.12.020]

[83] Zhang M, Zhao X. Alginate hydrogel dressings for advanced wound management. Int J Biol Macromol 2020; 162: 1414-28.
[http://dx.doi.org/10.1016/j.ijbiomac.2020.07.311] [PMID: 32777428]

[84] Koehler J, Wallmeyer L, Hedtrich S, Goepferich AM, Brandl FP. pH-modulating poly(ethylene glycol)/alginate hydrogel dressings for the treatment of chronic wounds. Macromol Biosci 2017; 17(5): 1600369.
[http://dx.doi.org/10.1002/mabi.201600369] [PMID: 27995736]

[85] Ertesvåg H, Valla S. Biosynthesis and applications of alginates. Polym Degrad Stabil 1998; 59(1-3):

85-91.
[http://dx.doi.org/10.1016/S0141-3910(97)00179-1]

[86]   Echave MC, Saenz del Burgo L, Pedraz JL, Orive G. Gelatin as biomaterial for tissue engineering. Curr Pharm Des 2017; 23(24): 3567-84.
[PMID: 28494717]

[87]   Avérous L. Avérous L. Polylactic acid: synthesis, properties and applications. Monomers, polymers and composites from renewable resources. Belgacem MN, Gandini A, eds. Elsevier 2008; pp. 433-50.

[88]   Tyler B, Gullotti D, Mangraviti A, Utsuki T, Brem H. Polylactic acid (PLA) controlled delivery carriers for biomedical applications. Adv Drug Deliv Rev 2016; 107: 163-75.
[http://dx.doi.org/10.1016/j.addr.2016.06.018] [PMID: 27426411]

[89]   Abdelsalam AM, Somaida A, Ayoub AM, *et al.* Surface-tailored zein nanoparticles: strategies and applications. Pharmaceutics 2021; 13(9): 1354.
[http://dx.doi.org/10.3390/pharmaceutics13091354] [PMID: 34575430]

[90]   Luo Y, Wang Q. Zein-based micro- and nano-particles for drug and nutrient delivery: A review. J Appl Polym Sci 2014; 131(16): app.40696.
[http://dx.doi.org/10.1002/app.40696]

[91]   Kumari S, Yadav BS, Yadav RB. Synthesis and modification approaches for starch nanoparticles for their emerging food industrial applications: A review. Food Res Int 2020; 128: 108765.
[PMID: 31955738]

[92]   Gaur N, Mishra S, Srivastava S, Parvez N. Naturapolyceutics-emerging science & technology in drug delivery system. Int J Pharm Res 2020.
[http://dx.doi.org/10.31838/ijpr/2020.SP1.224]

[93]   Castillo-Henríquez L, Castro-Alpízar J, Lopretti-Correa M, Vega-Baudrit J. Exploration of bioengineered scaffolds composed of thermo-responsive polymers for drug delivery in wound healing. Int J Mol Sci 2021; 22(3): 1408.
[http://dx.doi.org/10.3390/ijms22031408] [PMID: 33573351]

[94]   Ojha S, Tripathi S, Tripathi SM, Mishra S. Formulation and evaluation of glipizide-loaded mucoadhesive microparticle using salvia hispanica seeds mucilage as co-polymer. Curr Bioact Compd 2024; 20(8): e220124225836.
[http://dx.doi.org/10.2174/0115734072282524240101065517]

[95]   VM R, Edison LK. Safety issues, environmental impacts, and health effects of biopolymers. In: Thomas S, AR A, Chirayil CJ, Thomas B, eds. Handbook of Biopolymers. Singapore: Springer; 2023.
[http://dx.doi.org/10.1007/978-981-19-0710-4_54]

[96]   Luo Y, Wang Q, Zhang Y. Biopolymer-based nanotechnology approaches to deliver bioactive compounds for food applications: a perspective on the past, present, and future. J Agric Food Chem 2020; 68(46): 12993-3000.
[http://dx.doi.org/10.1021/acs.jafc.0c00277] [PMID: 32134655]

[97]   Mishra S, Shah H, Patel A, Tripathi SM, Malviya R, Prajapati BG. Applications of bioengineered polymer in the field of nano-based drug delivery. ACS Omega 2024; 9(1): 81-96.
[http://dx.doi.org/10.1021/acsomega.3c07356] [PMID: 38222544]

CHAPTER 3

Various Synthetic Pathways and Properties of Biopolymers

Piyush Anand[1]**, Deepak Kumar**[1]**, Juhi Tiwari**[1] **and Shashi Kant Singh**[1,*]

[1] *Faculty of Pharmaceutical Sciences, Mahayogi Gorakhnath University, Gorakhpur, Uttar Pradesh 273007, India*

Abstract: Biopolymers are naturally occurring macromolecules, such as proteins, nucleic acids, and polysaccharides, which are produced by living organisms. Over time, interest developed in both their natural synthesis and various synthetic pathways due to their importance in a variety of applications. Enzymatic reactions within organisms synthesize biopolymers through intricate biochemical processes known as natural biosynthesis. Nowadays, microbes may be engineered to produce unique biopolymers with specialized functions, indicating developments in synthetic biology. Moreover, synthetic variations that retain desirable capabilities can be carried out by using chemical synthesis techniques to mimic the architectures of genuine biopolymers. The combination of biopolymers with clay can enhance mechanical properties, leading to the development of new materials known as biopolymer–clay nanocomposites. These nanocomposites may represent a significant innovation in the development of biopolymers with enhanced features. The functionality of biopolymers depends on their features, which include good mechanical properties, biocompatibility, and biodegradability. These features facilitate their use in areas ranging from environmental sustainability to medicine. For example, polylactic acid (PLA) is a well-known artificial biopolymer that is being used in biomedical equipment and packaging because of its strength and biodegradability. As research advances, biopolymers are becoming increasingly attractive as alternatives to traditional petroleum-based materials, addressing environmental challenges and enabling innovative solutions across diverse industries.

Keywords: Biopolymers, Bioavailability, Bio-synthesis, Biocompatibility, Biodegradability, Enzymatic reactions.

INTRODUCTION

Natural materials often contain organic compounds known as biopolymers. The term "biopolymer" is derived from the Greek word "*bio*", meaning life, and "*polymer*", referring to many units. Biopolymers are long, repeating

* **Corresponding author Shashi Kant Singh:** Faculty of Pharmaceutical Sciences, Mahayogi Gorakhnath University, Gorakhpur, Uttar Pradesh 273007, India; E-mail: shashikantsingh59@gmail.com

Sudhanshu Mishra, Smriti Ojha, Shashi Kant Singh, Rishabha Malviya & Saurabh Kumar Gupta (Eds.)

macromolecules composed of several components. As biopolymers are both biocompatible and biodegradable, they can be used in a wide range of applications. These applications include their uses in food and pharmaceutical industries, such as "wound healing" and "tissue scaffolding", edible films, emulsions, drug transport materials, and implants. The most common macro-level compounds include biopolymers, such as proteins, carbohydrates, lipids, and nucleic acids, as well as larger non-polymeric molecules like macrocycles and lipids. Genetic manipulation of microorganisms facilitates the biotechnological production of various biopolymers with distinct properties, making them suitable for highly valuable medicinal applications, such as medicine delivery and tissue engineering [1].

Biopolymers are synthetic materials with superior qualities, including flexibility, tensile strength, stability, and reusability. They originate from biological sources and occur naturally, particularly during the growth cycle of various living organisms. The combination of two or more biopolymers yields the creation of "biocomposites", which have numerous innovative applications. Several processes have been discovered that enable the efficient synthesis of biopolymers from various life forms, including microorganisms, plants, and animals. The structure and functionality of biopolymers vary depending on their source. The environmental benefits and biodegradability of biopolymers make them a preferable alternative to chemically manufactured polymers. A key component of the pharmaceutical industry is biopolymers. Biopolymers have the potential to be utilized in both regenerative medicine and drug delivery, offering optimal therapeutic performance and minimal immunogenicity in treated patients [2].

The Role of Biopolymers in Nature and Industry

Directly derived from biomass sources, such as animal proteins and polysaccharides, biopolymers are most commonly used in the production of food packaging materials. The use of biopolymers from renewable resources can help solve the worldwide plastic pollution problem. Researchers and scientists have been working on creating packaging materials derived from biopolymers for a long time. However, due to their chemical makeup and structure, animal proteins and natural polysaccharides have certain unfavorable characteristics [3].

Biopolymers play a definite role in nature and industry. Natural biopolymers are macromolecules derived from microbes, plants, or animals. To expand the spectrum of applications for different polymers, they can be developed through particular chemical alterations or used directly. Since proteins are inexpensive and uncommon, materials derived from them have been used in a wide range of industries, including the food, cosmetic, textile, and biomedical sectors.

Biopolymers are biologically degradable and biocompatible. Typically, biomaterials are composed of natural proteins, such as collagen, keratin, silk, and gelatin. The primary protein in connective tissue is gelatin, a biopolymer created when collagen is heated to a low temperature. It possesses several remarkable qualities, including the ability to enhance cell attachment and growth, biodegradability, low antigenicity, and excellent biocompatibility. It is used in various industries, including the food industry as an emulsifier and gelling agent, the pharmaceutical industry for capsules and ointments, and the cosmetics industry as a cosmetic ingredient. Fig. (**1**) illustrates how the biopolymer-clay nanocomposites are made and processed using the melt insertion method [4].

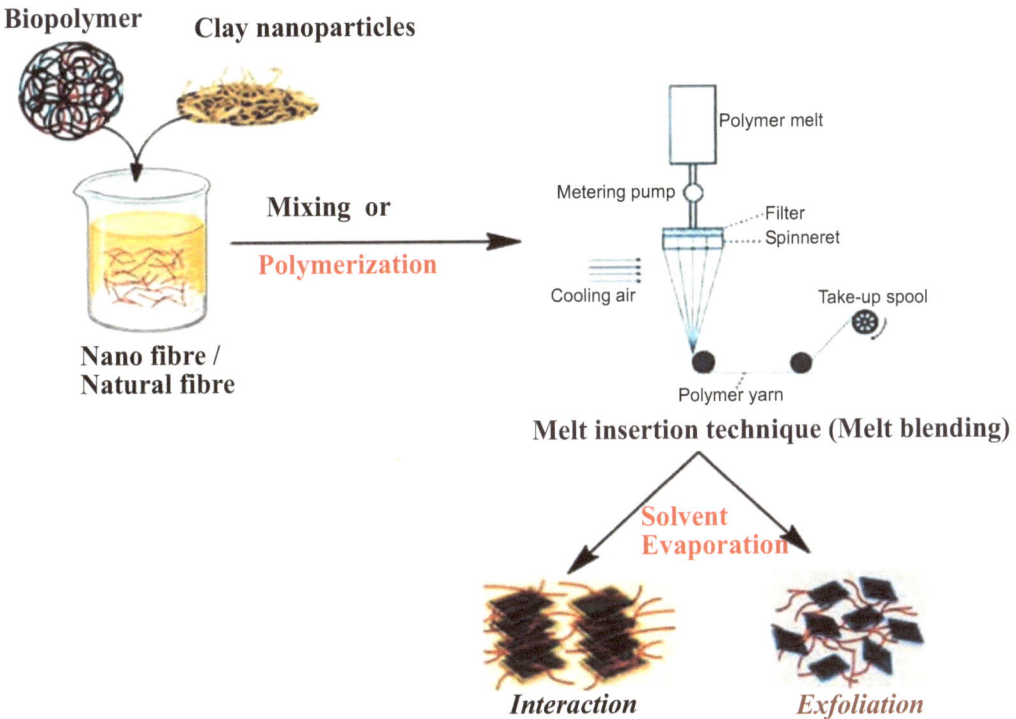

Fig. (1). Illustration of how the biopolymer-clay nanocomposites are made and processed using the melt insertion method.

In the pharmaceutical and medical sectors, alginate is one of the most commonly utilized biopolymers. This is due to its unique encapsulating properties and its role in wound healing. Since its initial isolation in the 1980s, it has evolved into a multifaceted compound with a broad range of applications. Starch has been utilized in the medical and pharmaceutical sectors as a plasma volume expander, controlled-release polymer, pharmaceutical excipient, and superdisintegrant (for immediate-release tablets), and is useful for patients undergoing cancer treatment or suffering from trauma [5].

Types of Biopolymers and Their Different Classifications

Biopolymers can be categorized into two major types: natural and synthetic biopolymer materials. The terms "diverse", "abundant", and "essential to life" aptly describe biopolymers and their primary biological products. For various reasons, biopolymers enhance the desirable properties of existing or emerging materials. For instance, protein molecules, polysaccharides, and monosaccharides, such as cellulose, starch, fructose, and glucose, can all be extracted from these biomaterials.

Major types of polymers produced by living organisms can be categorized based on their chemical structures. These classes include polysaccharides, proteins, and polyesters. Additionally, based on advancements in biotechnology, natural polymers can be produced through microbial fermentation. Fig. (**2**) illustrates the classification of biopolymers, accompanied by suitable examples [6].

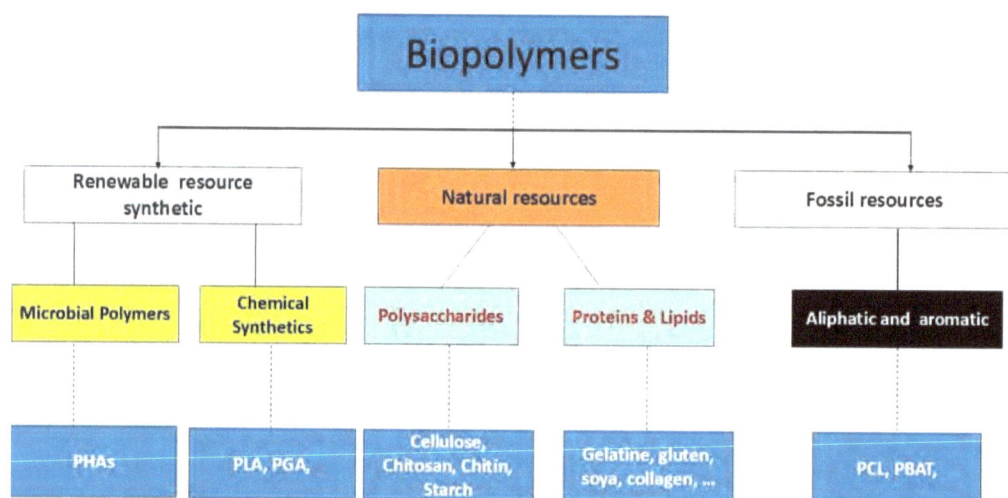

Fig. (2). Classification of biopolymers with suitable examples.

Natural Biopolymer

Natural biopolymers are macromolecules derived from plant, animal, or microbial sources. These polymers can be utilized immediately or modified chemically to increase their range of applications. Natural biopolymers can be produced from substrates that are high in fat, protein, or carbohydrates. Table **1** presents the type of natural biopolymers, their sources, and extraction methods [7].

Table1. Natural biopolymers, their sources, and extraction methods.

Natural Biopolymers	Sources	Extraction Methods
Starch	Cereal grains (corn, wheat, rice, barley, yams), legumes, roots, and starchy tuberous roots.	Centrifugal filtration method
Cellulose	Wood, flax, mushroom mycelium, and algae.	Alkaline pulping, chemo-mechanical method
Pectin	Fruits (apples, oranges) and vegetables (carrots, peas, potatoes).	Hydrolysis method, microwave-assisted extraction, ultrasonic, and enzymatic extraction
Chitosan	Shrimp shells, lobster shells, and insects' exoskeletons.	Demineralization, deproteinization, deacetylation
Alginate	Algae are brown, green, and red.	Acidification, alkaline extraction, drying
Casein	Milk and dairy products.	Precipitation
Collagen	Animal sources (cow bone, pig skin, sheep and goat skin), human bone, and connective tissues.	Fractional precipitation
Gelatin	Porcine, bovine, fish bones, cartilages.	Microwave heating

Protein-derived Biopolymers

Proteins are inexpensive and plentiful, and they have been utilized in a wide range of industries, including the food, cosmetic, textile, and biomedical sectors. They are naturally biocompatible and biodegradable [8].

They exhibit a diverse range of genetically encoded structures and properties. Many researchers have examined the structural and functional interactions of different proteins. Protein-based biopolymers include elastin, collagen, gelatin, and silk [9].

Polysaccharide-based Biopolymers

Materials composed of natural polymers are called biopolymers. Examples include polysaccharides (polymers of carbohydrates), polynucleotides or nucleic acids, such as RNA and DNA, and polypeptides. They consist of long chains of repeating units, such as monosaccharides, nucleotides, or amino acids, linked by covalent bonds. Polysaccharide-based biopolymers include alginate, chitin, and chitosan [10].

Gum-Based Biopolymers

Even at lesser concentrations, the cohesiveness of the solution can be effectively increased by natural gums, which are polysaccharides. These gums are inexpensive, odorless, chemically stable, non-toxic, and widely available in nature. They dissolve easily in water and are often referred to as hydrocolloids. They serve as structural components, nutrient reserves, and agents that bind water. Gum-based biopolymers include xanthan gum, gum tragacanth, dextran, carrageenan, and guar gum [11].

Synthetic Biopolymers

Petroleum is the source of biopolymers based on synthetic components that are also utilized to create biodegradable polymers, such as aliphatic-aromatic co-polyesters. Despite being prepared from synthetic materials, they are fully biodegradable and compostable. Table **2** presents the synthetic biopolymers, their molecular weights or masses, and glass transition temperatures (Tg) [12].

Natural vs. Synthetic Biopolymers

Polymeric materials are widely used in the biomedical industry. While synthetic polymers are often favored for their tailored properties, natural polymers are also well known for their inherent biocompatibility and biodegradability. Blending synthetic and natural polymers is another way to prepare polymeric materials for biomedical applications. Over the past three decades, there has been a growing interest in newly formed materials that are based on the blending of two or more polymers. While synthetically derived polymers may contain impurities or initiators that inhibit cell growth, natural polymers are generally biocompatible. However, compared to many naturally occurring polymers, synthetic polymers exhibit substantially higher mechanical and thermal stability. Furthermore, several natural polymers exhibit limited performance compared to synthetic polymers. Whereas natural polymers are difficult to form into a variety of shapes, synthetically derived polymers may be designed into a vast array of materialistic shapes. For use in biomedical applications, newly developed polymeric materials based on blends of natural and synthetic polymers must be biocompatible and exhibit good mechanical and thermal properties [13].

Polycaprolactone

Research on the biomaterial application of semi-crystalline polymers with lower melting points (59–64°C) and improved solubility, such as polycaprolactone, has been increasing recently.

Polydioxane

Polydioxanone (PDS) is a semi-crystalline, colorless, and non-toxic polymer. In the 1980s, it was the preferred material for commercially manufactured monofilament sutures, with a Tg of -10 to 0°C and approximately 55% crystallinity.

Polyurethane

Generally, polyurethanes are synthesized through polycondensation reactions between diisocyanates and alcohols or amines. Biologically stable polyurethanes and poly(ether urethanes), primarily used in cardiovascular system disorders, such as pacemakers and vascular grafts, are being thoroughly studied for their potential applications in future medical implants due to their mechanical strength and flexibility. They are biocompatible materials used in various medical devices [14].

Table 2. Synthetic biopolymers, their molecular weight/mass, and glass transition temperature (Tg).

Synthetic Biopolymer	Molecular weight	Glass transition temperature (Tg)
Polylactic Acid (PLA)	208,900 g mol^{-1}	55–65 °C [15]
Polyglycolic Acid (PGA)	91,000 g/mol	35–45°C [16]
Polyglutamic Acid (PGA)	5000 – 50000 g/mol	20 °C [17]
Polyurethane	548.589 g/mol	10 to 180 °C [18]
Polyacrylic Acid (PAA)	100,000 g/mol	126.5°C [19]
Polyvinyl Alcohol (PVA)	26,300–30,000 g/mol	75–85 °C [20]
Polyvinyl Chloride (PVC)	30,000 and 200,000 g/mol	70 – 125 °C [21]
Polyethylene Oxide (PEO)	100,000 g/mol	−17°C [22]
Polyorthoesters (POE)	10,000 – 500,000 g/mol	80–100°C [23]

VARIOUS SYNTHETIC PATHWAYS FOR BIOPOLYMERS

Various synthetic pathways are involved in the formation of different types of biopolymers. Fig. (3) illustrates different synthetic pathways of biopolymers with examples. Biopolymers can be chemically altered through various techniques, including graft copolymerization, which is a promising method that can result in a vast array of molecular designs through different types of chemical reactions. The structures of chemically generated polymeric materials are similar to those of naturally occurring polymeric materials. Biodegradation of polymeric materials with ester, amide, and peptide bond chains occurs easily. Polymerization is the process by which a vast number of tiny molecules, known as monomeric units, combine to form a network or chain that is covalently bonded. Throughout the process, each monomer may lose a few chemical groups. Using organic materials like glucose and starch as food sources, microorganisms produce a variety of complex polymeric materials, such as polyesters, silk, and polysaccharides. They also produce polyester-containing poly(hydroxyalkanoates) (PHAs), including poly(hydroxybutyrate) (PHB) and poly(hydroxybutyrate-hydroxyvalerate) (PHBV). A new strategy that can compete with physical modification and traditional chemical synthesis methods is enzyme-mediated polymerization. Enzyme-catalyzed polymerizations exhibit distinct characteristics. Due to the high specificity of the enzymes, these reactions produce no byproducts, simplifying product separation [15].

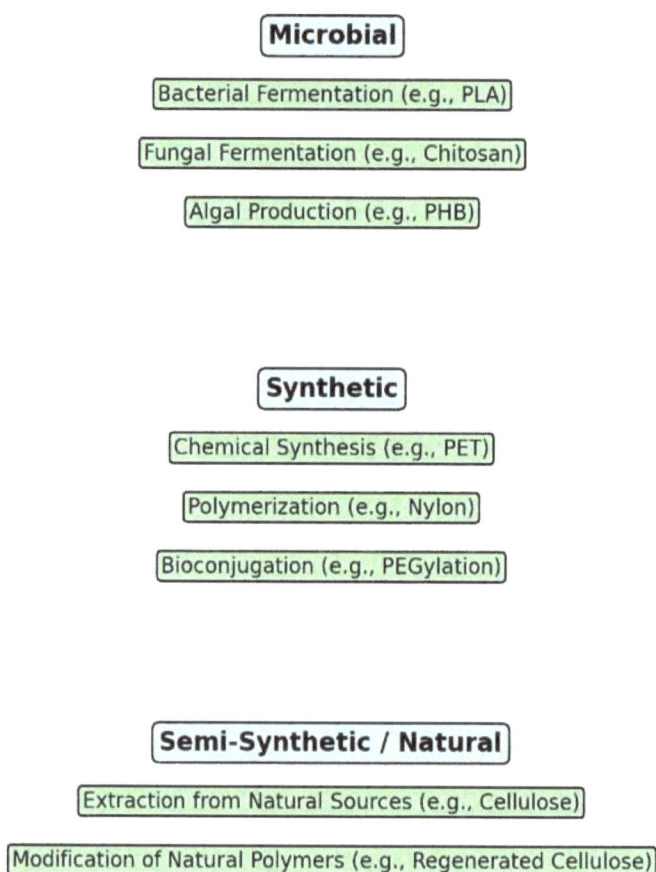

Microbial

Bacterial Fermentation (e.g., PLA)

Fungal Fermentation (e.g., Chitosan)

Algal Production (e.g., PHB)

Synthetic

Chemical Synthesis (e.g., PET)

Polymerization (e.g., Nylon)

Bioconjugation (e.g., PEGylation)

Semi-Synthetic / Natural

Extraction from Natural Sources (e.g., Cellulose)

Modification of Natural Polymers (e.g., Regenerated Cellulose)

Fig. (3). Different synthetic pathways of biopolymers with an example.

Biopolymer Synthesis by Enzymatic Polymerization

Enzymatic polymerization has also emerged as a greener and eco-friendly method for synthesizing biopolymers, particularly polyesters, due to its mild reaction conditions and the absence of toxic metal catalysts. Enzymatic polymerization aligns with the growing demand for bio-derived and biodegradable materials, which aim to reduce fossil fuel utilization and mitigate environmental degradation. Among the biopolymers, aromatic and aliphatic polyesters, including poly(lactic acid) (PLA), poly(butylene succinate) (PBS), and derivatives of poly(furanoate), have been effectively synthesized through enzymatic polymerization. The most widely applied biocatalysts are lipases, particularly Candida antarctica lipase B (CALB), due to their high selectivity and efficiency as catalysts. Two main mechanisms are used: enzymatic ring-opening polymerization (eROP) and direct enzymatic polycondensation. eROP, particularly for lactides, is the preferred route to PLA synthesis since it enables easier control of molecular weight and polymer properties. Enzyme catalysis entails the production of an acyl-enzyme intermediate followed by the growth of the chain by nucleophilic attack. Enzyme polycondensation, as used in the case of PBS and other polyesters, involves the condensation of monomeric hydroxyl and carboxyl groups with the release of water as a by-product. Optimized enzyme immobilization, reaction conditions, and choice of solvent facilitate efficient enzymatic polymerization. Ionic liquids (ILs) and supercritical fluids (SCFs) have improved monomer solubility and reaction rates. The process, however, must be upscaled to industrial levels without compromising cost-effectiveness and efficiency [16].

Biopolymer Synthesis *via* Microbial Production

Microbial production of biopolymers is of significant interest due to their sustainability, biodegradability, and potential to replace petroleum-based plastics. Of the biopolymers, polyhydroxyalkanoates (PHAs) are most widely studied as microbial polyesters with various applications ranging from medicine and packaging to agriculture. PHAs are produced by several microorganisms, including *Halomonas*, *Ralstonia eutropha* (also referred to as *Cupriavidus necator), Escherichia coli,* and *Pseudomonas* species. These microorganisms can store PHAs as intracellular storage granules in conditions of carbon-source excess and nutrient limitation. The process of synthesis is *via* metabolic pathways, including glycolysis, β-oxidation, and de novo fatty acid biosynthesis. Both genetic and metabolic engineering strategies have optimized the yield of PHA. For example, recombinant *E. coli* carrying the phaCAB gene cluster of *R. eutropha* can rapidly produce PHB (polyhydroxybutyrate) from both glucose and glycerol carbon sources. Also, *Pseudomonas* strains have been genetically modified to synthesize medium- and long-chain PHAs by modulating β-oxidation processes. Commercial-level production of PHAs has been demonstrated by companies, such as MedPHA, Danimer Scientific, and Tianan, with yields exceeding 100 g/L dry cell mass. Notwithstanding this, reducing production costs is a challenge, largely due to substrate and downstream processing expenses. Measurements such as high-cell-density fermentation, utilization of low-cost feedstocks, and chemical co-production of value-added chemicals have been employed to make it more economical. With advancements in synthetic biology and metabolic engineering, biopolymer production by microorganisms is a viable alternative for sustainable and environmentally friendly materials, leading to the possibility of mass-producing biopolymers [17].

Synthesis of Biopolymers Using Electrospinning Techniques

Electrospinning is a powerful method for the production of nanofibrous biopolymer scaffolds with properties like high porosity, high surface-area-to-volume ratio, and biomimetic structure. With its capability to generate nanostructured fibers that mimic the extracellular matrix (ECM), electrospinning has been extensively applied in tissue engineering. The electrospinning technique involves applying a high-voltage electric field to a polymer solution or melt, which produces thin fibers that solidify as they are deposited on a collector. Biopolymers of both synthetic and natural types, such as collagen, chitosan, polycaprolactone (PCL), and polylactic acid (PLA), have also been electrospun into fibrous scaffolds for biomedical applications. The primary parameters affecting fiber morphology are polymer concentration, viscosity, applied voltage, flow rate, and environmental conditions, including humidity and temperature. Technological advancements in electrospinning methods, including coaxial electrospinning, emulsification electrospinning, and dynamic water flow electrospinning, have boosted the functional properties of biopolymer-based nanofibers. Coaxial electrospinning, for example, enables the production of core-shell nanofibers, providing controlled delivery of bioactive agents. Similarly, emulsification electrospinning produces multicore nanofibers, which enhance the delivery of drugs and growth factors. Electrospun biopolymer scaffolds have demonstrated extensive potential in regenerative medicine, particularly for the engineering of bone, cartilage, cardiovascular, and skin tissues. The scaffolds provide cell attachment, growth, and differentiation along with mechanical stability. Optimizing fiber orientation, mechanical properties, and scale-up manufacturing for clinical use remains a challenge. Despite these challenges, electrospinning is a versatile and promising method for producing next-generation biopolymer-based scaffolds, which is propelling tissue engineering and biomedical research [18].

Some Conventional Synthetic Methods

Synthesis of Poly Lactic Acid (PLA)

Polylactic acid (PLA), produced from non-hazardous natural resources, is one of the most capable biopolymers that can be utilized for the creation of different biomedical materials. PLA has many beneficial properties, such as biocompatibility, biodegradability, mechanical strength, and process capability, which make it a very crucial and emerging biopolymeric material in medical fields.

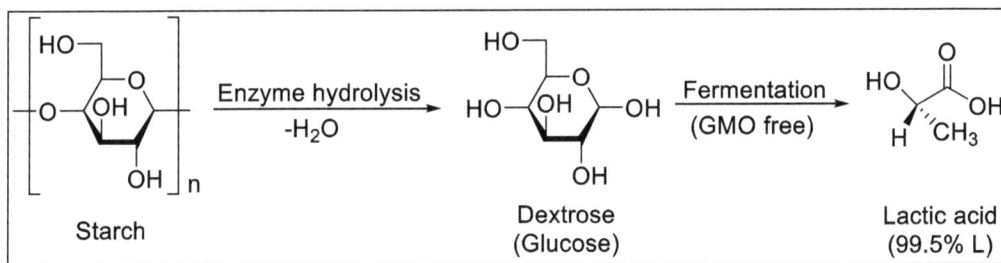

Scheme (1). Synthesis of lactic acid from starch.

Lactic acid (LA) is obtained through the enzymatic hydrolysis and fermentation of carbohydrates, such as starch, derived from regenerative resources like maize, corn, and sugarcane. Scheme **1** depicts the synthesis of lactic acid from starch. PLA is a naturally derived and environmentally friendly, nontoxic biopolymeric material with characteristics that allow it to be used in the human body. Some limitations are also associated with the widespread applications of PLA in various fields, such as delayed degradation rate, lipophilicity, and a short-term effect on its toughness, depending on the utilization purpose. PLA can be easily blended with many polymers, offering suitable possibilities to enhance its properties or produce novel PLA polymeric blends for various target applications. Numerous PLA mix types have been used in different fields, including tissue engineering, implants, medication delivery, and sutures [19]. Due to advanced and outstanding biocompatibility and mechanical strength, PLA and its derivatives are becoming widely used in tissue engineering for the functional repair of impaired or damaged tissues. LA manufacturing and PLA blends, as well as their implementation in biomedical and related sectors, are also explained in this chapter [20].

PLA is a biopolymer that can be synthesized from various natural products like rice, sugarcane, and starch. First, in 1932, Carothers polymerized PLA with a low molecular weight. Scheme **2** depicts the general synthetic scheme for the synthesis of PLA isomers. The first higher molecular weight PLA was created and patented in 1954 due to DuPont's subsequent efforts. It was also used for the first time in the biomedical industry as fiber materials for resorbable sutures. LA is the essential monomeric unit for the synthesis of different stereoisomers of PLA, as mentioned in Scheme (**1**). The two most common synthetic pathways available for the synthesis of PLA are:

1. DPC of lactic acid (direct condensation of poly-lactic acid)
2. ROP of lactic acid (mostly ROP- ring opening polymerization employed in industries for the production of PLA)

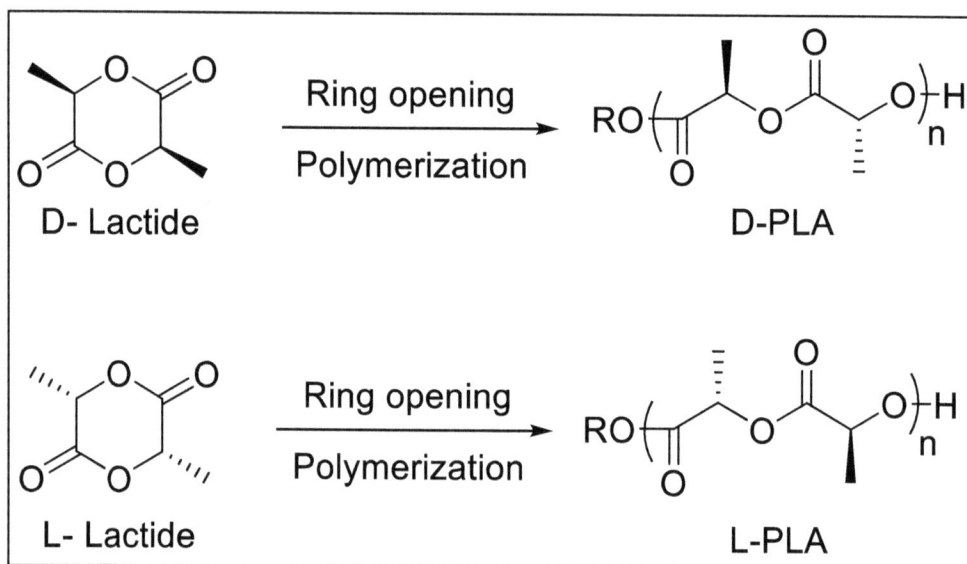

Scheme (2). General synthetic scheme for the synthesis of PLA isomers.

The chiral nature of PLA molecules is due to the availability of its three different isomeric forms, which are:

1. D, D- PLA
2. L, L- PLA
3. D, L- PLA

Polymerization of D- or L-lactides is used to make the isotactic (denoting a polymer in which all the repeating units have the same stereochemical configuration) homopolymers of L-PLA, D-PLA, and racemic D, L-PLA [21].

Poly(3-hydroxybutyrate) (PHB)

Various methods are available for the extraction, production, and retrieval of PHB-based biopolymers and their derivatives using several bacterial species. They increase the interest of researchers in the biosynthesis of PHB biopolymers, leading to the generation of different derivatives of PHB materials, which rely on the types of microorganisms used and the method of production adopted. According to the research, the presence of stereospecificity by the polymerizing enzyme PHA synthase makes it relatively simple to create stereospecific R- and S-configurations. Three main methods are mainly reported for synthesizing PHB compounds. Ring-opening polymerization (ROP) of β-butyrolactone (BL) is the first technique. Acetyl-CoA, the main substrate in PHA production, is available in the cells of natural or transgenic plants used in the second mode of biosynthesis. The third method for creating PHB compounds is bacterial fermentation. In the best-case scenario, PHA components are believed to comprise more than 90% of the dry weight of bacterial or microbial cells. The production of PHB is another common application for this technique [22]. Synthesis of PHB depends on the sequence of three enzymatic reactions, using the main carbon metabolite of acetyl-CoA.

1. First, the condensation reaction between two acetyl-CoA entities leads to the formation of acetoacetyl-CoA, catalyzed by β-keto thiolase. This step of the reaction is reversible.
2. Acetoacetyl-CoA reductase (PhaB) is an enzyme that enzymatically reduces acetoacetyl-CoA to (R)-3-hydroxybutyryl-CoA.
3. In the final step, the polymerization of (R)-3-hydroxybutyryl-CoA occurs to produce PHB, catalyzed by the enzyme PHB synthase (phbC gene).

These steps constitute the biosynthetic pathway of PHB from acetyl-CoA.

Poly(3-hydroxybutyrate), commonly known as the poly (hydroxy alkanoates) (PHAs), was the first PHA to be isolated and described. Scheme **3** depicts the enzyme-dependent biosynthesis of PHB. PHB has a linear chain structure that contains both crystalline and amorphous phases, but it is predominantly highly crystalline. It is available in the form of a polymer or can be found as part of copolymers or blends. It is commercially produced through a wide variety of bacterial strains through bacterial fermentation. PHB offers greater benefits than other synthetic polymers in the production of specific packaging materials, such as polyethylene (PE) and polypropylene (PP). Additionally, PHB has better barrier qualities than polyvinyl chloride (PVC) and polyethylene terephthalate (PET). In addition to this, another key property of PHB material is its biodegradability, which occurs in suitable biologically active environments, including soils, water, and both aerobic and anaerobic conditions, after a specified time interval, making it a desirable eco-friendly substitute for synthetic polymers [23].

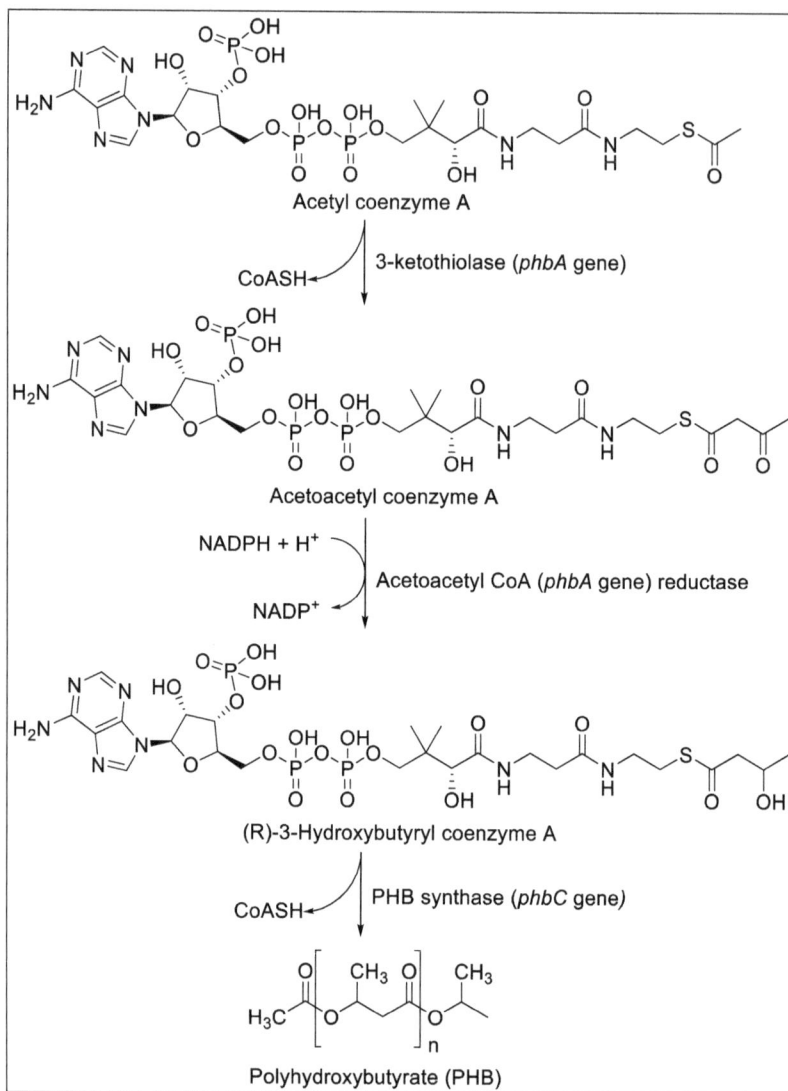

Scheme (3). Enzyme-dependent biosynthesis of PHB.

Polycaprolactone

Poly (caprolactone) (PCL) is composed of nonpolar methylene groups and a semi-polar ester group and is a hydrophobic, semi-crystalline, biodegradable polyester with more durability and biocompatibility. These characteristics have led to their widespread usage in biomedical sectors, specifically for scaffolding for tissue regeneration, drug delivery, tissue engineering, and drug carriers. Despite being highly crystalline (~50%), the biodegradability of PCL is supported by its significantly lower glass transition temperature (~-60°C) compared to other biodegradable polymers. Numerous studies have reportedly been conducted to enhance PCL's bioactivity by creating composites or introducing functional groups, including amino, carboxyl, and hydroxyl groups [24].

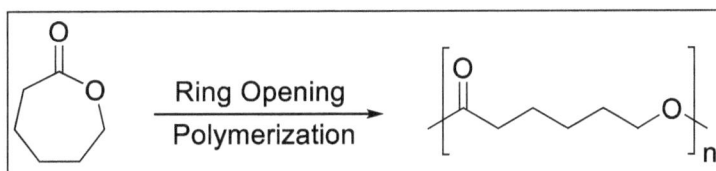

Scheme (4). General synthetic scheme for PCL.

PCL is a linear polyester synthesized by ROP of a seven-membered lactone, ε-caprolactone. Scheme **4** depicts a general synthetic scheme for PCL. The polymerization reaction is catalyzed by using catalysts, such as stannous octoate. Organic solvents with low molecular weight alcohols can be employed to regulate the polymer's molecular weight throughout the polymerization process [25].

Polyvinyl Alcohol (PVA)

PVA is an artificial crystalline structure made from polyvinyl acetate *via* hydrolysis and is easily biodegradable by biological organisms and water-soluble polymers. Lacquers, resins, surgical threads, and food packaging materials that frequently come into contact with food are just a few of the final products that have been produced using it in the commercial, industrial, medical, and food sectors. Polyvinyl alcohol (PVA) is a biodegradable synthetic polymer that mimics some properties of natural polymers and is commonly used for textile sizing and paper coating. It is utilized by blending with other polymers for several industrial applications to enhance the mechanical properties of films due to its compatible crystalline structure and hydrophilic (water-soluble) properties. Biodegradable polymers can be derived from either petroleum-based synthetic materials or renewable resources, and they break down naturally in either anaerobic (landfill) or aerobic (composting) environments. PVA is a widely used thermoplastic polymer that is non-toxic, safe, and kind to biological tissues. The hydroxyl groups on the carbon atoms in PVA facilitate hydrolysis, increasing its biodegradability and making the polymer more biodegradable. PVA is one of the most widely used synthetic polymers and has been commercially available for many years [26].

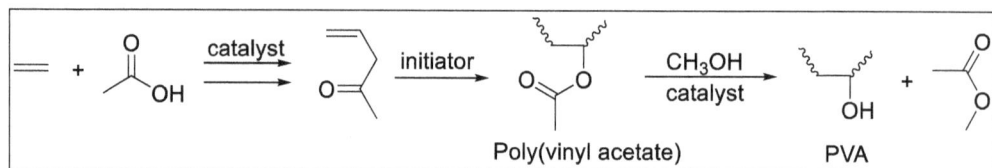

Scheme (5). General synthetic scheme for the synthesis of PVA.

Due to its ability to form films, PVA, which is produced by saponifying poly(vinyl acetate), has been utilized for a long time in combination with other natural polymers. Scheme 5 depicts a general synthetic scheme for the synthesis of PVA. Due to its inherent hydrophilic nature, PVA completely dissolves in water, requiring a water temperature of approximately 100 °C and a 30-minute holding period. Every PVA derivative is hydrophilic and is influenced by some variables, including particle crystal structure, molecular weight, and element dimensions of distribution [27].

Polytrimethylene Terephthalate (PTT)

One of the three polyesters initially created by Whinfield and Dickson in 1941, PTT, is an aromatic polyester produced by the melt polycondensation of 1,3-propanediol (PDO) with either terephthalic acid (TPA) or dimethyl terephthalate (DMT). Scheme 6 depicts a general synthetic scheme for the synthesis of PTT. It has been commercially accessible since 1998. Biopolymers with an even number of methylene groups in their structure, such as PET and PBT, are well-established, high-volume polymers. However, it has been considered that odd-numbered biopolymers, such as PTT, remain an ambiguous polymer. Trimethylene glycol, also known as 1,3-propanediol, is a clear, colorless liquid with a boiling point of 214°C and is used as a step in the PTT synthesis process. The conventional synthesis method involves hydrating acrolein to 3-hydroxypropyl (3-HPA) at high pressure in the presence of an acid catalyst, such as an acidic ion-exchange resin. A Raney nickel catalyst is used to hydrogenate the aqueous solution into PDO. In a newly developed commercial pathway by Shell, ethylene oxide (EO) serves as the initial raw material. First, ethylene oxide is hydroformylated into 3-HPA with a cobalt catalyst and a mixture of CO and H^2 gas. PDO is created by hydrogenating and concentrating HPA [28]. PTT is produced by a melt-polymerized procedure in which PDO is either trans-esterified with DMT or directly esterified with TPA. Instead of producing cyclic trimers as reaction side products, it generates cyclic dimers. It releases allyl alcohol and acrolein rather than the gaseous by-products of acetaldehyde. Acrolein needs to be handled and disposed of carefully. Since direct esterification of PDO with TPA is more cost-effective than using DMT, it is the commercially recommended method for polymerizing PTT [29].

Scheme (6). General synthetic scheme for the synthesis of PTT.

Some Emerging and Novel Synthetic Techniques

The synthesis of biopolymers has been revolutionized by new techniques, such as nanocatalysis, genetic engineering, and enzyme-catalyzed polymerization. These techniques provide more control over polymer structure, increased reaction efficiency, and improved material properties, making them highly useful for biomedical, industrial, and environmental applications.

Nanocatalysts for the Synthesis of Biopolymers

Nanocatalysts have emerged as a transformative reagent in the synthesis of polymers, allowing enhanced catalytic efficiency, molecular governance, and cost-saving procedures. The catalysts, like metal nanoparticles, immobilized enzyme-based systems, and semiconductor-derived compounds, allow rapid reaction kinetics as well as a controlled process of polymerization that leads to modified biopolymer architectures. For instance, gold and silver nanoparticles were utilized to facilitate the polymerization of polylactic acid (PLA), resulting in high-molecular-weight polymers with enhanced thermal stability. The use of nanocatalysts offers high selectivity and reactivity with reduced unwanted byproducts and impurities in the product. They enable energy-efficient synthesis with mild reaction conditions, making the process more sustainable. Nonetheless, issues, such as nanoparticle agglomeration, cytotoxicity, and long-term stability in biomedical applications, are still under investigation. Overcoming these constraints through surface modification and biocompatible coatings can further enhance their usability [30].

Genetic Engineering to Produce Precision Biopolymers

Genetic engineering, particularly through CRISPR-Cas9 and recombinant DNA technology, has provided unparalleled precision for modifying metabolic pathways in microbes, thereby paving the way for the potential production of optimized biopolymers with enhanced properties. Through genetic modifications, microbes, such as *Escherichia coli* and *Cupriavidus necator*, have been genetically modified to optimize polyhydroxyalkanoate (PHA) production and enhance the mechanical strength of the polymers (Chen *et al.*, 2023). PHAs are highly biodegradable and elastically tunable and hence find applications that range from biodegradable plastics to biomedical implants. The primary advantages of genetic engineering for biopolymer synthesis include direct control over polymer composition, maximum yield efficiency, and sustainability through microbial fermentation. Nevertheless, this technology has disadvantages, including elevated costs associated with genetic modification, scalability issues, and regulatory barriers to genetically modified biomaterials. Ongoing advancements in synthetic biology and metabolic engineering are aimed at addressing these challenges, paving the way for more cost-effective and efficient production methods [31].

High-performance Biopolymers by Enzyme-directed Polymerization

Enzyme-catalyzed polymerization is a green, environmentally friendly, and biocompatible process compared to traditional chemical synthesis, and it enables the controlled synthesis of polymers with specific molecular weights and architectures. Enzyme-catalyzed polymerization is particularly beneficial in the synthesis of biodegradable polyesters, where enzymatic catalysts ensure efficient and environmentally friendly polymerization. A notable example is lipase-catalyzed polymerization, which has been widely applied for synthesizing polycaprolactone (PCL). It produces very biocompatible and tunable polymers, which are utilized to their maximum potential in controlled drug delivery systems and tissue engineering. Although enzyme-directed polymerization offers numerous benefits, including high specificity and mild reaction conditions, it also presents challenges, such as the requirement for high enzyme stability and the development of cost-effective immobilization methods for large-scale applications. Latest advances in enzyme engineering and immobilization technologies are countering these limitations, taking another step toward the practicality of enzymatic polymerization for the bulk-scale production of biopolymers.

PROPERTIES OF BIOPOLYMERS

Biopolymers are a broad class of materials derived from natural sources. They have special physical characteristics that enable them to be used in a variety of ways. This section provides an in-depth discussion of the mechanical and thermal characteristics of biopolymers. Biodegradable polymers possess unique characteristics, and biopolymers, by nature, are inherently biodegradable. Due to their unique surface chemistry, excellent biocompatibility, anisotropic shape, remarkable mechanical properties, renewable nature, and intriguing optical features, biopolymers are gaining significant attention in the domains of materials science and biomedical engineering. Biopolymers have unexpected intrinsic characteristics that make them valuable in an expanded range of optical-electronic devices, such as being chemically inert, amphiphilic in nature, having good mechanical properties, being tough, and having a low density. Table **3** presents the mechanical properties of various biopolymers, including specific numerical values for tensile strength, elastic modulus, synthesis methods, and corresponding applications.

Table 3. Mechanical properties of various biopolymers, their tensile strength, elasticity modulus, synthesis methods, and their applications.

S. No	Biopolymer	Tensile Strength (GPa /MPa)	Elasticity Modulus (GPa /KPa/MPa)	Synthesis Method	Application	References
1.	Cellulose (Crystalline)	7.5–7.7 GPa	110–220 GPa	Synthesized in plants by cellulose synthase complexes (CSCs).	Used in food, construction, medicine, and other industries.	[32]
2.	Starch	45 MPa	15 MPa to 294.98 MPa	A multi-enzyme and substrate-based complicated process in plants regulated by phosphorylation, redox conditions, and gene transcription.	Used in the food, pharmaceutical, and paper industries.	[33]
3.	Glycogen	1.08 MPa	10–70 kPa	Glycogenesis	Used in the food, cosmetic, and pharmaceutical sectors as a thickener.	[34, 35]
4.	Chitin	1.6–3.0 GPa	1.2 GPa to 191.5 GPa	Synthesized by a family of enzymes called chitin synthases, found in the cell wall of fungi.	Used in wound dressing, agricultural material, and wastewater treatment.	[36, 37]
5.	Hyaluronic acid	0.1-200 MPa	200 kPa to 22 kPa	Synthesized by enzymes called hyaluronan synthases (HAS).	Used in the cosmetic and aesthetic industries for skincare products like moisturizers, serums, and dermal fillers.	[38 - 40]
6.	Collagen	47–580 MPa	6.1 GPa.	Synthesis inside the cell by various processes, including transcription, hydroxylation.	Used in wound healing, tissue engineering, and drug delivery.	[41 - 43]
7.	Keratin	200–260 MPa (Human Hair)	10 MPa to 2.5 GPa	The process of keratin synthesis involves nucleation, elongation, fusion, and bundling of keratin precursors.	Used in wound healing, tissue engineering, and drug delivery.	[44, 45]

(Table 3) cont.....

S. No	Biopolymer	Tensile Strength (GPa /MPa)	Elasticity Modulus (GPa /KPa/MPa)	Synthesis Method	Application	References
8.	Elastin	2 MPa	1 MPa	Synthesized by fibroblasts through a process called neoelastinogenesis.	Used in wound healing, tissue engineering, drug delivery, skin care products, and hair care.	[46, 47]
9.	Myosin	3-4 GPa	0.71 ± 0.16 GPa.	Synthesized in skeletal muscle by electrical stimulation, creatine, and other factors.	Applications in medicine, agriculture, and biological sciences.	[48, 49]
10.	Casein	110 MPa	1 – 10 MPa	Synthesized by acidifying skim milk or treating it with rennet.	Used in the cosmetic, food, and medicine industries.	[50, 51]
11.	Lignin	40 MPa	2.5–3.7 GPa	Plants undergo a complex biochemical process that creates lignin monomers, which are subsequently polymerized to form lignin.	Used in biomedicine, energy production, and manufacturing.	[52, 53]
12.	Agarose	100 kPa	200 kPa	Synthesized by applying an alkali treatment to red marine algae, then removing the seaweed and isolating the resulting agar.	Applications in life sciences, including gel electrophoresis, chromatography, and cell culture.	[54, 55]
13.	Alginate	70 MPa	1-100 KPa	Synthesized by enzymatic processes, chemical modifications, and reactions with calcium ions.	Used in medicine, dentistry, and food.	[56, 57]
14.	Pectin	4.99 MPa to 10.84 MPa	1-100 KPa	Synthesized in the Golgi lumen of plant cells by glycosyltransferases (GTs).	Used in food, health, and other industries.	[58, 59]
15.	Zeatin	10-100 MPa	0.1 -10 GPa	Synthesized by two pathways: the tRNA pathway and the AMP pathway.	Used in plant tissue culture, agriculture, and cosmetics.	[60, 61]

Physical Properties

The physical properties of biopolymers must be understood to comprehend how they react in different environments and applications. These attributes may be broadly classified into two groups based on their mechanical and thermal characteristics [62].

Mechanical Properties

The mechanical properties of biopolymers are essential for their ability to respond to external stimuli. Important components include:

Tensile Strength

This refers to the highest tensile stress (pulling stress) that a material is capable of withstanding before breaking down. The content and structure of biopolymers can cause them to have variable tensile strengths. When compared to synthetic polymers, the tensile strength of certain biopolymers, such as those derived from starch, may be inferior; however, these properties can be improved by alterations or the inclusion of plasticizers [63].

Elasticity

The capacity of a substance to regain its original shape following deformation is measured by its elasticity. Biopolymers exhibit a high degree of elasticity, which is advantageous in applications where flexibility is required, such as in packaging materials. The molecular makeup of biopolymers and the addition of additives can affect their flexibility [64].

Durability

One important factor influencing the performance of biopolymers, especially in applications exposed to environmental stressors, is their durability. Although many biopolymers degrade naturally, they can be made more durable by mixing or cross-linking them with other materials, thereby maintaining their structural integrity over time [65].

Thermal Properties

It is essential to understand the thermal properties of biopolymers to comprehend how they behave at various temperatures. Some of the essential thermal characteristics include:

Melting Point

The point at which a solid turns into a liquid is known as the melting temperature. When it comes to high-temperature applications, biopolymers may not be as useful as conventional plastics because of their generally lower melting temperatures. For processing methods, such as extrusion and molding, knowledge of the melting point is crucial [66].

Thermal Stability

This characteristic indicates the duration a material can withstand heat exposure without undergoing degradation. The chemical composition of biopolymers and the additives they contain can affect their stability at different temperatures. It is possible to acquire increased heat stability by changing the polymer structure or adding stabilizers [67].

Glass Transition Temperature (Tg)

The temperature range in which a polymer changes from a hard, glassy state to a softer, more rubbery form is known as the glass transition temperature. Tg has a significant impact on the mechanical characteristics and suitability of biopolymers for various applications. Greater stiffness may be suggested by a higher Tg, whilst greater flexibility may be indicated by a lower Tg [68].

Chemical Properties

Reactivity: Relationships with Bases, Acids, and Solvents

The chemical reactivities of marine polymers vary greatly based on the structural makeup of the polymer. Different marine biopolymers interact differently with solvents; for instance, chitosan dissolves in acidic solutions but is insoluble in basic or neutral ones. This tendency is primarily caused by the protonation of amine groups at lower pH values, which enhances solubility. Alginates, on the other hand, interact with multivalent cations in solution with a unique reactivity that promotes the creation of gel-like formations. Certain marine polymers hydrolyze, that is, break down into smaller units when exposed to acids. Comparably, in unfavorable circumstances, the reactivity of these polymers may result in deacetylation or depolymerization, which can compromise their structural integrity and limit their use in various contexts, including medication delivery systems [69].

Degradation: Biodegradation Mechanisms and Factors Affecting Degradation

Marine polymers naturally biodegrade due to their natural origins. Their primary mode of breakdown is enzymatic, typically facilitated by marine bacteria or other microorganisms that produce specific enzymes. For example, Alginate lyases function on alginates, whereas chitinases degrade chitin. The rate of degradation is influenced by the kind of polymers, temperature, pH, and other environmental factors, in addition to the concentration of microbial populations. The polymer's structural complexity, which influences the rate of breakdown, includes factors, such as polymer branching and acetylation degree. Understanding these routes is essential to maximize the use of marine polymers in biological applications where controlled breakdown is crucial [70].

Biological Properties

Biocompatibility: Interaction with Biological Tissues and Cells

The marine biopolymers, chitosan and alginate, are ideal for biomedical applications due to their exceptional biocompatibility, which results in a generally positive interaction with biological tissues, promoting cell adhesion and proliferation. For example, the positive charge of chitosan improves interactions with negatively charged cell membranes, facilitating cell attachment and promoting tissue regeneration. Alginates, when used in gel form, create a supportive environment that mimics extracellular matrices, thereby encouraging cell growth and differentiation. Nevertheless, the purity of the polymers and any potential residual contaminants must be carefully considered, as they may affect their biocompatibility [71].

Immunogenicity: Immune Response and Potential for Allergic Reactions

While most marine polymers do not elicit an immune response, certain substances may trigger an immune response. A polymer's molecular makeup, degree of purity, and presence of residual proteins can all impact its immunogenicity. For instance, chitosan has been shown to exhibit specific immunomodulatory effects, which are dependent on its molecular weight and degree of deacetylation. The capacity of chitosan to activate macrophages and enhance cytokine release may be the cause of its therapeutic benefits in wound healing and immune control. However, since shellfish is a typical source of chitosan, it is essential to assess the likelihood of allergic reactions, especially for individuals who are sensitive to it [72].

Environmental Properties

Biodegradability: Breakdown Processes in Natural Environments

Marine polymers decompose in both terrestrial and aquatic environments due to microbial degradation, a natural process. Enzymatic activity from bacteria, fungi, and other microorganisms initiates the process by secreting enzymes that hydrolyze the polymeric chains into smaller, easier-to-manage fragments. The rate and efficiency of this process depend on variables, including moisture content, temperature, oxygen availability, and the presence of particular microbial populations. For example, in marine environments, chitosan and alginate gradually decompose to contribute to the carbon cycle without leaving behind toxic residues. Due to this characteristic, marine polymers can be utilized in biomedicine and packaging as environmentally friendly alternatives [73].

Sustainability: Environmental Impact and Sustainability Aspects

Due to their biodegradability and renewable nature, marine polymers are widely used in accordance with sustainability standards. Unlike synthetic polymers made from petrochemicals, these biopolymers are harvested from abundant marine sources, such as seaweed or crab shells, and therefore have a comparatively smaller environmental impact. The environmental effect of marine biopolymers is further minimized by the fact that their extraction and processing frequently require less energy-intensive techniques. In terms of sustainability, promoting the use of marine polymers can facilitate the development of more environmentally friendly technologies in industries, such as biomedicine, agriculture, and environmental remediation, while also helping to mitigate the growing problem of plastic pollution [74].

APPLICATION OF BIOPOLYMERS

Biopolymers are used for a variety of purposes. Different biopolymers possess distinct properties, and their suitable properties determine their applications.

Biomedical Application

Biomedical polymer architectures have applications across various areas, including artificial tissues, joint replacements, sutures, maxillofacial implants, dentistry, cardiovascular devices, and gene- and drug-delivery systems. Biopolymers have been used in many well-defined biomedical applications, such as artificial hearts, breast implants, intraocular lenses, soft-tissue replacement, bone cement, dialyzers, liver, pancreas, bladder, kidney, coverings for capsules and tablets, pacemakers, implanted pumps, cardiac assist devices, encapsulations, replacements of joints, artificial skin, dental care, drug delivery, and tumor or inflammatory targeting [75].

Food Industry Application

Water-soluble gums with unique characteristics produced by a variety of microorganisms are known as polysaccharides. The low cost of these biopolymers has allowed them to develop into distinctive and useful polymeric compounds for industry. The structural and physicochemical diversity of microbial polysaccharides makes them useful for a variety of applications in the food industry. They serve as coagulants, gelling agents, binders, suspending agents, stabilizers, and emulsifiers. These biopolymers are ideal for the food sector due to their regular structure and excellent purity, which confer upon them their unique rheological properties.

Wastewater Treatment Application

Chitosan is renowned for its superior capacity to adsorb heavy metals from wastewater, including Pb, Cd, and Cu. In wastewater treatment facilities, biopolymers can be used as flocculants and coagulants to enhance the settling and filtration processes, thereby reducing turbidity and suspended particles. The use of biopolymers in wastewater treatment procedures can help reduce the need for synthetic chemicals and mitigate the negative environmental effects associated with them [75].

CASE STUDIES AND EXAMPLES

Case Study: Biodegradable Polymers: Examples like PLA (Polylactic Acid) and PHA (Polyhydroxyalkanoates)

The potential of biodegradable polymers, including PHA and PLA, to mitigate the environmental impact of traditional plastics has generated considerable interest. Polylactic acid (PLA) has extensive use in packaging, agriculture, and biomedical devices. It is obtained from sustainable sources, such as sugarcane or maize starch. Its main benefit is that, through hydrolytic breakdown, it can be broken down into non-toxic byproducts. For this reason, it is a common material for biodegradable packaging and medical implants that eventually dissolve after serving their purpose. Nevertheless, the brittleness and other mechanical characteristics of PLA could prevent it from being used in some situations [76].

On the other hand, bacteria ferment carbohydrates or lipids to form polyhydroxyalkanoates (PHA). PHAs are especially well-suited for use in maritime conditions, where traditional plastics can persist for decades, due to their exceptional biodegradability in both aerobic and anaerobic environments. PHAs are stronger and more flexible than PLA, and they are being investigated for a range of uses, including biodegradable packaging and medical equipment. Both PLA and PHA represent significant advancements in the creation of sustainable biopolymers; however, to increase their use, issues with affordability and scalability still need to be addressed [77].

Case Study: Biomedical Applications: Examples like Collagen-based Scaffolds and Chitosan in Drug Delivery

Natural polymers, such as collagen and chitosan, have become essential tools in the biomedical field for applications in regenerative and therapeutic medicine. As collagen-based scaffolds resemble the extracellular matrix (ECM), which promotes cell proliferation and tissue regeneration, they are often utilized in tissue engineering. Moreover, collagen scaffolds can be designed to break down at a controlled rate and are highly biocompatible, enabling them to promote tissue repair without harming the body. In wound healing, bone repair, and reconstructive surgery, these scaffolds are commonly used to provide a temporary matrix that promotes native tissue regeneration [78].

The biopolymer chitosan, produced by deacetylating chitin, is crucial in drug delivery systems due to its gel-forming and mucoadhesive properties. Due to its positive charge, it can interact with negatively charged cell membranes to enhance the transport of proteins, DNA, and medications. Furthermore, chitosan is a suitable option for targeted drug administration, particularly in cancer therapy and wound healing applications, due to its biocompatibility and ability to form nanoparticles. As a bioactive substance, it may also boost the immune system, hastening the healing process and providing a versatile platform for cutting-edge drug delivery systems [79].

Case Study: Synthetic *vs.* Natural Polymers: Comparative Analysis of Properties and Applications

Synthetic and natural polymers differ significantly in their characteristics and applications. Petrochemical techniques are commonly used to create synthetic polymers, such as polyethylene and polypropylene, which offer a high degree of control over the molecular structure and physical properties of the material. These polymers are ideal for industrial applications, including building materials, packaging, and automotive components, due to their strength, resilience, and resistance to environmental degradation. However, the harm that synthetic polymers cause to the environment is becoming increasingly significant, especially since they are not biodegradable and contribute to the persistent pollution caused by plastic [80].

On the other hand, natural polymers with higher biocompatibility and biodegradability, such as chitosan, collagen, and alginate, come from renewable biological sources. In biomedical applications where contact with biological tissues and ultimate degradation within the body are crucial, these materials are preferred. Natural polymers are frequently utilized in tissue engineering, medication delivery, and the cure and healing of infections and wounds. They cannot be as versatile or strong mechanically as synthetic polymers, which makes them less suitable for high-stress situations.

One important field of current study is the synthesis of hybrid materials, which combine the best features of natural and artificial polymers. Through the addition of synthetic elements, these hybrid systems seek to make much better mechanical strength, consistency, and tunability of natural polymers while preserving their biocompatibility and sustainability [81].

FUTURE DIRECTIONS AND CHALLENGES

In this chapter, we discussed various synthetic pathways and their role in the formulation of biopolymers. Biopolymeric materials have been utilized in a wide range of bioengineering applications, offering superior qualities, such as physical and mechanical strength, durability, biocompatibility, and eco-friendliness. In the future, the most critical challenges in human life are expected to revolve around energy and resource management, healthcare, food security, communication, transportation, and infrastructure maintenance. Biopolymers can be utilized as an efficient tool for addressing these serious issues and problems. Biopolymers are anticipated to serve as key materials for the next era of development, where their synthesis and synthetic pathways will play a crucial role in advancing sustainable and biodegradable solutions. They could provide the optimal solution to the widespread use of non-degradable, environmentally harmful, and pollution-causing plastics and thermoplastics employed across various essential and developmental sectors. They could be the best options for use in medicines, the creation of medical devices, and may also lead to the production of artificial organs. In the future, pharmaceuticals may increasingly adopt various synthetic and natural biopolymers to replace toxic and non-degradable materials currently used in the synthesis and packaging of pharmaceutical products. Biopolymeric materials help conserve energy and enhance renewable resources. Based on the current scenario and evolving conditions, it is assumed that fossil fuels and petroleum-based resources will be unable to meet global needs. Biopolymeric materials are projected to play a crucial role in addressing future challenges. They may also be utilized as a replacement for batteries and electric generators. Furthermore, modified biopolymeric features may enhance the utilization of these materials in electrical devices and components. Biopolymers are expected to become more cost-effective as innovative solid materials, leading to a reduction in manufacturing expenses for large-scale production. Biopolymeric materials have the potential to replace conventional plastics in numerous applications, including the manufacture of plates, mugs, cutlery, food packaging, and storage, thereby contributing to the reduction of pollution caused by plastic waste. Many electrical industries have progressively adopted plastic materials for making and utilizing plugs, sockets, wire and cable insulators, bulbs, LEDs, household electricals, and electronic gadgets. The majority of polymer-based materials find applications in diverse fields such as ceramics, stem cell biology, pharmaceutical formulation and packaging, automotive components, aerospace materials, and various electronic devices. Some researchers also assume that biopolymeric packaging may advance and improve the accuracy of the expiry date and time of packed materials. Studies on biopolymers offer a ray of hope for nutritionally rich food packaging and extended expiry dates for both packed items and packaging materials. In the coming era, the development of biopolymer-based materials with tailored properties will be essential to support diverse applications in biomedical fields, bioengineering, and biochemistry. Advances in molecular biology are expected to further enhance the biological activity of biopolymer-based pharmaceutical and medical formulations. For example, cellulose, a sustainable biomedical material, has inspired various research groups to develop advanced cellulose-based products for use in tissue engineering platforms, wound and burn dressings, infection treatments, therapeutic transplants, and drug delivery systems due to its superior physical and biological properties, biocompatibility, biodegradability, and low cytotoxicity. Additionally, from a commercial perspective, this represents an exciting evolution and a promising new area of research. It offers benefits through the development of various biopolymer groups and derivatives derived from natural or synthetic sources, including artificial, organic, and inorganic materials. Biopolymers, derived from renewable resources, represent promising alternatives to conventional plastics, but they currently face several limitations and challenges. Major challenges may include increased manufacturing and production costs, difficulties in achieving large-scale production, and restricted or limited mechanical and physical properties. Furthermore, their biodegradability depends on the specific environmental conditions that vary for different types of biopolymers [82].

CONCLUSION

In this chapter, we discussed various natural, synthetic, and semi-synthetic pathways for the synthesis of biopolymers, which are utilized as therapeutic adjuvants. In the semi-synthetic domain, specific bacteria have been utilized for the stereospecific and thermoregulated synthesis of biopolymers. Biopolymers derived from natural resources have been replacing synthetic polymers, which are neither biodegradable nor eco-friendly. Biopolymers can be obtained through various extraction methods using waste management systems from agricultural, industrial, and natural resources. Their properties are more advantageous and comparable to other existing conventional petroleum-based materials. PHB plays a significant role as a biodegradable biopolymer in appropriate environmental conditions. The cost-efficient production of such biopolymers plays a crucial role in reducing the reliance on thermoplastic materials. Some primary properties of biopolymers, such as water solubility (hydrophilicity) and biocompatibility, have enhanced their utilization in various fields, including biomedicals, the food industry, and tissue engineering. Some synthetic biopolymers, such as PLA, are considered valuable and biodegradable materials that can be used extensively in various industries, including the medical field and healthcare systems. Various studies suggest that synthetic biopolymers are more advantageous than petroleum-based biopolymers, which depend on their characteristics, including physical, chemical, mechanical, biocompatibility, and biodegradability properties. These parameters enable their use in various applications, ranging from eco-friendly products to synthesis, drug delivery systems, foodstuffs, and packaging materials. Additionally, various derivatives are serving as models for future research groups, functioning as versatile biopolymers with properties that render them eco-friendly and sustainable. Consequently, biopolymers are regarded as a promising alternative to conventional petroleum-based materials.

LIST OF ABBREVIATIONS

BL   β-Butyrolactone

DPC   Direct polycondensation

LA   Lactic acid

PLA   Polylactic acid

PHB   Poly(3-hydroxybutyrate)

PHAs Polyhydroxyalkanoates

PCL   Poly(caprolactone)

PVA   Polyvinyl alcohol

PTT   Poly(trimethylene terephthalate)

PU   Polyurethane

PhaA β-ketothiolase

PhaB Acetoacetyl-CoA reductase

PhbC PHB synthase

ROP   Ring-opening polymerization

ACKNOWLEDGEMENTS

The authors would like to thank the Mahayogi Gorakhnath University, Gorakhpur, for providing the necessary facilities for the study.

AUTHORS' CONTRIBUTIONS

All authors contributed equally to the development of the chapter's framework and conceptualization for this book. This collaborative effort ensured a comprehensive and balanced perspective throughout the study.

REFERENCES

[1]   Panchal SS, Vasava DV. Biodegradable polymeric materials: synthetic approach. ACS Omega 2020; 5(9): 4370-9.
[http://dx.doi.org/10.1021/acsomega.9b04422] [PMID: 32175484]

[2]   Das A, Ringu T, Ghosh S, Pramanik N. A comprehensive review on recent advances in preparation, physicochemical characterization, and bioengineering applications of biopolymers. Polym Bull 2023; 80(7): 7247-312.
[http://dx.doi.org/10.1007/s00289-022-04443-4] [PMID: 36043186]

[3]   Gkountela CI, Vouyiouka SN. Enzymatic polymerization as a green approach to synthesizing bio-based polyesters. Macromol 2022; 2(1): 30-57.
[http://dx.doi.org/10.3390/macromol2010003]

[4]   Gao Q, Yang H, Wang C, *et al.* Advances and trends in microbial production of polyhydroxyalkanoates and their building blocks. Front Bioeng Biotechnol 2022; 10: 966598.
[http://dx.doi.org/10.3389/fbioe.2022.966598] [PMID: 35928942]

[5]   Owida HA, Al-Nabulsi JI, Alnaimat F, *et al.* Recent applications of electrospun nanofibrous scaffold in tissue engineering. Appl Bionics Biomech 2022; 2022(1): 1-15.
[http://dx.doi.org/10.1155/2022/1953861] [PMID: 35186119]

[6]   Singhvi MS, Zinjarde SS, Gokhale DV. Polylactic acid: synthesis and biomedical applications. J Appl Microbiol 2019; 127(6): 1612-26.
[http://dx.doi.org/10.1111/jam.14290] [PMID: 31021482]

[7]   Mishra S, Shah H, Patel A, Tripathi SM, Malviya R, Prajapati BG. Applications of bioengineered polymer in the field of nano-based drug delivery. ACS Omega 2024; 9(1): 81-96.
[http://dx.doi.org/10.1021/acsomega.3c07356] [PMID: 38222544]

[8]   McAdam B, Brennan Fournet M, McDonald P, Mojicevic M. Production of polyhydroxybutyrate (PHB) and factors impacting its chemical and mechanical characteristics. Polymers (Basel) 2020; 12(12): 2908.
[http://dx.doi.org/10.3390/polym12122908] [PMID: 33291620]

[9]   Rahmani S, Maroufkhani M, Mohammadzadeh-Komuleh S, Khoubi-Arani Z. Polymer nanocomposites for biomedical applications. Fundamentals of bionanomaterials. Elsevier 2022; pp. 175-215.
[http://dx.doi.org/10.1016/B978-0-12-824147-9.00007-8]

[10]   Baranwal J, Barse B, Fais A, Delogu GL, Kumar A. Biopolymer: A Sustainable Material for Food and Medical Applications. Polymers. 2022; 14(5): 983.
[http://dx.doi.org/10.3390/polym14050983]

[11]   Gaaz T, Sulong A, Akhtar M, Kadhum A, Mohamad A, Al-Amiery A. Properties and applications of polyvinyl alcohol, halloysite nanotubes, and their nanocomposites. Molecules 2015; 20(12): 22833-47.
[http://dx.doi.org/10.3390/molecules201219884] [PMID: 26703542]

[12]   Chuah HH. Crystallization kinetics of poly(trimethylene terephthalate). Polym Eng Sci 2001; 41(2):

308-13.
[http://dx.doi.org/10.1002/pen.10730]

[13]   Flórez M, Cazón P, Vázquez M. Selected biopolymers' processing and their applications: A review. Polymers (Basel) 2023; 15(3): 641.
[http://dx.doi.org/10.3390/polym15030641] [PMID: 36771942]

[14]   Rana H, Sharma A, Dutta S, Goswami S. Recent approaches on the application of agro waste-derived biocomposites as green support matrix for enzyme immobilization. J Polym Environ 2022; 30(12): 4936-60.
[http://dx.doi.org/10.1007/s10924-022-02574-3]

[15]   Holyavka MG, Goncharova SS, Redko YA, Lavlinskaya MS, Sorokin AV, Artyukhov VG. Novel biocatalysts based on enzymes in complexes with nano- and micromaterials. Biophys Rev 2023; 15(5): 1127-58.
[http://dx.doi.org/10.1007/s12551-023-01146-6] [PMID: 37975005]

[16]   Eskandar K. Revolutionizing biotechnology and bioengineering: unleashing the power of innovation. J Appl Biotechnol Bioeng 2023; 10(3): 81-8.
[http://dx.doi.org/10.15406/jabb.2023.10.00332]

[17]   Chen C, Chen X, Liu L, Wu J, Gao C. Engineering microorganisms to produce bio-based monomers: progress and challenges. Fermentation (Basel) 2023; 9(2): 137.
[http://dx.doi.org/10.3390/fermentation9020137]

[18]   Safdar A, Ismail F, Hussain A, *et al.* Implementation of nanobiocatalysis in food industry. In Bionanocatalysis: From Design to Applications 2023; 223-48.

[19]   Baranwal J, Barse B, Fais A, Delogu GL, Kumar A. Biopolymer: a sustainable material for food and medical applications. Polymers 2022; 14: 983.
[http://dx.doi.org/10.3390/polym14050983]

[20]   Jorda J, Kain G, Barbu MC, Köll B, Petutschnigg A, Král P. Mechanical properties of cellulose and flax fiber unidirectional reinforced plywood. Polymers (Basel) 2022; 14(4): 843.
[http://dx.doi.org/10.3390/polym14040843] [PMID: 35215756]

[21]   Lehrhofer AF, Goto T, Kawada T, Rosenau T, Hettegger H. The *in vitro* synthesis of cellulose – A mini-review. Carbohydr Polym 2022; 285: 119222.
[http://dx.doi.org/10.1016/j.carbpol.2022.119222] [PMID: 35287852]

[22]   Fatema N, Ceballos RM, Fan C. Modifications of cellulose-based biomaterials for biomedical applications. Front Bioeng Biotechnol 2022; 10: 993711.
[http://dx.doi.org/10.3389/fbioe.2022.993711] [PMID: 36406218]

[23]   Żołek-Tryznowska Z, Kałuża A. The influence of starch origin on the properties of starch films: packaging performance. Materials (Basel) 2021; 14(5): 1146.
[http://dx.doi.org/10.3390/ma14051146] [PMID: 33671033]

[24]   Rovira-Truitt R, Patil N, Castillo F, White JL. Synthesis and characterization of biopolymer composites from the inside out. Macromolecules 2009; 42(20): 7772-80.
[http://dx.doi.org/10.1021/ma901324b]

[25]   Maleki H, Azimi B, Ismaeilimoghadam S, Danti S. Poly (lactic acid)-based electrospun fibrous structures for biomedical applications. Appl Sci (Basel) 2022; 12(6): 3192.
[http://dx.doi.org/10.3390/app12063192]

[26]   Singh AK, Sharma L, Mallick N, Mala J. Progress and challenges in producing polyhydroxyalkanoate biopolymers from cyanobacteria. J Appl Phycol 2017; 29(3): 1213-32.
[http://dx.doi.org/10.1007/s10811-016-1006-1]

[27]   Kaur R, Pathak L, Vyas P. Biobased polymers of plant and microbial origin and their applications - a review. Biotechnol Sustain Mater 2024; 1(1): 13.
[http://dx.doi.org/10.1186/s44316-024-00014-x]

[28] Gaur N, Mishra S, Srivastava S, Parvez N. Naturapolyceutics-emerging science & technology in drug delivery system. Int J Pharm Res. 2020.
[http://dx.doi.org/10.31838/ijpr/2020.SP1.224]

[29] Mishra S, Shah H, Patel A, Tripathi SM, Malviya R, Prajapati BG. Applications of bioengineered polymer in the field of nano-based drug delivery. ACS omega. 2023 Dec 18; 9(1): 81-96.
[http://dx.doi.org/10.1021/acsomega.3c07356]

[30] Singh A, Kumari K, Kundu PP. Synthesis of biopolymer-polypeptide conjugates and their potential therapeutic interests. Phys Sci Rev 2024; 9(9): 2947-64.
[http://dx.doi.org/10.1515/psr-2022-0185]

[31] Prochon M, Dzeikala O, Szczepanik S. High-performance structures of biopolymer gels activated with scleroprotein crosslinkers. Molecules 2025; 30(3): 627.
[http://dx.doi.org/10.3390/molecules30030627] [PMID: 39942731]

[32] Velásquez, E., Guerrero Correa, M., Garrido, L., Guarda, A., Galotto, M.J., López de Dicastillo, C. 2021. Food Packaging Plastics: Identification and Recycling. In: Parameswaranpillai, J., Mavinkere Rangappa, S., Gulihonnehalli Rajkumar, A., Siengchin, S. (eds) Recent Developments in Plastic Recycling. Composites Science and Technology. Springer, Singapore.
[http://dx.doi.org/10.1007/978-981-16-3627-1_14]

[33] Pooja N, Banik S, Chakraborty I, *et al.* Comparative analysis of biopolymer films derived from corn and potato starch with insights into morphological, structural and thermal properties. Discov Sustain 2024; 5(1): 467.
[http://dx.doi.org/10.1007/s43621-024-00626-3]

[34] Murray B, Rosenbloom C. Fundamentals of glycogen metabolism for coaches and athletes. Nutr Rev 2018; 76(4): 243-59.
[http://dx.doi.org/10.1093/nutrit/nuy001] [PMID: 29444266]

[35] Adeva-Andany MM, González-Lucán M, Donapetry-García C, Fernández-Fernández C, Ameneiros-Rodríguez E. Glycogen metabolism in humans. BBA Clin 2016; 5: 85-100.
[http://dx.doi.org/10.1016/j.bbacli.2016.02.001] [PMID: 27051594]

[36] Chen P, Zhao C, Wang H, *et al.* Quantifying the contribution of the dispersion interaction and hydrogen bonding to the anisotropic elastic properties of Chitin and Chitosan. Biomacromolecules 2022; 23(4): 1633-42.
[http://dx.doi.org/10.1021/acs.biomac.1c01488] [PMID: 35352926]

[37] Lenardon MD, Munro CA, Gow NAR. Chitin synthesis and fungal pathogenesis. Curr Opin Microbiol 2010; 13(4): 416-23.
[http://dx.doi.org/10.1016/j.mib.2010.05.002] [PMID: 20561815]

[38] Varma S, Jaber M, Fanas S, Desai V, Al Razouk A, Nasser S. Effect of hyaluronic acid in modifying tensile strength of nonabsorbable suture materials: An *in vitro* study. J Int Soc Prev Community Dent 2020; 10(1): 16-20.
[http://dx.doi.org/10.4103/jispcd.JISPCD_343_19] [PMID: 32181217]

[39] Papakonstantinou E, Roth M, Karakiulakis G. Hyaluronic acid: A key molecule in skin aging. Dermatoendocrinol 2012; 4(3): 253-8.
[http://dx.doi.org/10.4161/derm.21923] [PMID: 23467280]

[40] Liu Y, Ballarini R, Eppell SJ. Tension tests on mammalian collagen fibrils. Interface Focus 2016; 6(1): 20150080.
[http://dx.doi.org/10.1098/rsfs.2015.0080] [PMID: 26855757]

[41] Wenger MPE, Bozec L, Horton MA, Mesquida P. Mechanical properties of collagen fibrils. Biophys J 2007; 93(4): 1255-63.
[http://dx.doi.org/10.1529/biophysj.106.103192] [PMID: 17526569]

[42] Shenoy M, Abdul NS, Qamar Z, *et al.* Collagen structure, synthesis, and its applications: A systematic

review. Cureus 2022; 14(5): e24856.
[http://dx.doi.org/10.7759/cureus.24856] [PMID: 35702467]

[43]   Bonser RHC, Purslow PP. The Young's modulus of feather keratin. J Exp Biol 1995; 198(4): 1029-33.
[http://dx.doi.org/10.1242/jeb.198.4.1029] [PMID: 9318836]

[44]   Taichman LB, Prokop CA. Synthesis of keratin proteins during maturation of cultured human keratinocytes. J Invest Dermatol 1982; 78(6): 464-7.
[http://dx.doi.org/10.1111/1523-1747.ep12510153] [PMID: 6177798]

[45]   Guthold M, Liu W, Sparks EA, *et al.* A comparison of the mechanical and structural properties of fibrin fibers with other protein fibers. Cell Biochem Biophys 2007; 49(3): 165-81.
[http://dx.doi.org/10.1007/s12013-007-9001-4] [PMID: 17952642]

[46]   Mehta-Ambalal S. Neocollagenesis and neoelastinogenesis: From the laboratory to the clinic. J Cutan Aesthet Surg 2016; 9(3): 145-51.
[http://dx.doi.org/10.4103/0974-2077.191645] [PMID: 27761083]

[47]   Kreplak L, Herrmann H, Aebi U. Tensile properties of single desmin intermediate filaments. Biophys J 2008; 94(7): 2790-9.
[http://dx.doi.org/10.1529/biophysj.107.119826] [PMID: 18178641]

[48]   Ingwall JS, Morales MF, Stockdale FE. Creatine and the control of myosin synthesis in differentiating skeletal muscle. Proc Natl Acad Sci USA 1972; 69(8): 2250-3.
[http://dx.doi.org/10.1073/pnas.69.8.2250] [PMID: 4506094]

[49]   Wagh YR, Pushpadass HA, Emerald FME, Nath BS. Preparation and characterization of milk protein films and their application for packaging of Cheddar cheese. J Food Sci Technol 2014; 51(12): 3767-75.
[http://dx.doi.org/10.1007/s13197-012-0916-4] [PMID: 25477643]

[50]   Homareda H, Komine S. Casein synthesis by mouse polysomes and their messenger ribonucleic acid extracts. J Dairy Sci 1982; 65(6): 915-9.
[http://dx.doi.org/10.3168/jds.S0022-0302(82)82291-1] [PMID: 7108010]

[51]   Cave ID. The anisotropic elasticity of the plant cell wall. Wood Sci Technol 1968; 2(4): 268-78.
[http://dx.doi.org/10.1007/BF00350273]

[52]   Vanholme R, Demedts B, Morreel K, Ralph J, Boerjan W. Lignin biosynthesis and structure. Plant Physiol 2010; 153(3): 895-905.
[http://dx.doi.org/10.1104/pp.110.155119] [PMID: 20472751]

[53]   Normand V, Lootens DL, Amici E, Plucknett KP, Aymard P. New insight into agarose gel mechanical properties. Biomacromolecules 2000; 1(4): 730-8.
[http://dx.doi.org/10.1021/bm005583j] [PMID: 11710204]

[54]   Jarosz A, Kapusta O, Gugała-Fekner D, Barczak M. Synthesis and characterization of agarose hydrogels for release of diclofenac sodium. Materials (Basel) 2023; 16(17): 6042.
[http://dx.doi.org/10.3390/ma16176042] [PMID: 37687735]

[55]   West E, Xu M, Woodruff T, Shea L. Physical properties of alginate hydrogels and their effects on *in vitro* follicle development. Biomaterials 2007; 28(30): 4439-48.
[http://dx.doi.org/10.1016/j.biomaterials.2007.07.001] [PMID: 17643486]

[56]   Hay ID, Rehman ZU, Moradali MF, Wang Y, Rehm BHA. Microbial alginate production, modification and its applications. Microb Biotechnol 2013; 6(6): 637-50.
[http://dx.doi.org/10.1111/1751-7915.12076] [PMID: 24034361]

[57]   Mohnen D. Pectin structure and biosynthesis. Curr Opin Plant Biol 2008; 11(3): 266-77.
[http://dx.doi.org/10.1016/j.pbi.2008.03.006] [PMID: 18486536]

[58]   Freitas CMP, Coimbra JSR, Souza VGL, Sousa RCS. Structure and applications of pectin in food, biomedical, and pharmaceutical industry: A review. Coatings 2021; 11(8): 922.

[http://dx.doi.org/10.3390/coatings11080922]

[59]   Jameson PE. Zeatin: The 60th anniversary of its identification. Plant Physiol 2023; 192(1): 34-55.
        [http://dx.doi.org/10.1093/plphys/kiad094] [PMID: 36789623]

[60]   Großkinsky D, Edelsbrunner K, Pfeifhofer H, Van der Graaff E, Roitsch T. Cis- and trans-zeatin
        differentially modulate plant immunity. Plant Signal Behav 2013; 8(7): e24798.
        [http://dx.doi.org/10.4161/psb.24798] [PMID: 23656869]

[61]   Jung G, Qin Z, Buehler MJ. Mechanical properties and failure of biopolymers: atomistic reactions to
        macroscale response. Top Curr Chem. 2015, 369, 317-43.
        [http://dx.doi.org/10.1007/128_2015_643]

[62]   Das, A., Ringu, T., Ghosh, S. et al. A comprehensive review on recent advances in preparation,
        physicochemical characterization, and bioengineering applications of biopolymers. Polym. Bull. 2023;
        80: 7247–7312 .
        [http://dx.doi.org/10.1007/s00289-022-04443-4]

[63]   Baranwal J, Barse B, Fais A, Delogu GL, Kumar A. Biopolymer: A sustainable material for food and
        medical applications. Polymers (Basel) 2022; 14(5): 983.
        [http://dx.doi.org/10.3390/polym14050983] [PMID: 35267803]

[64]   Samir A, Ashour FH, Hakim AAA, Bassyouni M. Recent advances in biodegradable polymers for
        sustainable applications. npj Mater Degrad 2022; 6(1): 68.
        [http://dx.doi.org/10.1038/s41529-022-00277-7]

[65]   Chattopadhyay DK, Webster DC. Thermal stability and flame retardancy of polyurethanes. Prog
        Polym Sci 2009; 34(10): 1068-133.
        [http://dx.doi.org/10.1016/j.progpolymsci.2009.06.002]

[66]   McNamara JT, Morgan JL, Zimmer J. A molecular description of cellulose biosynthesis. Annu Rev
        Biochem. 2015; 84: 895-921.
        [http://dx.doi.org/10.1146/annurev-biochem-060614-033930] [PMID: 26034894]

[67]   Aranaz I, Alcántara AR, Civera MC, *et al.* Chitosan: An overview of its properties and applications.
        Polymers (Basel) 2021; 13(19): 3256.
        [http://dx.doi.org/10.3390/polym13193256] [PMID: 34641071]

[68]   Mohanan N, Montazer Z, Sharma PK, Levin DB. Microbial and enzymatic degradation of synthetic
        plastics. Front Microbiol 2020; 11: 580709.
        [http://dx.doi.org/10.3389/fmicb.2020.580709] [PMID: 33324366]

[69]   Desai N, Rana D, Salave S, *et al.* Chitosan: a potential biopolymer in drug delivery and biomedical
        applications. Pharmaceutics 2023; 15(4): 1313.
        [http://dx.doi.org/10.3390/pharmaceutics15041313] [PMID: 37111795]

[70]   Dintzis HM, Dintzis RZ, Vogelstein B. Molecular determinants of immunogenicity: the immunon
        model of immune response. Proc Natl Acad Sci USA 1976; 73(10): 3671-5.
        [http://dx.doi.org/10.1073/pnas.73.10.3671] [PMID: 62364]

[71]   Bher A, Mayekar PC, Auras RA, Schvezov CE. Biodegradation of biodegradable polymers in
        mesophilic aerobic environments. Int J Mol Sci 2022; 23(20): 12165.
        [http://dx.doi.org/10.3390/ijms232012165] [PMID: 36293023]

[72]   Tennakoon P, Chandika P, Yi M, Jung WK. Marine-derived biopolymers as potential bioplastics, an
        eco-friendly alternative. iScience 2023; 26(4): 106404.
        [http://dx.doi.org/10.1016/j.isci.2023.106404] [PMID: 37034997]

[73]   Gough CR, Callaway K, Spencer E, *et al.* Biopolymer-based filtration materials. ACS Omega 2021;
        6(18): 11804-12.
        [http://dx.doi.org/10.1021/acsomega.1c00791] [PMID: 34056334]

[74]   Naser AZ, Deiab I, Darras BM. Poly(lactic acid) (PLA) and polyhydroxyalkanoates (PHAs), green

alternatives to petroleum-based plastics: a review. RSC Advances 2021; 11(28): 17151-96.
[http://dx.doi.org/10.1039/D1RA02390J] [PMID: 35479695]

[75]   Pinaeva LG, Noskov AS. Biodegradable biopolymers: Real impact to environment pollution. Sci Total Environ 2024; 947: 174445.
[http://dx.doi.org/10.1016/j.scitotenv.2024.174445] [PMID: 38981547]

[76]   Kim Y, Zharkinbekov Z, Raziyeva K, *et al.* Chitosan-based biomaterials for tissue regeneration. Pharmaceutics 2023; 15(3): 807.
[http://dx.doi.org/10.3390/pharmaceutics15030807] [PMID: 36986668]

[77]   Satchanska G, Davidova S, Petrov PD. Natural and synthetic polymers for biomedical and environmental applications. Polymers (Basel) 2024; 16(8): 1159.
[http://dx.doi.org/10.3390/polym16081159] [PMID: 38675078]

[78]   Zhao L, Zhou Y, Zhang J, Liang H, Chen X, Tan H. Natural polymer-based hydrogels: From polymer to biomedical applications. Pharmaceutics 2023; 15(10): 2514.
[http://dx.doi.org/10.3390/pharmaceutics15102514] [PMID: 37896274]

[79]   Ojha S, Sharma S, Mishra S. Hydrogels as potential controlled drug delivery system: drug release mechanism and applications. Nanoscience & Nanotechnology-Asia. 2023 Jun 1;13(3):42-50.

[80]   Flury M, Narayan R. Biodegradable plastic as an integral part of the solution to plastic waste pollution of the environment. Curr Opin Green Sustain Chem 2021; 30: 100490.
[http://dx.doi.org/10.1016/j.cogsc.2021.100490]

[81]   Manfra L, Marengo V, Libralato G, Costantini M, De Falco F, Cocca M. Biodegradable polymers: A real opportunity to solve marine plastic pollution? J Hazard Mater 2021; 416: 125763.
[http://dx.doi.org/10.1016/j.jhazmat.2021.125763] [PMID: 33839500]

[82]   Gross RA, Kalra B. Biodegradable polymers for the environment. Science 2002; 297(5582): 803-7.
[http://dx.doi.org/10.1126/science.297.5582.803] [PMID: 12161646]

CHAPTER 4

Biopolymer Mechanism: Pharmacokinetics and Pharmacodynamics

Ajay Pandey[1,*] and **Bharat Mishra**[2]

[1] *Department of Pharmaceutical Engineering and Technology, Indian Institute of Technology (Banaras Hindu University), Varanasi, India*

[2] *Institute of Pharmacy DR. Shakuntala Misra National Rehabilitation University, Mohan Road, Lucknow 226017, India*

Abstract: Biopolymers have become apparent as a potential therapeutic material, which shows an important role in treating different diseases like cancer, neurological disorders, cardiovascular disorders, infectious diseases, diabetes, and drug delivery. Due to its biocompatible and biodegradable properties, it is widely used for therapeutic purposes. This chapter provides detailed information regarding the pharmacokinetic and pharmacodynamic properties of biopolymers. It involves a comprehensive analysis of how these materials are involved in drug absorption, distribution, metabolism, excretion, and their interaction with physiological systems. The pharmacokinetic properties of biopolymers depend on various factors such as molecular weight, degradation behavior, and structure of biopolymers, which are important for therapeutic efficacy. On the other hand, pharmacodynamic properties provide detailed information regarding mechanisms like cellular uptake and internalization, modulation in immunological responses, and physical and chemical interactions by which these materials show therapeutic responses by targeting specific cells or tissues, extending beyond the target consequences, and receptor-associated interactions. Recent advancements in the field of biopolymers involve their utilization for specific therapeutic effects, enhancing patient compliance and outcomes, with various mechanisms of action involved for showing therapeutic responses by biopolymers mentioned in this chapter. By gathering information regarding biopolymers' pharmacokinetic and pharmacodynamic properties, this chapter aims to provide detailed information on various mechanisms involving the effectiveness and efficacy of biopolymer-based therapeutics in clinical utilization.

Keywords: Applications, Biopolymers, Mechanism of action, Pharmacodynamics, Pharmacokinetics.

---

* **Corresponding author Ajay Pandey:** Department of Pharmaceutical Engineering and Technology, Indian Institute of Technology (Banaras Hindu University), Varanasi, India; E-mail: pandeyajay.1633@gmail.com

INTRODUCTION

Biopolymers are macromolecular compounds that are obtained from resources found in nature and can be synthesized chemically with the help of biological elements or can be fully biosynthesized by living things [1]. The structural composition of these compounds is carbon, nitrogen, and oxygen, which makes biopolymers susceptible to degradation. The broken parts of biopolymers, like organic macromolecules, water, and some other natural products, do not cause any harm to nature. Biopolymers combine bio and polymers, respectively, meaning life and nature [2].

Biopolymers are utilized for their emerging properties like biocompatibility, no immunogenic reactions, lack of toxicity, extended systemic circulation, and biodegradability. Apart from these properties, these are non-carcinogenic and non-thrombogenic, which makes them eco-friendly and excellent materials for the delivery of larger as well as smaller drugs [3, 4]. A comparison of different biopolymers for drug delivery is mentioned in Table **1**. Biopolymers consist of recurring sequences with various recurring functional moieties (such as carboxyl, hydroxyl, and amino) and a large range of chemical constituents, which makes biopolymers vulnerable to cross-linking [5]. Applications of biopolymers are mentioned in Fig. (**1**).

Table 1. Characteristics and applications of biopolymers [4, 6, 7].

Biopolymer	Advantages	Disadvantages	Suitable Drug Types	Delivery Routes A
PLGA (Poly(lactic-co-glycolic acid))	Biodegradable, biocompatible, controlled drug release, FDA-approved	Acidic degradation byproducts, burst release possible	Small molecules, peptides, and proteins	Injectable, implantable, oral
PLA (Polylactic Acid)	Biocompatible, slow degradation, good mechanical strength	Hydrophobic, low protein adsorption, acidic degradation	Hydrophobic drugs, proteins, and peptides	Injectable, implantable, transdermal
Polysaccharides	Natural, biocompatible, mucoadhesive, controlled drug release	Variable properties, stability concerns, batch-to-batch variations	Small molecules, peptides, proteins, and vaccines	Oral, injectable, transdermal
Collagen	Biodegradable, promotes cell adhesion, widely used in tissue engineering	Expensive, potential immunogenicity	Proteins, peptides, growth factors	Injectable, implantable, transdermal

(Table 1) cont.....

Biopolymer	Advantages	Disadvantages	Suitable Drug Types	Delivery Routes A
Gelatin	Biocompatible, thermosensitive gelation, widely used in drug delivery	Enzymatic degradation, poor mechanical strength	Hydrophilic drugs, proteins, and vaccines	Injectable, transdermal
Starch	Biodegradable, easily modifiable, widely available	Swelling issues, limited mechanical strength	Small molecules, peptides	Oral, transdermal, injectable
Hyaluronic Acid	Biodegradable, enhances drug permeability, high water retention	Rapid degradation, expensive	Hydrophilic drugs, peptides, and proteins	Ocular, injectable, transdermal
Polynucleotides	Biocompatible, essential in gene therapy, highly specific	Stability issues, enzymatic degradation	Gene-based drugs, oligonucleotides	Injectable, transdermal
DNA	Highly specific, enables gene therapy, stable when encapsulated	Prone to enzymatic degradation, immune activation risk	Gene-based drugs, plasmids	Injectable, oral
RNA	High specificity, potential for mRNA vaccines, and gene silencing	Unstable, rapid enzymatic degradation	mRNA vaccines, siRNA, gene therapy	Injectable, nasal
Proteins and Peptides	High specificity, biocompatible, diverse functionality	Short half-life, requires stabilization	Therapeutic peptides, enzymes, and hormones	Injectable, oral, transdermal
Zinc-containing Proteins	Important for enzyme activation, immune modulation	Potential toxicity at high doses, stability concerns	Enzyme-based drugs, protein therapies	Injectable, oral
Bovine Serum Albumin (BSA)	High drug binding capacity, stabilizes drugs, biocompatible	Potential immunogenicity, batch variability	Small molecules, proteins, and peptides	Injectable, oral, transdermal
Chitosan	Biodegradable, mucoadhesive, enhances drug permeation	Limited solubility at physiological pH, possible immunogenicity	Peptides, proteins, and hydrophilic drugs	Oral, nasal, transdermal, ocular

Chemically, biopolymers comprise different functional groups such as amino, hydroxyl, and carboxyl. They become reactive and vulnerable to cross-linking as a result. Thus, large molecular weight molecules with repetitive sequences are known as biopolymers, and they may have a high potential for chemical reactions with different substances [5].

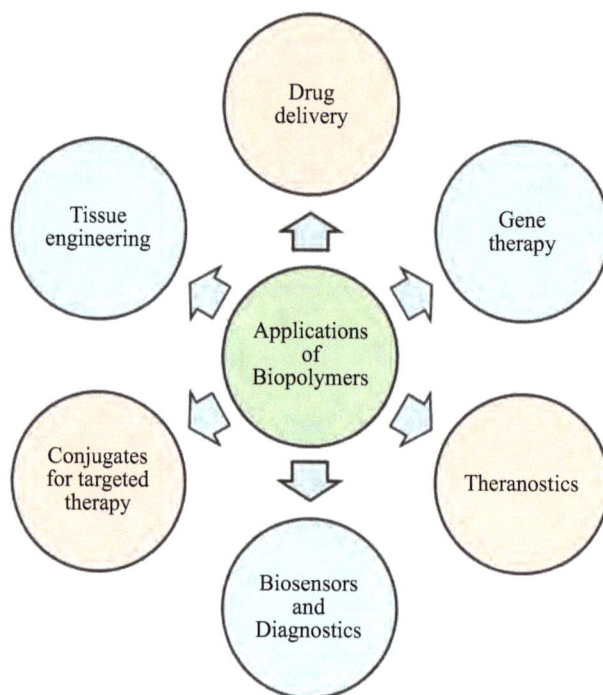

Fig. (1). Applications of biopolymers.

Pharmacokinetics

"Pharmacokinetics" refers to the action of physiological systems on administered pharmaceutical ingredients, which involves four phases such as absorption, distribution, metabolism and elimination. When medications are taken into the body and move from the point of incorporation into the circulatory system, this is referred to as absorption and is the initial stage of pharmacokinetics. Distribution is the next stage in pharmacokinetics. As the third phase in pharmacokinetics, it results in the inactivation of medications, and excretion, the final stage of pharmacokinetics, explains how inactivated metabolites are eliminated by the body's organs [8].

Pharmacodynamics

The word "Pharmacodynamics" refers to the way that drugs function in the physiological system. As a medication passes through the bloodstream, its particular affinity—the extent to which it binds to a drug-receptor site is explained by the process of pharmacodynamics. The way medicines work and whether they stay in circulation after delivery is influenced by lock and key processes linking the drugs and receptor sites. In medicine, the term "mechanism of action" describes how a medication works inside the body. The way a medicine associates with a certain receptor can be used to determine its mode of action.

Many medications attach to certain receptors on the surface of cells to work. Medications can act on receptor sites in an agonistic or antagonistic manner. An agonist forms a strong binding with a receptor to accomplish a desired result. By competing with other chemicals, an antagonist stops a specific activity or response at a receptor region [8].

PHARMACOKINETICS OF BIOPOLYMERS

As mentioned above, pharmacokinetics involves the interaction of medicinal products with physiological systems. The term pharmacokinetics comprises two Greek words, "*Pharmakon*" and "*kinetikos*," meaning putting a drug in motion. Pharmacokinetics is also explained as a quantitative analysis of absorption, distribution, metabolism, and excretion [9]. Various pharmacokinetic parameters are mentioned in Fig. (**2**). These four factors collectively affect drug concentrations and the rate at which drugs are exposed to tissues, which in turn affects the efficacy and pharmacological action of medicine and other pharmaceutical inactive excipients. Toxicological testing and ADME profiling are carried out throughout the development and discovery of new drugs. A novel physiologically active compound's ADME/Tox characteristics will ascertain whether it may be turned into a beneficial medication item. Pharmaceutical dosage forms, including a tablet or parenteral solution, contain several inactive ingredients in addition to the active pharmaceutical ingredient (API). Components that are referred to as pharmaceutical excipients or just excipients, which derive from the Latin word "*to except*" or "*other*," compared to the active pharmaceutical ingredients. Even when an excipient is thought to be an inert pharmaceutical component, it makes it possible to provide patient access to the API.

Various Parameters of Pharmacokinetics

Whenever an active pharmaceutical ingredient or excipients are administered into the physiological environment, they get eliminated from the body by following various pharmacokinetic processes depending on the route of administration. ADME is characterized by several pharmacokinetic parameters, including volume of distribution (V_d), elimination half-life ($t_{1/2}$), clearance (Cl), and bioavailability (F), and mostly pharmacokinetic mechanisms involve first-order kinetics. V_d explains the drug's internal distribution throughout the body. Smaller compounds (*e.g.*, 0.2L/kg) are mostly found in blood plasma, but bigger compounds (*e.g.*, considerably greater as compared with physiological volume) may be heavily attached to tissue. The elimination half-life of chemical substances is the duration needed to lower their plasma concentration in the circulatory system by 50%.

Another measure that shows how quickly a medication leaves the body is called clearance, which can be separated into two primary elimination mechanisms, hepatic clearance (ClH) and renal clearance (ClR). Whereas ClR is linked to the excretion of unmetabolized compounds in the urine, ClH is linked to the compound's metabolism in the liver. ClH plus ClR adds up to the total body clearance (ClT), which may be expressed as ClT=ClH+ClR. The rate and extent of an API or excipients that reaches the systemic circulation are known as its bioavailability. Although following PO administration, high bioavailability (*i.e.*, almost total absorption [F_1]) of API is preferred, with the majority of the excipient being processed from the stomach or expelled undigested in feces [10]. Pharmacokinetic parameters are mentioned in Fig. (**2**).

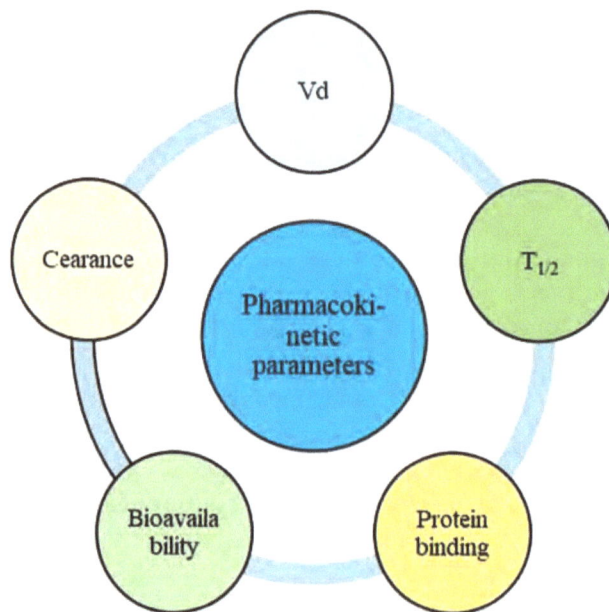

Fig. (2). Pharmacokinetic parameter.

Drugs and other inactive excipients are mostly eliminated *via* the circulatory system by the process of filtration by the glomerulus in the kidneys. Drugs that have been globularly filtered may be reabsorbed passively in the peritubular capillaries [9, 10].

ABSORPTION

Absorption is one of the crucial components of pharmacokinetics that impacts the bioavailability and therapeutic efficacy of formulations, especially in the context of drug administration. Because of their higher molecular size, intricate structural makeup, and interactions with biological surroundings, polymers display distinct absorption properties in contrast to tiny molecules. The absorption method is greatly influenced by the polymer structural changes, hydrophilicity, molecular weight, and mode of administration. For macromolecular compounds and drugs, the oral route is seen to be among the best ways to provide medication. The absorption, distribution, and removal of medication from the body require it to cross several biological membranes [11].

Bioavailability

Bioavailability is the extent to which a substance or drug meets its designated biological target (s). More specifically, bioavailability is the proportion and pace at which the initially administered amount of a chemical penetrates the site of effect or physiological fluid region.

The bioavailability (F) of a drug administered through non-intravenous routes can be determined by dividing the quantity that enters the bloodstream by the entire dosage of the medication.

F = Drug concentration delivered to the circulatory system ÷ The cumulative amount of the drug administered

Various Mechanisms are Involved in the Absorption of Biopolymers

Passive Diffusion

The most common method of drug or macromolecular absorption is passive diffusion, which is based on Fick's rule of diffusion, which suggests a drug or macromolecule moves through a concentration gradient from higher drug concentrations to lower amounts until the state of equilibrium is reached. The association of medications or macromolecules to carriers occurs during active diffusion. This complex dissociates on the opposite side of the membrane after helping the medication get through it. Potentially very particular to the medication molecule is the carrier molecule. Drugs with comparable structures may contend with one another for the carrier in absorption regions. Since there are only a limited number of carrier molecules accessible if the drug concentration is sufficiently large, the binding sites on the carrier may become saturated, at which point an increase in dosage does not affect the drug concentration. Certain transporters can significantly hinder drug absorption, whereas others, including P-glycoprotein (P-gp), actually help with absorption. Energy-dependent P-gp (MDR1) is an efflux transporter that helps secrete molecules back towards the intestinal lumen, which limits the overall absorption [12].

Active Transport

Active transport is characterized by the energy-mediated process where proteins present on the membrane help to transfer macromolecules into the cells. As the sodium/potassium-ATPase and hydrogen-ATPase pumps show, the mode of action through a chemical compound, including ATP hydrolysis, propels the immediate progression of the substance to produce specific variations in concentration. It occurs when substances, similar to glucose or amino acids, diffuse from areas of high concentration to low concentration through protein carriers or membrane openings.

Several compounds cannot pass across cell membranes without the help of transmembrane proteins, since their mobility would be hindered by the phospholipid bilayer or electrochemical gradient. One way that cells achieve this movement is by active transport, which involves acting to prevent the establishment of an equilibrium. This is done by concentrating molecules based on different demands required for the cell, such as ions, carbohydrates, and amino acids. Transmembrane ATPases are mostly used in primary active transport, which also frequently uses ion pumps and channels to move metal ions such as sodium, potassium, magnesium, and calcium. Secondary active (coupled) transport moves other molecules against their respective gradients by utilizing the energy within electrochemical potential differences resulting from direct active transport, which is primarily produced by sodium ions with the sodium-potassium ATPase. Notably, this occurs without directly coupling to ATP [13].

Endocytosis

This delicate transport mechanism works by swallowing external components within a section of the cell membrane to create a vesicle, which is subsequently pinched off intracellularly. This process is also known as corpuscular or vesicular transport. This is the sole transport method that allows a substance or medication to be absorbed without needing to be in an aqueous phase. The cellular absorption of macromolecular nutrients like lipids and starches, water-soluble vitamins like B12, oil-soluble vitamins like A, E, K, and D, and medications like insulin is caused by this process. The fact that the medication avoids first-pass hepatic metabolism by being absorbed into the lymphatic system adds further relevance to this process [14].

DISTRIBUTION

The passage of a medication or macromolecules across the body's extravascular (intracellular & extracellular) and intravascular (blood/plasma) compartments is referred to as the distribution mechanism. A pharmaceutical product exists in equilibrium across every physiological compartment, either in its free or protein-bound form. Active macromolecules in the bloodstream will eventually be processed by the liver and kidneys [15].

Compartment Models for Distribution

A drug first reaches the "central" compartment when an intravenous (IV) bolus is administered. This compartment contains the plasma, organs with good blood perfusion, such as the liver and kidneys, and additional tissues where the drug is rapidly absorbed. The "peripheral" compartment, which consists of tissues wherein drug distribution happens at a slower pace, may eventually begin to replace the central compartment as the site for certain drug distribution.

One-compartment Model

Certain medications and macromolecules disperse "instantaneously. These medications don't spread to the peripheral compartments but rather stay in the central compartment. As a consequence, a reduction in drug plasma concentration occurs due to the removal of compounds from the body. Since these medications do not migrate to peripheral compartments, they are considered to exhibit single-compartment models of distribution. One measure, the V_d of the central compartment (Vc), may be used to indicate the V_d of these medications [16].

$$Vc~(L) = \text{Dose administered (mg)} / Co~(mg/L)~[17].$$

For medications exhibiting a straight line graph on plasma *vs.* time curves for single-compartment distribution kinetics, it is difficult to determine the drug's initial plasma concentration at time = 0 (Co) since the medication is believed to disperse instantly. As a result, the initial plasma concentration of the drug is calculated by extrapolating to time = 0 on a plasma concentration *vs.* time curve [17, 18].

Multi-compartment Model

The majority of medications or macromolecules have delayed distribution kinetics, meaning that an early period of dispersion follows a later phase of elimination. Drugs with multi-compartment distribution models move from their central compartment into outer compartments before being removed. Phases associated with models of multi-compartment placement include:

Phase of distribution: After injection, the body's overall drug concentration stays constant, while the concentration of the drug or macromolecules within plasma first decreases. A single medication may have many time-dependent V_d levels as a result of this phenomenon [18].

Phase of terminal elimination: The medication will be removed from the central compartment (excretory systems) after the distribution phase, which will alter the drug's body content and plasma concentration. As a result, during the terminal elimination phase (Vbeta), additional V_d values can be calculated and depend on drug clearance.

Steady-state: There is a transitional state between the distribution and elimination phases that is referred to as the "steady state". When a medicine has reached "dynamic equilibrium" and is fully distributed throughout the body's central and peripheral compartments, it is said to be in a steady state. The drug's net flow among the central and peripheral compartments is zero in a steady state. A different value (Vss) for V_d can be computed in the steady-state phase. Since the observation will be used to calculate a drug's loading dosage, it is often the most therapeutically significant one [18].

$$Vd \ (L) = A_{(t)} \ (mg) \ / \ C_{(t)} \ (mg/L)$$

A (t) represents the amount of drug in the body at time = t

C (t) represents the plasma concentration of the drug at time = t

Volume of Distribution

One pharmacokinetic metric that shows how a medication tends to either remain in circulation or disseminate into other tissue compartments is the volume of distribution (V_d). In essence, V_d connects the drug's overall amount in the human system to its plasma concentration at a given moment in time.

Volume of Distribution (L) = Amount of drug in the body (mg) / Plasma concentration of drug (mg/L)

High V_d needs a greater dosage to reach the specified plasma concentration because it is more likely to exit the plasma and enter the body's extravascular compartments. (High V_d -> Greater dispersion to adjacent tissues) .On the other hand, low V_d tends to stay in the plasma, which means that a lower medication dosage is needed to reach a certain plasma concentration. Reduced distribution to other tissues due to low V_d [18].

Half-life

The term "half-life" ($t_{1/2}$) describes how long it takes for a drug's plasma concentration to drop by 50%. The rate constant (k), which is connected to V_d and clearance, determines $t_{1/2}$ [19].

Half-life (hours)= 0.693 x (Volume of distribution (L)/ Clearance (L/hr))

At a constant clearance rate, the compound having a higher V_d will have a longer elimination half-life than one with a lower V_d [19].

ELIMINATION

The process of metabolism and excretion together is considered the elimination process, where macromolecules that are hydrophobic need to be metabolically changed to become more polar to be eliminated. Conversely, hydrophilic medications do not require metabolic modifications to their molecular structures to be excreted directly [20].

Metabolism

The majority of medications are xenobiotics, or macromolecules that are detoxified by the body through a variety of mechanisms, which lessen their toxicity and make them easily excreted. These mechanisms, sometimes referred to as drug metabolism or metabolic biotransformation, enable the chemical conversion of pharmaceutical macromolecules into their metabolites. These metabolites, which come in three different forms—toxic, inactive, and active—are the end products of metabolism. Inactive metabolites are biochemically inert substances with neither poisonous nor beneficial properties, whereas active metabolites are chemically active substances with therapeutic effects. Like active metabolites, toxic metabolites are biochemically active substances with various negative consequences [21, 22].

Pharmaceutical compounds or macromolecules are mostly eliminated through the body by the kidneys; despite this, lipophilic substances easily pass through the renal tubules' cell membrane and get reabsorbed back into the blood. Therefore, before the compounds can be excreted, lipophilic substances have to be processed in the liver. Phase I (alteration), Phase II (conjugation), and occasionally Phase III (further modification and excretion) are the different processes that might occur during drug metabolism.

Phase I alterations include oxidation, reduction, hydrolysis, cyclization/ decyclization, and addition or subtraction of oxygen from more polar molecules to change the chemical composition of lipophilic drugs. This mechanism can occasionally turn an inert prodrug into a metabolically active drug. Metabolites produced by oxidation usually have some residual pharmacological action. Methylation, acetylation, sulphation, glucuronidation, and glutathione or glycine conjugation are types of conjugation mechanisms. A macromolecular compound interacts with a different substance in a conjugation process during phase II modifications [23]. A molecule that has undergone conjugation is often water soluble and pharmacologically inactive, making it easy to eliminate. Following the excretion of conjugates and metabolites from the cells during phase II metabolism, phase III metabolism could happen as well. The enzymatic catalysis of phase I and II activities is a crucial component of drug metabolism. Effective drug metabolism depends on the kind and quantity of liver enzymes. Cytochrome P450 and monoamine oxidase are the two enzymes that are most frequently utilized in medicine. Several hundred biomolecules and xenobiotic compounds are metabolized by these two enzymes [24].

Excretion

The method by which the macromolecular compounds are removed from the body is called excretion. Excretion of some compounds usually happens through the kidneys, although it can also happen through the gastrointestinal system, skin, or lungs. Drugs can be eliminated from the body by extrusion in the tubules or passive filtration in the glomerulus, which can be hampered by some substances' reabsorption [25].

Clearance

It may be defined as the ratio of a drug's rate of removal to its concentration in plasma. This is dependent upon the medication, blood flow, and organ function (typically the kidneys) of the patient. The total blood flow across the organ would constitute the limiting factor in the ideal extraction organ, where blood would be free of drugs. Comprehending clearance enables professionals to choose the proper dosage of pharmaceuticals. The goal of maintained dosing is to replenish the dosage that has been removed from the body since the last delivery. Maintenance dose can be determined by multiplying the plasma concentration by clearance and total divided by bioavailability [26].

$$\text{Clearance (Cl)} = \text{Elimination rate / Plasma drug concentration}$$

Renal clearance (Cl_R): Volume of plasma or blood getting free from the unchanged drug by the kidney per unit time.

$$CL_R = \text{Rate of urinary excretion / Plasma drug concentration}$$

Renal clearance can also be defined as the ratio of "addition of the rate of glomerular filtration and active secretion minus the rate of reabsorption" to concentration in plasma.

Half-life

The amount of time needed for a 50% decrease in blood medication concentrations is known as the half-life. The formula t = (0.693xVd)/Clearance, which defines a drug's half-life, demonstrates a strong correlation with the volume of distribution as well as an inverse association with clearance. Modifications in clearance parameters brought on by disease or age commonly impact medicinal or macromolecule half-lives [11, 27].

Drug Kinetics

This illustrates the half-life of a macromolecular compound and a graphical representation of metabolism and excretion by considering zero-order and first-order kinetics. Regardless of the medication's concentration, the zero-order kinetics process demonstrates a constant rate of drug metabolism and/or removal. When the overall serum concentrations decrease, the half-life also decreases. Conversely, the proportion of the drug's plasma concentration determines first-order kinetics. First-order plasma clearance decreases with time, given a constant. For the majority of drugs, this is the main elimination model. Estimating steady states and total drug removal may be done using these kinetic models. When a drug's or macromolecule's administration and clearance are in equilibrium, a steady state is reached, resulting in a constant plasma concentration throughout time. When an active agent or macromolecule is supplied *via* continuous infusion, optimal therapeutic outcomes are reached four to five half-lives into the course of treatment. The system is considered to be in a steady state at this stage. Only adjustments to the dosage, timing of doses, or drug clearance can affect this steady-state concentration. Similarly, half-lives can be used to quantify complete elimination. If a medicine exhibits first-order elimination kinetics, it may be expected that it is removed fully after four to five half-lives, meaning that by then, ninety-four to ninety-seven percent of the medication has left the body [28]. The pharmacokinetic properties of various biopolymers are mentioned in Table **2**. Types of biopolymers are mentioned below in Fig. (**3**).

Table 2. Pharmacokinetics of various biopolymers.

Biopolymers	Absorption	Distribution	Metabolism	Excretion	Half-life	References
Polyesters						
PLGA	Slow hydrolytic degradation. Drug release *via* matrix erosion (first- or zero-order kinetics).	Primarily local distribution, systemic for smaller nanoparticles.	Degrades to lactic acid and glycolic acid. Lactic acid enters the citric acid cycle; glycolic acid is converted to glyoxylate.	Metabolized into CO_2 and H_2O, excreted *via* lungs (CO_2) and kidneys (H_2O).	1–2 weeks for 50:50 PLGA, up to 6 months for 85:15 PLGA.	[29, 30]
PLA	Hydrolytic degradation is slower than PLGA. Drug release is dependent on matrix breakdown.	Localized at the application site, with minimal systemic distribution.	Degrades to lactic acid, converted into pyruvate, and enters the citric acid cycle.	Excreted as CO_2 (*via* respiration) and H_2O (*via* kidneys).	Months to over a year, depending on molecular weight and formulation.	[31]
Polysaccharides						
Collagen	Degraded by collagenases and absorbed as peptides.	Localized in the tissue where applied. Peptides are absorbed into the bloodstream and distributed systemically.	Collagen peptides are further broken down into amino acids and utilized for new protein synthesis or metabolized for energy.	Excreted as nitrogenous waste *via* the kidneys (urea) and urine.	Varies with crosslinking; crosslinked collagen degrades more slowly.	[32]
Gelatin	Absorbed after enzymatic degradation into peptides, followed by absorption of amino acids.	Local distribution: Peptides are absorbed into the bloodstream and distributed to target tissues.	Degraded by proteases into amino acids, which are used in new protein synthesis or metabolized for energy.	Excreted as urea through the kidneys and urine.	Varies based on formulation; typically shorter half-life compared to collagen.	[33]

(Table 2) cont.....

Biopolymers	Absorption	Distribution	Metabolism	Excretion	Half-life	References
Polyesters						
Starch	Absorbed as glucose after enzymatic breakdown by amylases in the digestive system.	Glucose is rapidly distributed to tissues throughout the body *via* the bloodstream.	Starch is metabolized into glucose, which is used for energy or stored as glycogen in the liver and muscles.	Excreted as CO_2 and water following the complete metabolism of glucose.	Rapid degradation in the digestive system; short half-life.	[34]
Hyaluronic Acid	Absorbed through subcutaneous, intravenous, or topical administration.	Distributed primarily in the skin, joints, and eyes.	Degraded by hyaluronidases into oligosaccharides, further broken down into glucuronic acid and N-acetylglucosamine.	Excreted primarily through the kidneys in the form of urine.	Half-life in tissues is 12-24 hours; in the bloodstream, around 3-5 minutes.	[35]
Polynucleotides						
DNA	Absorption is limited when administered directly and often requires delivery systems like viral vectors or liposomes.	Viral vectors or nanoparticles can enhance systemic distribution. DNA fragments may be localized in the nucleus after cellular uptake.	DNA is enzymatically degraded by nucleases into nucleotides.	Degraded nucleotides are further metabolized or excreted *via* the kidneys and urine.	Short half-life unless encapsulated in delivery systems (*e.g.,* liposomes, viral vectors).	[6]
RNA	Absorption is limited due to rapid degradation by ribonucleases. Can be delivered *via* nanoparticles or liposomes for enhanced absorption.	Systemic distribution is possible using delivery systems, with uptake into cells where RNA can exert its therapeutic effect.	RNA is degraded by ribonucleases into nucleotides and further metabolized.	Degraded nucleotides are excreted *via* the kidneys and urine.	Very short half-life due to rapid degradation by ribonucleases; extended with delivery systems.	[36]

(Table 2) cont.....

Biopolymers	Absorption	Distribution	Metabolism	Excretion	Half-life	References
Polyesters						
Proteins and Peptides						
Zinc-containing Proteins	Absorbed through the small intestine when administered as a supplement. Endogenous zinc proteins are regulated by zinc homeostasis.	Bound to proteins like albumin; widely distributed in tissues, especially in the liver, bones, and muscles.	Zinc-containing proteins are involved in gene regulation and immune response. Zinc is not metabolized but is incorporated into proteins and enzymes.	Excreted mainly *via* the gastrointestinal tract, with smaller amounts *via* urine and sweat.	Dependent on zinc protein function, the half-life varies based on protein turnover.	[37]
Bovine Serum Albumin (BSA)	Absorption *via* parenteral routes acts as a drug carrier and stabilizer.	Primarily confined to the vascular system and extracellular spaces. Serves as a carrier for drugs and endogenous substances (*e.g.*, fatty acids).	Broken down by proteolytic enzymes into peptides and amino acids for new protein synthesis or energy metabolism.	Excreted as urea and other nitrogenous waste products *via* the kidneys (urine).	Varies depending on formulation and use (as a drug carrier or nanoparticle stabilizer).	[38]
Chitosan	Poorly absorbed in the gastrointestinal tract unless chemically modified.	Local distribution in the GI tract; modified chitosan can achieve systemic distribution when used as a drug carrier.	Metabolized by lysozymes into glucosamine, which can enter metabolic pathways.	Degraded glucosamine is excreted *via* the kidneys and urine.	Varies based on formulation; deacetylated chitosan degrades more slowly than acetylated forms.	[39]

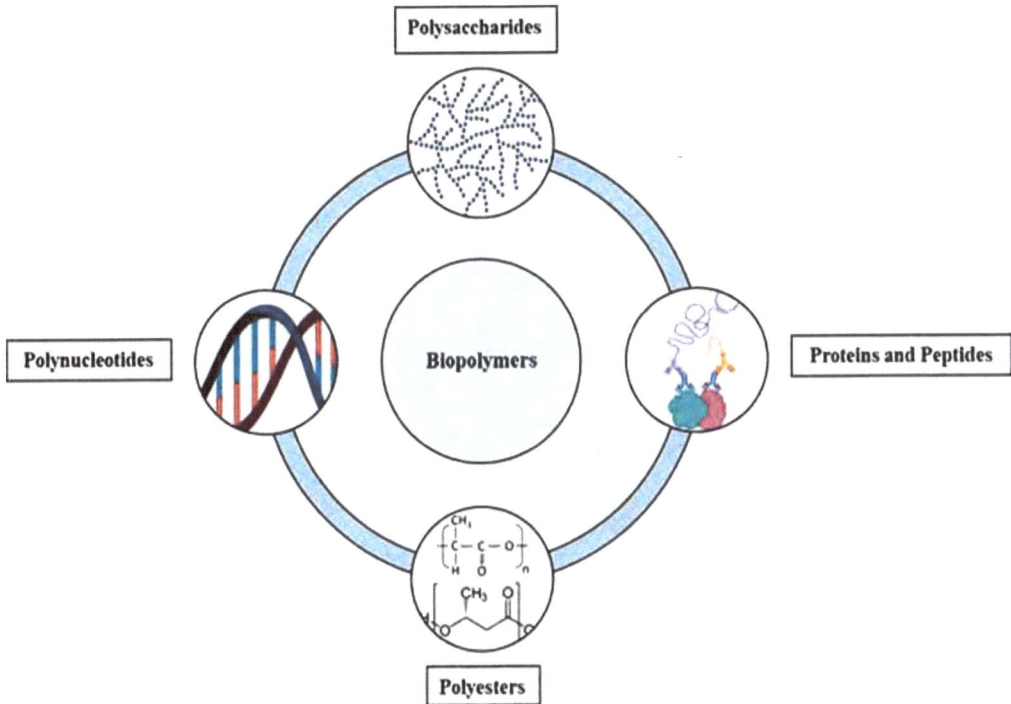

Fig. (3). Types of biopolymers.

PHARMACODYNAMICS

Pharmacodynamics explores a drug's or macromolecule's physiological, biochemical, and pharmacological consequences. It comes from the Greek words "*pharmakon*," indicating "drug," and "*dynamikos*," which signifies "power." All drugs change the nature of the target component concerning succeeding molecule relationships by interacting at the molecular level with biological frameworks or targets. These interactions include receptor binding, post-receptor implications, and chemical relationships. Examples of these interactions include macromolecules that connect to the functioning domain of an enzyme, those that connect with signaling proteins on the cell surface to block downstream signaling, and those that function by binding components like tumor necrosis factor (TNF) [40].

Pharmacodynamics activities are:

- Stimulating action by effectively blocking a receptor and its subsequent consequences.
- The effects of direct receptor inhibition and its consequences.
- Antagonistic, meaning it binds to a receptor and stops it from activating.

- Stabilizing effect, meaning the drug doesn't appear to have either an agonistic or antagonistic impact.
- Direct chemical reactions, including therapeutic side effects, are both favorable and unfavorable.

Theories of Pharmacodynamics

Pharmacodynamics is defined as the extent and length of a drug's effect using several important phrases and concepts. Emax is the greatest effect that a medication can have on the variable that is being measured. This might indicate the biggest drop in blood pressure or, in the case of *ex vivo* assessments, platelet suppression. The EC50 of a substance is the steady-state concentration at which half of its maximum action is generated.

The slope of the relationship involving medication activity and concentration is known as the "hill coefficient". A steep connection is indicated by a hill coefficient value above 2, meaning that tiny variations in concentration result in considerable changes in effect. An almost immediate "all or none" impact is indicated by a hill coefficient value above 3 [40].

General Mechanism of Action of Biopolymers

Drugs or macromolecules work by interacting with biological targets to create their effects, but the target's mechanism and biochemical route determine how long the pharmacodynamic impact lasts. It is possible to categorize effects as instantaneous, delayed, direct, or indirect. Macromolecules that interact with an enzyme or receptor that is essential to the effect's route typically have direct effects. Drugs that combine with proteins and receptors in other biological structures, which are situated much upstream from the final biochemical reaction that generates the drug action, might have unforeseen consequences. In the cytoplasm of the cell, corticosteroids attach to nuclear transcription factors, which go to the nucleus and prevent DNA from being transcribed into mRNA that codes for several inflammatory proteins [40].

Pharmacodynamics of Polysaccharides

Polysaccharides are composed of a continuous chain of carbohydrates. Because of their hydrophilicity and gel-forming properties, they have a variety of uses in medication administration and immunomodulation. Polysaccharides' pharmacodynamic properties are determined by how they interact with immune cells, biological barriers, and their capacity to either stimulate or moderate immunological responses.

Mechanism of Action: Chitosan is a polysaccharide that functions as a drug delivery system and immunomodulator. By opening tight junctions, chitosan can improve medication absorption by increasing the permeability of biological barriers such as mucosal membranes. When used in immunotherapy, chitosan boosts the production of pro-inflammatory cytokines by immune cells like dendritic and macrophages, strengthening the body's defenses against infections or cancerous cells.

a) Chitosan:

The delivery of antigens in vaccine formulations has been accomplished *via* chitosan-based nanoparticles, which serve as an adjuvant and a delivery vehicle. Antigen-presenting cells (APCs) absorb chitosan more readily because of its positive charge, which is associated with cell membranes that have a negative charge [41].

Human enzymes break down chitosan, which promotes hemostasis and accelerates tissue regeneration to aid in the healing of wounds. Furthermore, the production of this biopolymer comes from reusable resources, and chitosan has emerged as a key focus in research for diverse applications. It has been blended with various inorganic biologically active compounds and other polymeric biomaterials, demonstrating significant potential in orthopedics for applications such as bone replacements, annulus fibrosus and cartilage repair, and bone tissue engineering. Renowned for its excellent antibacterial properties, chitosan and its nanoparticles are extensively utilized in numerous biological fields, encompassing food safeguarding. Furthermore, their immunostimulatory mechanisms on fish and crustaceans make them highly beneficial to the aquaculture sector. Furthermore, fish and other animal species' illnesses have been treated with chitosan nanoparticles. Because of CS-NPs' incredibly potent biological qualities, including their biodegradability, biocompatibility, and antibacterial ability, they are appealing candidates for a variety of fish medicine applications [42, 43].

b) Hyaluronic Acid:

The anionic homopolymer known as hyaluronic acid is comprised of beta-1,4-d-glucuronic acid and beta-1,3-N-acetyl-d-glucosamine. Because of this copolymer's many beneficial characteristics, like its harmonious interaction with biological environments and natural decomposition, it is employed in medicine, for instance, to treat a variety of pathological conditions, including arthritis. It has been extensively utilized in medication transferable systems and tissue engineering. Hyaluronic acid is a linear polysaccharide made up of N-acetyl-β-d-glucosamine disaccharides and β-d-glucuronic acid that are connected in both 1-3 and 1-4 directions [44].

Another useful polysaccharide is hyaluronic acid (HA). HA binds to cell surface receptors, including CD44, which is involved in cell migration and proliferation. Drug delivery systems that utilize HA improve the ability of medications to target certain tissues, particularly in cancer treatment. Example: Targeted delivery of anticancer medications to tumors where CD44 is overexpressed is made possible by HA's interaction with the CD44 receptor in HA-based drug delivery systems. Reduced systemic toxicity and higher medication accumulation in the tumor are the outcomes [45].

c) Starch:

Amylose and amylopectin combine to form starch, a polysaccharide that is mostly broken down by enzymes in the gastrointestinal (GI) tract. The release of Active drug components can be regulated by using starch-based drug delivery systems because of their capacity to create hydrogels, microparticles, and nanoparticles. In the stomach's acidic environment, starch matrices can prevent the medication from breaking down, allowing the medicine to be released gradually into the intestine.

In the GI tract, starch is hydrolyzed by enzymes such as α-amylase, which dissociates its glycosidic bonds into glucose units. Because starch-based formulations are frequently intended for oral medication administration, this enzymatic breakdown enables regulated drug release [46].

d) Gelatin:

Gelatin is a polymeric protein obtained through the processing of collagen and has good biodegradability and biocompatibility. Gelatin's capacity to produce hydrogels, microspheres, and nanoparticles, which can contain both hydrophilic and hydrophobic drugs, is the basis for its pharmacodynamic properties. The body breaks down gelatin by the action of proteolytic enzymes like collagenase and gelatinase, which makes the medicine that has been encapsulated easier to release.

Because of its capacity to create gels and increase the bioavailability of poorly soluble medicines, gelatin has found widespread application in drug administration. Drugs can diffuse out of gelatin matrices because they swell in watery settings. By altering the degree of cross-linking in the gelatin, one may regulate the release rate by changing the pace at which the polymer degrades [48].

Pharmacokinetics of Polynucleotides

The monomers of nucleotides, such as DNA and RNA, combine to form polynucleotides. As therapeutic macromolecules or carriers for gene therapy and vaccine administration, polynucleotides' pharmacodynamic properties are determined by their functions.

a) DNA:

DNA-based medicines act by inserting a particular gene sequence into cells, where it is translated into a useful protein after being transcribed into messenger RNA (mRNA). Usually, plasmids or viral vectors are used to transfer DNA. The DNA is moved to the nucleus of the target cell, where the biological machinery transcribes the gene into mRNA. The therapeutic protein is then translated from this mRNA by ribosomes located in the cytoplasm. The protein that is generated can carry out many therapeutic tasks, such as inducing an immune response or substituting for a protein that is damaged.

DNA vaccines: DNA vaccines infect host cells with a plasmid that has a gene that encodes an antigen. The antigen is expressed inside the cells, whereupon the immune system views it as alien and mounts an attack. DNA vaccines have the potential to induce humoral immunity mediated by antibodies and cellular immunity driven by T cells [47, 48].

Genes that are defective and lead to genetic illnesses can be replaced or repaired using DNA. The therapeutic gene is inserted into the patient's cells using a viral or non-viral vector. Once the gene is expressed inside the nucleus, the defective gene is fixed, or normal cellular function is restored by the functioning protein. For the treatment of lipoprotein lipase deficiency (LPLD), a rare inherited condition, Glybera was the first gene therapy licensed in Europe. An effective version of the LPL gene is introduced into muscle cells as part of the treatment using an adeno-associated virus (AAV) vector. The amplified LPL protein lowers the risk of pancreatitis in those with triglyceride overload by aiding in its breakdown [49].

b) RNA:

Two main processes underlie the effectiveness of RNA-based therapeutics: the production of therapeutic proteins by messenger RNA (mRNA) or the silencing of particular gene expression by small interfering RNA (siRNA). The capacity of RNA to either translate into protein (mRNA) or break down complementary mRNA molecules (siRNA) is what gives it its pharmacologic qualities, controlling the amount of protein produced in cells.

Messenger RNA (mRNA): mRNA-based medicines insert synthetic mRNA into cells, where ribosomes transform it into a particular protein. Through its ability to modulate cellular pathways, function as a vaccine antigen, and replace damaged proteins, this protein can have therapeutic benefits.

Messenger RNA (mRNA) vaccines are particularly effective against infectious diseases because of their potent immunological responses. mRNA vaccines allow cells to manufacture an antigen by directly delivering the antigen's coding (Similar to the spike protein of SARS-CoV-2) to the cells. When the immune system perceives an antigen as being foreign, it mounts an immunological response that shields the body from the infection [50, 51].

Pharmacodynamics of Proteins and Peptides

Proteins and peptides are essential for therapeutic applications as they can regulate physiological processes and provide treatment for a wide range of illnesses. It is necessary to investigate their mechanisms of action, interactions with cellular targets, and methods of exerting therapeutic effects to comprehend their pharmacodynamics. Two prominent examples of proteins having a variety of medicinal uses are collagen and proteins containing zinc (BSA). Through amide bonds, amino acid subunits combine to form peptides and proteins, which are types of natural macromolecules. Proteins and peptides consist of different amino acids that form complex secondary structures (such as loops, chains, turns, β-sheets, and helices) and tertiary structures (involving multiple folds), in contrast to the uniform repeat sequences in homopolymers or the random arrangements of different monomers in copolymers. When a protein has several chains and consequently takes on a certain quaternary structure (*i.e.*, the three-dimensional organization of individual chains), the complexity of its structure rises [52].

a) Collagen:

Collagen, the predominant protein in the human body, plays a key role in supporting the structure of connective tissues, skin, and bones. It is essential for tissue regeneration, wound healing, and preserving the integrity of different tissues. In regenerative medicine, collagen treatments are often employed, especially as biomaterials for skin transplants, wound dressings, and tissue scaffolding.

The way collagen-based treatments work is by supplying a matrix that encourages cell adhesion, growth, and differentiation. Additionally, it accelerates the healing of wounds by stimulating the formation of new vascular networks, or angiogenesis, and by improving the deposition of extracellular matrix (ECM). Collagen can influence inflammatory responses and tissue healing mechanisms through interactions with growth factors and cytokines [53].

b) Zinc-containing proteins:

The structure and operation of several proteins, enzymes, and transcription factors depend on iron, an important trace element. Zinc metalloenzymes, such as superoxide dismutase (SOD), have an impact on antioxidant defense mechanisms, while zinc-containing proteins, such as zinc fingers, control DNA binding and gene transcription. Zinc ions can influence the biological activity of different proteins by functioning as structural or catalytic elements.

The pharmacodynamic effects of zinc include its capacity to regulate several cellular processes, such as antioxidant defense, immunological response, cell division, and apoptosis. Proteins containing zinc have an extensive selection of therapeutic effects, involving immunological regulation, wound healing, as well as neuroprotection [54].

c) Bovine Serum Albumin:

Bovine Serum Albumin (BSA) is a massive, globular protein obtained from cows that is widely utilized in several biochemical and medicinal applications. Numerous compounds, which are comprised of fatty acids, hormones, and medications, can be bound and transported by BSA. As a carrier protein, it improves the solubility and bioavailability of medicinal drugs, which contributes to its pharmacodynamic qualities. In circulation, BSA can prolong the half-life of therapies, stabilize medications, and stop them from aggregating.

BSA's capacity to bind tiny molecules makes it a perfect carrier for hydrophobic medicines at its binding sites, which facilitates the controlled release of these pharmaceuticals into the body and their transportation. Stabilizing vaccine formulations, recombinant proteins, and cell culture mediums are common uses of BSA in the biopharmaceutical industry [55].

Pharmacodynamics of Polyesters

One of the most popular types of biodegradable polymers in pharmaceutical and medical applications is polyesters, namely poly (lactic-co-glycolic acid) (PLGA) and poly(lactic acid) (PLA). Because of their biocompatibility and regulated biodegradation, these synthetic polymers have found widespread application in regenerative tissue engineering, implant technology, and medication delivery mechanisms. As they decompose into biocompatible monomers, which are eventually metabolized and expelled from the body, their pharmacodynamics are controlled.

a) Poly(lactic-co-glycolic acid) (PLGA):

Lactic acid and glycolic acid copolymer PLGA break down into their monomeric components by hydrolysis of their ester linkages. The body naturally uses the citric acid cycle to digest these monomers. Both glycolic acid and lactic acid are transformed into glyoxylate and pyruvate, which are then expelled as carbon dioxide and water. By varying the lactic acid to glycolic acid ratio, PLGA's disintegration rate may be customized to suit a variety of therapeutic purposes. Due to its hydrolytic instability, PLGA breaks down when water seeps through the polymer matrix. This causes bulk erosion, which frees the medication or biomolecule that is encapsulated.

The goal of PLGA-based drug delivery systems is to release encapsulated medications in a controlled or sustained manner for a predetermined amount of time. The drug's release rate is regulated by the breakdown of PLGA matrices, which allows consistent therapeutic effects over several weeks or months without requiring regular dosage adjustments. This is especially helpful for treating chronic illnesses, as sustained medication release is necessary.

Encapsulating drugs, proteins, and other bioactive compounds in PLGA nanoparticles, microparticles, or implants offers a controlled release method while shielding the molecules from deterioration. The physical and chemical attributes of the medication being administered, Particle dimensions, and the polymer's breakdown mechanism rates all affect the pharmacodynamics [56, 57].

b) Polylactic acid (PLA):

Lactic acid is an organic acid that occurs naturally and is used to make PLA, a biodegradable polyester. The hydrolysis of PLA's ester linkages causes the generation of lactic acid monomers, which the body then metabolizes through the citric acid cycle. The molecular weight, crystallinity, and variables in the environment (such as temperature and pH) all affect how quickly PLA degrades. Regarding biocompatibility and biodegradation, PLA's pharmacodynamic characteristics are comparable to those of PLGA; however, because of PLA's greater hydrophobicity and crystallinity, it degrades more gradually.

Applications involving tissue engineering have made considerable use of PLA. It gives tissues a rigid foundation and breaks down naturally in the body, letting cells proliferate and replacing the scaffolding with an extracellular matrix. The PLA's breakdown kinetics guarantee that the scaffold lasts sufficient time to facilitate tissue regeneration before being absorbed and broken down [30, 31]. The pharmacodynamic properties of various biopolymers are mentioned in Table 3.

BIOPOLYMER INTERACTIONS WITH IMMUNE CELLS AND IMMUNOGENICITY MITIGATION STRATEGIES

Biopolymers, widely utilized in drug delivery, regenerative medicine, and tissue engineering, inevitably interact with the immune system upon administration. The immune response to biopolymers is governed by various factors, including their physicochemical properties, surface chemistry, and degradation products. Understanding the mechanisms of immune recognition and response is crucial for designing biopolymers with minimal immunogenicity while maintaining therapeutic efficacy.

Table 3. Pharmacodynamics properties of various biopolymers.

Biopolymer	Mechanism of Action	Applications	Pharmacodynamics Properties	References
Polysaccharides				
Chitosan	Facilitating drug penetration *via* the disruption of tight junctions in epithelial cells. Form hydrogels for controlled release.	Drug delivery, wound healing, vaccine adjuvant	It enhances drug absorption, and mucoadhesive properties, promotes sustained release, and has bioadhesive, low immunogenicity.	[42, 58]
Starch	Biodegradable, can form microparticles for drug encapsulation, with slow enzymatic degradation.	Drug carriers, wound dressings	Good biocompatibility and controlled degradation act as a sustained release matrix for drugs.	[46, 59]
Gelatin	Forms gels and nanoparticles for drug delivery and degrades *via* enzymatic action.	Drug delivery, tissue scaffolds	Biocompatible and biodegradable, good for sustained release, and can trigger mild immune responses in some cases.	[47, 60]
Hyaluronic Acid	Binds to CD44 receptors, modulating cellular behavior. Acts as a carrier for medication administration.	Biotechnology for tissue repair, drug administration	High biocompatibility promotes wound healing, increases drug bioavailability, and targets CD44-positive cells.	[44, 45]
Polynucleotide				
DNA	Encodes proteins and vaccines, delivered *via* gene therapy systems, integrating into the host genome for expression.	Gene therapy, vaccine development	Directs protein expression, can induce an immune response, and needs precise targeting to avoid off-target effects.	[4]

Biopolymer	Mechanism of Action	Applications	Pharmacodynamics Properties	References
Polysaccharides				
RNA	Delivers mRNA for protein expression in cells and avoids integration into the genome.	mRNA vaccines, gene therapy	Transient protein expression, rapid degradation, low long-term risk, efficient immune activation (*e.g.*, in vaccines).	[50, 51]
Proteins and Peptides				
Zinc-containing Proteins	Zinc ions play a catalytic or structural role, impacting cellular pathways.	Drug delivery, enzyme carriers	Enhances the stability of protein structures, modulates immune responses, used in enzyme and drug stabilization.	[54]
Bovine Serum Albumin (BSA)	Binds to drugs and enhances stability, acts as a facilitator for drug transport.	Medication delivery, diagnostics, protein carriers	High binding capacity for drugs, improves drug stability, is biocompatible, low immunogenicity.	[55]
Polyesters				
PLGA (Poly(lactic-co-glycolic acid))	Biodegrades into lactic and glycolic acid, and acts as a matrix for sustained drug release.	Drug delivery, tissue engineering	Biodegradable, controlled release properties, low toxicity, and mild local immune response are possible.	[56, 57]
PLA (Polylactic acid)	Hydrolyzes into lactic acid, releasing encapsulated drugs over time.	Drug carriers, implants	Slow degradation, sustained drug release, low toxicity, and high mechanical strength.	[30]

Role of Innate Immune Cells in Biopolymer Recognition

The innate immune system serves as the first line of defense, where immune cells such as macrophages, dendritic cells (DCs), and neutrophils recognize and respond to biopolymers. Macrophages express pattern recognition receptors (PRRs), including toll-like receptors (TLRs) and scavenger receptors, which enable them to detect biopolymers and their degradation products. The interaction between biopolymers and macrophages can lead to the release of pro-inflammatory cytokines such as tumor necrosis factor-alpha (TNF-α) and interleukin-6 (IL-6), triggering an inflammatory response [61].

Adaptive Immune Response and Biopolymer Recognition

T cells and B cells mediate the adaptive immune response against biopolymers. CD4+ T helper cells play a critical role by releasing cytokines that guide immune responses. Some biopolymers can be processed into antigenic peptides that activate CD4+ T cells, leading to antibody production by B cells (Gupta *et al.*, 2022). Biopolymers that resist enzymatic degradation and remain intact for extended periods may exhibit prolonged immune activation, contributing to chronic inflammation. Biopolymer-based drug carriers, such as poly (lactic-co-glycolic acid) (PLGA), have been shown to stimulate humoral immune responses when conjugated with immunogenic molecules. Conversely, PEGylation reduces the formation of anti-polymer antibodies, mitigating immune-mediated clearance and enhancing circulation time [62].

Biopolymers' Mechanism Underlying Pathogenicity

a) Complement the Activation Pathway

Biopolymers can trigger the complement system through classical, alternative, or lectin pathways, leading to opsonization, inflammation, and cell lysis. For example, poly(ethylene imine) (PEI) nanoparticles have been reported to activate complement proteins, resulting in increased clearance by macrophages. Strategies such as surface modification with hydrophilic polymers, including PEG or zwitterionic materials, help evade complement activation [63].

b) Oxidative Stress and Inflammatory Pathways

Certain biopolymers, particularly those with high molecular weight or cationic charge, induce oxidative stress by stimulating reactive oxygen species (ROS) production in immune cells. This oxidative stress can activate nuclear factor-kappa B (NF-κB), a transcription factor responsible for inflammatory cytokine production. Chitosan-based nanoparticles, for instance, have been found to increase ROS levels in macrophages, leading to inflammatory responses [64].

Strategies to Mitigate Immunogenicity

To mitigate the immunogenicity of biopolymers, surface modification techniques like PEGylation and zwitterionic coatings reduce protein adsorption and immune recognition. Controlling degradation rates through crosslinking and hybrid biopolymers minimizes inflammatory responses. Immunomodulatory strategies, such as incorporating dexamethasone or anti-inflammatory peptides, suppress immune activation. Optimizing biopolymer charge and hydrophilicity can further reduce complement activation and macrophage uptake. Additionally, biomimetic modifications, including glycosylation or cell-membrane coating, enhance immune evasion. Rational polymer design focusing on non-immunogenic materials and controlled drug release ensures compatibility. These strategies collectively enhance the biocompatibility of biopolymers for safer and more effective biomedical applications [63, 65, 66].

CHALLENGES AND FUTURE PROSPECTIVE

Because of their inherent biocompatibility, capacity for biodegradation, and wide variety of chemical and physical characteristics, biopolymers are showing great promise as potential therapeutic materials. By increasing bioavailability, lowering toxicity, and facilitating targeted distribution, they function as therapeutic adjuvants, augmenting the effectiveness of medications and other therapeutic agents. Because of their adaptability, biopolymers are now important components of vaccine formulations, tissue engineering technologies, regenerative healthcare, and drug distribution systems. The capacity of biopolymers to act as transporters for medications, proteins, or other therapeutic compounds is one of their main benefits. Biopolymers provide regulated and prolonged administration of medications to certain tissues or organs by enclosing pharmaceuticals in protective matrices that delay their premature breakdown or release.

Furthermore, several biopolymers have been designed to react in response to environmental changes (like pH, temperature, or enzymes) to minimize side effects by releasing therapeutic chemicals in precisely the right proportions under predetermined circumstances. For instance, the potential of chitosan, a biopolymer derived from chitin, to create hydrogels and nanoparticles that carry medications in a regulated way has been extensively investigated. Similar to this, naturally occurring biopolymers like collagen and alginate are utilized in tissue scaffolds and wound dressings to promote healing while also delivering medications or growth hormones.

These materials' adaptability makes it possible to utilize them in a variety of therapeutic contexts, from the administration of vaccines to cancer therapies. Nevertheless, several obstacles are preventing the widespread use of biopolymers as medicinal adjuvants, despite their encouraging qualities. One of the most important hurdles is immunogenicity because some biopolymers, when ingested by the body, can cause immunological reactions. The polymer could be recognized by the immune system as a non-self component, which might cause inflammation, allergic responses, or even lower the effectiveness of the medication that is encapsulated. To counteract this, biopolymers are frequently chemically altered or covered in biocompatible materials (such as polyethylene glycol) to lessen immunological recognition. Although these tactics have shown promise, it is still difficult to strike a careful balance between immunogenicity and functioning.

FUTURE PROSPECTIVE

In the future, the fields of green chemistry, synthetic biology, and nanotechnology will propel major progress in the field of biopolymers as medicinal adjuvants. The creation of stimuli-responsive biopolymers is a promising future. These materials can alter their behavior in response to temperature, pH, light, and other environmental stimuli. For example, a polymer that reacts to the acidic environment of a tumor to release a cancer medicine might offer more targeted therapy without harming healthy cells.

Biopolymers provide interesting new opportunities in medicine delivery thanks to nanotechnology. Drugs can be encapsulated in biopolymer-based nanoparticles to prevent degradation while they are being circulated throughout the body. Therapeutic compounds can be delivered specifically to the intended tissue with the help of these designed nanoparticles, increasing effectiveness and lowering systemic toxicity. These systems hold great promise for treating illnesses like cancer, where it is necessary to use focused treatment to prevent harm to healthy cells. Biopolymers have the potential to be extremely important in customized medicine when it comes to individualized patient care. Through the creation of biopolymers with distinct molecular characteristics, scientists may be able to develop customized medication delivery systems that adjust to the individual biological circumstances of every patient. Treatments for complicated diseases such as cancer, autoimmune disorders, and hereditary abnormalities might be revolutionized by this method. Combination treatments provide biopolymer-based systems with a fascinating new direction. It is feasible to improve therapy results by co-delivering medications and adjuvants in a single biopolymer matrix, particularly in complicated conditions like cancer and infectious disorders. Combination therapy based on biopolymers may maximize pharmacological effectiveness while lowering dose requirements and minimizing negative effects.

CONCLUSION

Biopolymers' distinct pharmacokinetic and pharmacodynamic characteristics make them essential to contemporary treatments. Because of their capacity to form complex structures, biodegradability, and biocompatibility, they are perfect for drug delivery and other biological applications. Optimizing their therapeutic potential requires an understanding of the processes controlling their absorption, distribution, metabolism, and excretion (ADME). They are also important for achieving focused and long-lasting effects because of their interactions with biological systems, which include receptor binding and regulated drug release. Developments in the study of biopolymers have created new opportunities for developing novel delivery methods, enhancing effectiveness, and minimizing adverse effects. This chapter highlights the significance of customizing biopolymers to meet certain pharmacological requirements and highlights how they have revolutionized healthcare.

REFERENCES

[1]   Smith AM, Moxon S, Morris GA. Biopolymers as wound healing materials In: Ågren SM, ed, Wound Heal Biomater. Elsevier 2016; pp. 261-87.
[http://dx.doi.org/10.1016/B978-1-78242-456-7.00013-1]

[2]   Baranwal J, Barse B, Fais A, Delogu GL, Kumar A. Biopolymer: A sustainable material for food and medical applications. Polymers (Basel) 2022; 14(5): 983.
[http://dx.doi.org/10.3390/polym14050983] [PMID: 35267803]

[3]   Fazal T, Murtaza BN, Shah M, *et al.* Recent developments in natural biopolymer based drug delivery systems. RSC Advances 2023; 13(33): 23087-121.
[http://dx.doi.org/10.1039/D3RA03369D] [PMID: 37529365]

[4]   Opriş O, Mormile C, Lung I, Stegarescu A, Soran ML, Soran A. An overview of biopolymers for drug delivery applications. Appl Sci (Basel) 2024; 14(4): 1383.
[http://dx.doi.org/10.3390/app14041383]

[5]   Musa Y, Bwatanglang IB. Current role and future developments of biopolymers in green and sustainable chemistry and catalysis In: Mohammad F, Al-Lohedan HA, Jawaid M, eds Sustainable Nanocellulose and Nanohydrogels from Natural Sources. Elsevier 2020; pp. 131-54.
[http://dx.doi.org/10.1016/B978-0-12-816789-2.00006-7]

[6]   Hudry E, Vandenberghe LH. Therapeutic AAV gene transfer to the nervous system: A clinical reality. Neuron 2019; 101(5): 839-62.
[http://dx.doi.org/10.1016/j.neuron.2019.02.017] [PMID: 30844402]

[7]   Tören E, Buzgo M, Mazari AA, Khan MZ. Recent advances in biopolymer based electrospun nanomaterials for drug delivery systems. Polym Adv Technol 2024; 35(3): e6309.
[http://dx.doi.org/10.1002/pat.6309]

[8]   Ernstmeyer K CE. Pharmacokoinetics and Pharmacodynamics [Internet]. Available from: https://www.ncbi.nlm.nih.gov/books/NBK595006/

[9]   Singh S, Chunglok W. Pharmacokinetics and toxicology of pharmaceutical excipients In: Biopolymers Towards Green and Sustainable Development Bentham Science Publishers. 2022; pp. 168-81.
[http://dx.doi.org/10.2174/9789815079302122010011]

[10]   Loftsson T. Excipient pharmacokinetics and profiling. Int J Pharm 2015; 480(1-2): 48-54.
[http://dx.doi.org/10.1016/j.ijpharm.2015.01.022] [PMID: 25596414]

[11]   Sean Grogan; Charles V. Preuss. Pharmacokinetics [Internet]. 2023. Available from: https://www.ncbi.nlm.nih.gov/books/NBK557744/

[12]   Abdulrahman A. Alagga; Mark V. Pellegrini; Vikas Gupta. Drug absorption [Internet]. 2024. Available from: https://www.ncbi.nlm.nih.gov/books/NBK557405/

[13]   Neverisky DL, Abbott GW. Ion channel–transporter interactions. Crit Rev Biochem Mol Biol 2016; 51(4): 257-67.
[http://dx.doi.org/10.3109/10409238.2016.1172553] [PMID: 27098917]

[14]   Cooper GM. Endocytosis, The Cell: A Molecular Approach [Internet]. Available from: https://www.ncbi.nlm.nih.gov/books/NBK9831/

[15]   Asad Mansoor; Navid Mahabadi. DISTRIBUTION [Internent]. 2024. Available from: https://doi.org/https://www.ncbi.nlm.nih.gov/books/NBK545280/

[16]   Smith DA, Beaumont K, Maurer TS, Di L. Volume of distribution in drug design. J Med Chem 2015; 58(15): 5691-8.
[http://dx.doi.org/10.1021/acs.jmedchem.5b00201] [PMID: 25799158]

[17]   Asad Mansoor; Navid Mahabadi. Volume of distribution [Internet]. 2023. Available from: https://www.ncbi.nlm.nih.gov/books/NBK545280/

[18]   Fan J, de Lannoy IAM. Pharmacokinetics. Biochem Pharmacol 2014; 87(1): 93-120.
[http://dx.doi.org/10.1016/j.bcp.2013.09.007] [PMID: 24055064]

[19]   Toutain PL, Bousquet-Mélou A. Volumes of distribution. J Vet Pharmacol Ther 2004; 27(6): 441-53.
[http://dx.doi.org/10.1111/j.1365-2885.2004.00602.x]

[20]   Kocz   AZGSBPR.   Drug   Elimination   [Internet].   2023.   Available   from:
https://www.ncbi.nlm.nih.gov/books/NBK547662/

[21]   Simone Phang-Lyn; Valerie A. Llerena. Biochemistry, Biotransformation [Internet]. 2023. Available
from: https://www.ncbi.nlm.nih.gov/books/NBK544353

[22]   Judge A, Dodd MS. Metabolism. Essays Biochem 2020; 64(4): 607-47.
[http://dx.doi.org/10.1042/EBC20190041] [PMID: 32830223]

[23]   Mohsin NA, Farrukh M, Shahzadi S, Irfan M. Drug metabolism: Phase I and Phase II metabolic
pathways, 2024.

[24]   Lykkesfeldt J, Tveden-Nyborg P. The pharmacokinetics of vitamin C. Nutrients 2019; 11(10): 2412.
[http://dx.doi.org/10.3390/nu11102412] [PMID: 31601028]

[25]   Bonate PL, Cunningham CC, Gaynon P, *et al.* Population pharmacokinetics of clofarabine and its
metabolite 6-ketoclofarabine in adult and pediatric patients with cancer. Cancer Chemother Pharmacol
2011; 67(4): 875-90.
[http://dx.doi.org/10.1007/s00280-010-1376-z] [PMID: 20582417]

[26]   Rowland M, Benet LZ, Graham GG. Clearance concepts in pharmacokinetics. J Pharmacokinet
Biopharm 1973; 1(2): 123-36.
[http://dx.doi.org/10.1007/BF01059626] [PMID: 4764426]

[27]   Smith DA, Di L, Kerns EH. The effect of plasma protein binding on *in vivo* efficacy: misconceptions
in drug discovery. Nat Rev Drug Discov 2010; 9(12): 929-39.
[http://dx.doi.org/10.1038/nrd3287] [PMID: 21119731]

[28]   Chillistone S, Hardman JG. Modes of drug elimination and bioactive metabolites. Anaesth Intensive
Care Med 2023; 24(8): 482-5.
[http://dx.doi.org/10.1016/j.mpaic.2023.05.011]

[29]   Makadia HK, Siegel SJ. Poly lactic-co-glycolic acid (plga) as biodegradable controlled drug delivery
carrier. Polymers (Basel) 2011; 3(3): 1377-97.
[http://dx.doi.org/10.3390/polym3031377] [PMID: 22577513]

[30]   Blasi P. Poly(lactic acid)/poly(lactic-co-glycolic acid)-based microparticles: an overview. J Pharm
Investig 2019; 49(4): 337-46.
[http://dx.doi.org/10.1007/s40005-019-00453-z]

[31]   Anderson JM, Shive MS. Biodegradation and biocompatibility of PLA and PLGA microspheres. Adv
Drug Deliv Rev 2012; 64: 72-82.
[http://dx.doi.org/10.1016/j.addr.2012.09.004]

[32]   Kassam HA, Bahnson EM, Cartaya A, *et al.* Pharmacokinetics and biodistribution of a collagen-
targeted peptide amphiphile for cardiovascular applications. Pharmacol Res Perspect 2020; 8(6):
e00672.
[http://dx.doi.org/10.1002/prp2.672] [PMID: 33090704]

[33]   Iwao Y. Albumin Nanoparticles In: Otagiri M, Chuang VTG, eds Albumin in Medicine. Singapore:
Springer Singapore 2016; pp. 91-100.
[http://dx.doi.org/10.1007/978-981-10-2116-9_5]

[34]   Jungheinrich C, Neff TA. Pharmacokinetics of hydroxyethyl starch. Clin Pharmacokinet 2005; 44(7):
681-99.
[http://dx.doi.org/10.2165/00003088-200544070-00002] [PMID: 15966753]

[35]   Gupta RC, Lall R, Srivastava A, Sinha A. Hyaluronic acid: molecular mechanisms and therapeutic trajectory. Front Vet Sci 2019; 6: 192.
[http://dx.doi.org/10.3389/fvets.2019.00192] [PMID: 31294035]

[36]   Setten RL, Rossi JJ, Han S. The current state and future directions of RNAi-based therapeutics. Nat Rev Drug Discov 2019; 18(6): 421-46.
[http://dx.doi.org/10.1038/s41573-019-0017-4] [PMID: 30846871]

[37]   Costa MI, Sarmento-Ribeiro AB, Gonçalves AC. Zinc: From biological functions to therapeutic potential. Int J Mol Sci 2023; 24(5): 4822.
[http://dx.doi.org/10.3390/ijms24054822] [PMID: 36902254]

[38]   Roopenian DC, Low BE, Christianson GJ, Proetzel G, Sproule TJ, Wiles MV. Albumin-deficient mouse models for studying metabolism of human albumin and pharmacokinetics of albumin-based drugs. MAbs 2015; 7(2): 344-51.
[http://dx.doi.org/10.1080/19420862.2015.1008345] [PMID: 25654695]

[39]   Li H, Jiang Z, Han B, Niu S, Dong W, Liu W. Pharmacokinetics and biodegradation of chitosan in rats. J Ocean Univ China 2015; 14(5): 897-904.
[http://dx.doi.org/10.1007/s11802-015-2573-5]

[40]   Mark Marino. Zohaib Jamal; Patrick M. Zito. Pharmacodynamics. NCBI 2023.

[41]   Kravanja G, Primožič M, Knez Ž, Leitgeb M. Chitosan-based (nano)materials for novel biomedical applications. Molecules 2019; 24(10): 1960.
[http://dx.doi.org/10.3390/molecules24101960] [PMID: 31117310]

[42]   Fatullayeva S, Tagiyev D, Zeynalov N, Mammadova S, Aliyeva E. Recent advances of chitosan-based polymers in biomedical applications and environmental protection. J Polym Res 2022; 29(7): 259.
[http://dx.doi.org/10.1007/s10965-022-03121-3]

[43]   Kalpana Manivannan R, Sharma N, Kumar V, Jayaraj I, Vimal S, Umesh M. A comprehensive review on natural macromolecular biopolymers for biomedical applications: Recent advancements, current challenges, and future outlooks. Carbohydr Polym Technol Appl 2024; 8: 100536.
[http://dx.doi.org/10.1016/j.carpta.2024.100536]

[44]   Goh GD, Lee JM, Goh GL, Huang X, Lee S, Yeong WY. Machine learning for bioelectronics on wearable and implantable devices: challenges and potential. Tissue Eng Part A 2023; 29(1-2): 20-46.
[http://dx.doi.org/10.1089/ten.tea.2022.0119] [PMID: 36047505]

[45]   de Souza AB, Chaud MV, Santana MHA. Hyaluronic acid behavior in oral administration and perspectives for nanotechnology-based formulations: A review. Carbohydr Polym 2019; 222: 115001.
[http://dx.doi.org/10.1016/j.carbpol.2019.115001] [PMID: 31320101]

[46]   Tharanathan RN. Starch-value addition by modification. Crit Rev Food Sci Nutr 2005; 45(5): 371-84.
[http://dx.doi.org/10.1080/10408390590967702] [PMID: 16130414]

[47]   Mikhailov OV. Gelatin as it is: history and modernity. Int J Mol Sci 2023; 24(4): 3583.
[http://dx.doi.org/10.3390/ijms24043583] [PMID: 36834993]

[48]   Lu B, Lim JM, Yu B, *et al.* The next-generation DNA vaccine platforms and delivery systems: advances, challenges and prospects. Front Immunol 2024; 15: 1332939.
[http://dx.doi.org/10.3389/fimmu.2024.1332939] [PMID: 38361919]

[49]   Jablonka S, Hennlein L, Sendtner M. Therapy development for spinal muscular atrophy: perspectives for muscular dystrophies and neurodegenerative disorders. Neurol Res Pract 2022; 4(1): 2.
[http://dx.doi.org/10.1186/s42466-021-00162-9] [PMID: 34983696]

[50]   Jackson NAC, Kester KE, Casimiro D, Gurunathan S, DeRosa F. The promise of mRNA vaccines: a biotech and industrial perspective. NPJ Vaccines 2020; 5(1): 11.
[http://dx.doi.org/10.1038/s41541-020-0159-8] [PMID: 32047656]

[51]   Pardi N, Hogan MJ, Porter FW, Weissman D. mRNA vaccines — a new era in vaccinology. Nat Rev

Drug Discov 2018; 17(4): 261-79.
[http://dx.doi.org/10.1038/nrd.2017.243] [PMID: 29326426]

[52]   Shamblin SL, Hancock BC, Zografi G. Water vapor sorption by peptides, proteins and their formulations. Eur J Pharm Biopharm 1998; 45(3): 239-47.
[http://dx.doi.org/10.1016/S0939-6411(98)00006-X] [PMID: 9653628]

[53]   Rezvani Ghomi E, Nourbakhsh N, Akbari Kenari M, Zare M, Ramakrishna S. Collagen-based biomaterials for biomedical applications. J Biomed Mater Res B Appl Biomater 2021; 109(12): 1986-99.
[http://dx.doi.org/10.1002/jbm.b.34881] [PMID: 34028179]

[54]   Maywald M, Rink L. Zinc in human health and infectious diseases. Biomolecules 2022; 12(12): 1748.
[http://dx.doi.org/10.3390/biom12121748] [PMID: 36551176]

[55]   Shen X, Liu X, Li T, *et al.* Recent advancements in serum albumin-based nanovehicles toward potential cancer diagnosis and therapy. Front Chem 2021; 9: 746646.
[http://dx.doi.org/10.3389/fchem.2021.746646] [PMID: 34869202]

[56]   Zhang D, Liu L, Wang J, *et al.* Drug-loaded PEG-PLGA nanoparticles for cancer treatment. Front Pharmacol 2022; 13: 990505.
[http://dx.doi.org/10.3389/fphar.2022.990505] [PMID: 36059964]

[57]   Puricelli C, Gigliotti CL, Stoppa I, *et al.* Use of poly lactic-co-glycolic acid nano and micro particles in the delivery of drugs modulating different phases of inflammation. Pharmaceutics 2023; 15(6): 1772.
[http://dx.doi.org/10.3390/pharmaceutics15061772] [PMID: 37376219]

[58]   Hassan HAFM, Ali AI, ElDesawy EM, ElShafeey AH. Pharmacokinetic and pharmacodynamic evaluation of gemifloxacin chitosan nanoparticles as an antibacterial ocular dosage form. J Pharm Sci 2022; 111(5): 1497-508.
[http://dx.doi.org/10.1016/j.xphs.2021.12.016] [PMID: 34929155]

[59]   Sivamaruthi BS, Nallasamy P, Suganthy N, Kesika P, Chaiyasut C. Pharmaceutical and biomedical applications of starch-based drug delivery system: A review. J Drug Deliv Sci Technol 2022; 77: 103890.
[http://dx.doi.org/10.1016/j.jddst.2022.103890]

[60]   Verma D, Bhatia A, Chopra S, *et al.* Advancements on microparticles-based drug delivery systems for cancer therapy Adv Drug Deliv Syst Manag Cancer. Elsevier 2021; pp. 351-8.
[http://dx.doi.org/10.1016/B978-0-323-85503-7.00003-1]

[61]   Akira S, Uematsu S, Takeuchi O. Pathogen recognition and innate immunity. Cell 2006; 124(4): 783-801.
[http://dx.doi.org/10.1016/j.cell.2006.02.015] [PMID: 16497588]

[62]   Lee JH, Shin SJ, Lee JH, Knowles JC, Lee HH, Kim HW. Adaptive immunity of materials: Implications for tissue healing and regeneration. Bioact Mater 2024; 41: 499-522.
[http://dx.doi.org/10.1016/j.bioactmat.2024.07.027] [PMID: 39206299]

[63]   Kuriakose A, Chirmule N, Nair P. Immunogenicity of biotherapeutics: causes and association with posttranslational modifications. J Immunol Res 2016; 2016: 1-18.
[http://dx.doi.org/10.1155/2016/1298473] [PMID: 27437405]

[64]   Gambini J, Stromsnes K. Oxidative stress and inflammation: from mechanisms to therapeutic approaches. Biomedicines 2022; 10(4): 753.
[http://dx.doi.org/10.3390/biomedicines10040753] [PMID: 35453503]

[65]   Hoang Thi TT, Pilkington EH, Nguyen DH, Lee JS, Park KD, Truong NP. The importance of poly(ethylene glycol) alternatives for overcoming peg immunogenicity in drug delivery and bioconjugation. Polymers (Basel) 2020; 12(2): 298.
[http://dx.doi.org/10.3390/polym12020298] [PMID: 32024289]

[66]   Mazor R, King EM, Pastan I. Strategies to reduce the immunogenicity of recombinant immunotoxins. Am J Pathol 2018; 188(8): 1736-43.
[http://dx.doi.org/10.1016/j.ajpath.2018.04.016] [PMID: 29870741]

CHAPTER 5

Biopolymer-based Chemotherapeutics: Combination Therapies and Synergistic Effects

Rufaida Wasim[1], **Tarique Mahmood**[1,*], **Saba Parveen**[2], **Aamir Anwar**[1] and **Asad Ahmad**[1]

[1] *Department of Pharmacy, Integral University, Dasauli, Kursi Road, Lucknow, Uttar Pradesh, India*

[2] *Department of Pharmacy, Madan Mohan Malaviya University of Technology, Gorakhpur, Uttar Pradesh, India*

Abstract: Applications of nanotechnology have increased the importance of research and nanocarriers, which have revolutionized medication delivery in recent years to treat a range of diseases, including cancer. Due to its multidrug resistance to several chemotherapeutic treatments, cancer, one of the most dangerous diseases in the world, has drawn the attention of experts. Scientists have created a different way to deliver chemotherapeutic drugs to the desired location while reducing side effects and enhancing delivery efficacy on healthy cells by incorporating them into nanocarriers such as synthetic polymers, nanotubes, micelles, dendrimers, magnetic nanoparticles, Quantum Dots (QDs), lipid nanoparticles, nano-biopolymeric substances, *etc.* Nanotechnology applications have made research and nanocarriers—which have recently transformed drug delivery to treat a variety of illnesses, including cancer—even more crucial. One of the most deadly illnesses in the world, cancer, has caught the attention of scientists because of its multidrug resistance to several chemotherapeutic therapies. By integrating chemotherapeutic drugs into nanocarriers like synthetic polymers, nanotubes, micelles, dendrimers, magnetic nanoparticles, Quantum Dots (QDs), lipid nanoparticles, nano-biopolymeric substances, *etc.*, researchers have developed an alternative method of delivering these medications to the intended site while minimizing side effects and improving delivery efficacy on healthy cells. Preclinical and clinical research on cancer treatment has yielded promising results. Biopolymers stand out as viable options for anticancer nano drug delivery systems due to their exceptional biocompatibility. Moreover, the presence of ligands in some biopolymers that are naturally present on the surface of human cells enables active targeting.

Keywords: Biopolymers, Chemotherapeutics, Drugs, Nanocarrier, Nanotechnology.

* **Corresponding author Tarique Mahmood:** Department of Pharmacy, Integral University, Dasauli, Kursi Road, Lucknow, Uttar Pradesh, India; E-mail: tmahmood@iul.ac.in

Sudhanshu Mishra, Smriti Ojha, Shashi Kant Singh, Rishabha Malviya & Saurabh Kumar Gupta (Eds.)

INTRODUCTION

The dramatic rise in cancer as the world's top cause of death highlights the pressing need for scientific research and healthcare initiatives. Ten million people died from cancer in 2020 alone, according to estimates, and the World Health Organization estimates that number might triple by 2040 [1 - 3]. A wide range of adverse effects are linked to modern cancer treatments, including hormone therapy, immunotherapy, radiation, chemotherapy, and surgery [4 - 6]. Chemotherapy's infamous side effects, which include myelotoxicity and cardiotoxicity, are caused by non-specific delivery and the exacerbation of unintentional cellular damage [7 - 11]. This makes it abundantly evident how urgently anticancer medications are needed. In light of this, nanomedicine has become more than just a substitute; it is a revolutionary tool in pharmaceutical research, particularly in the field of drug delivery [12 - 14].

Cancer therapy is being revolutionized by nanomaterials, which provide new and improved therapeutic alternatives. Their special qualities—such as their compact size, high surface area-to-volume ratio, and capacity to be surface-engineered for precise targeting—make them perfect for use in cancer therapy applications. By restricting drug exposure to healthy cells, they are used to deliver pharmaceuticals directly to tumor cells, significantly reducing the adverse consequences linked to conventional chemotherapy. Furthermore, thermal treatment makes use of nanomaterials, which are designed to absorb particular light wavelengths to heat up and kill cancer cells with the least amount of harm to the surrounding tissues. They are also essential in diagnostic applications, where they improve imaging techniques to increase the sensitivity and specificity of cancer detection procedures. The use of nanomaterials in cancer therapy has enormous potential to enhance therapeutic results, lessen adverse effects, and open the door to more individualized and effective cancer care [15].

Because of their inherent biomimetic qualities, design flexibility, and biocompatibility, biopolymers (which have particle sizes ranging from 10 to 1000 nm) have emerged as the focal point of this research revolution [16, 17]. Because of their increased specificity and bioavailability, recent research has highlighted the potential of naturally derived nanomedicines, ranging from complex composite nanocarriers to lipid nanocarriers for targeted drug delivery [18 - 21]. However, due to their possible cytotoxicities, their synthetic equivalents raise doubts [22 - 24].

Research to enable precision-targeted medication delivery has traditionally focused on the combination of polymer conjugates and nanomedicine [25 - 29].

Innovative developments like polylactic acid conjugates, which are intended for better medication penetration, highlight this story [30 - 41].

PRODUCTION AND CHARACTERIZATION OF BIOPOLYMERIC NANOPARTICLES (NCS) UTILIZED IN DRUG DELIVERY SYSTEMS (DDS)

Anticancer medication-loaded NCs provide several benefits over free medication. They guarantee target-specific tumor cell death by preventing medicines from degrading too quickly and interacting with nonspecific substances [42 - 44]. One of the most crucial characteristics of NCs is their biocompatibility, which increases their effectiveness and prolongs the drug's shelf life in use [45].

Natural materials with nontoxic, biocompatible, and biodegradable qualities are called biopolymers. For the creation of NC formulations, only polymers have been studied thus far [46 - 49]. Both natural and synthetic polymers can be used for encapsulation; natural polymers, such as chitosan, silk, alginate, albumin, starch, carbohydrates, proteins, and lipid materials, can be used without the medication having to be chemically altered [50]. Over the past few decades, several biopolymeric NP formulations and methodologies have produced effective nanotransporting capabilities with outstanding anticancer effects [51 - 53].

LIPID-BASED BIOPOLYMER

In the past few decades, organic polymers have been used [54]. Proteins, liposomes, and solid lipid nanoparticles are examples of organic polymers that have been considered appropriate nanocarriers for Drug Delivery Systems (DDSs) [55, 56]. The widespread use of lipid polymers may be attributed to their effective ability to encapsulate both hydrophilic and hydrophobic medications [57 - 60]. Water-insoluble medications have recently been administered *via* lipid nanoparticles (LPNs) [61]. Stearic acid-modified polyglycerol adipate (PGAS) is a lipid substance that has been demonstrated by Weiss *et al.* to be a potential drug delivery vehicle that does not require a surfactant [62]. Only N-(2-hydroxypropyl) methacrylamide (HPMA) copolymer coating, either covalently or non-covalently, was carried out [62]. Despite having comparable particle sizes, these nanoparticles can show lower or negative zeta-potentials [62]. The fluorescent dyes DiR and DYOMICS, which were covalently bonded to an HPMA copolymer, were used to double-label the NPs in a non-covalent fashion [63]. Using optical imaging based on various spectra, the biodistribution was examined noninvasively. The pharmacokinetics and biodistribution of cancer-bearing and healthy mice were altered by coating (Fig. 1).

With the introduction of recent biopolymer techniques in DDSs, the nanocarriers provide a half-burst release before building up at the tumor site, which results in toxicity from material circulation [64]. As a result, scientists are now concentrating on carriers that respond to inputs. In the presence of PEGylated lipids (PEG = polyethylene glycol) and a mildly acidic pH environment, Michelle Stollzoff *et al.* described a unique lipid-coated nanoparticle that is pH sensitive and expands in size by 100–1000 nm [65]. Folic Acid (FA) and folate receptor targeting can be introduced by altering the surface of PEG-L-eNPs [65]. Consequently, when loaded with paclitaxel *in-vitro* and in comparison to nontargeted PEGL-eNPs, these resulting polymer/lipid hybrid nanocarriers, FA-PEG-LeNPs, offer improved efficacy and uptake [65]. Biao Kang *et al.* also published an investigation on surface modification of the nanocarrier for improved targeting [66]. Various researchers prepared the PEGylated nanocarrier using hydroxyethyl starch and added mannose to the outer PEG layer to target dendritic cells [66]. Better targeting behavior with dendritic cells is ensured by the human plasma contact, a unique pattern of protein adsorption, and increased affinity (Fig. **2**) [66].

Liposome preparation via Thin Film Hydration

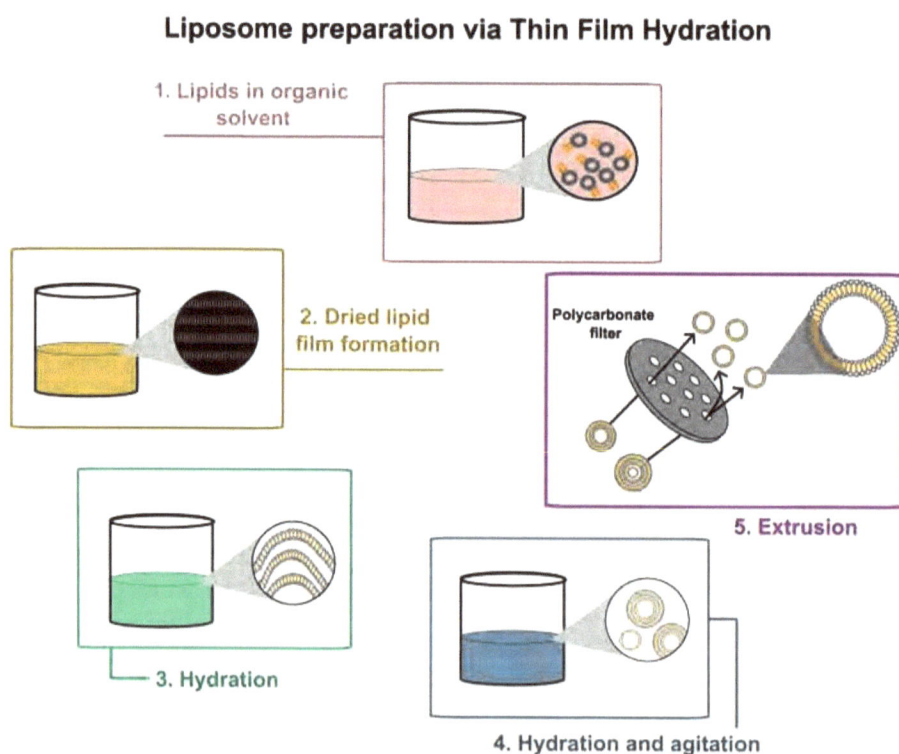

Fig. (1). Liposome preparation and nanocarrier production using film hydration.

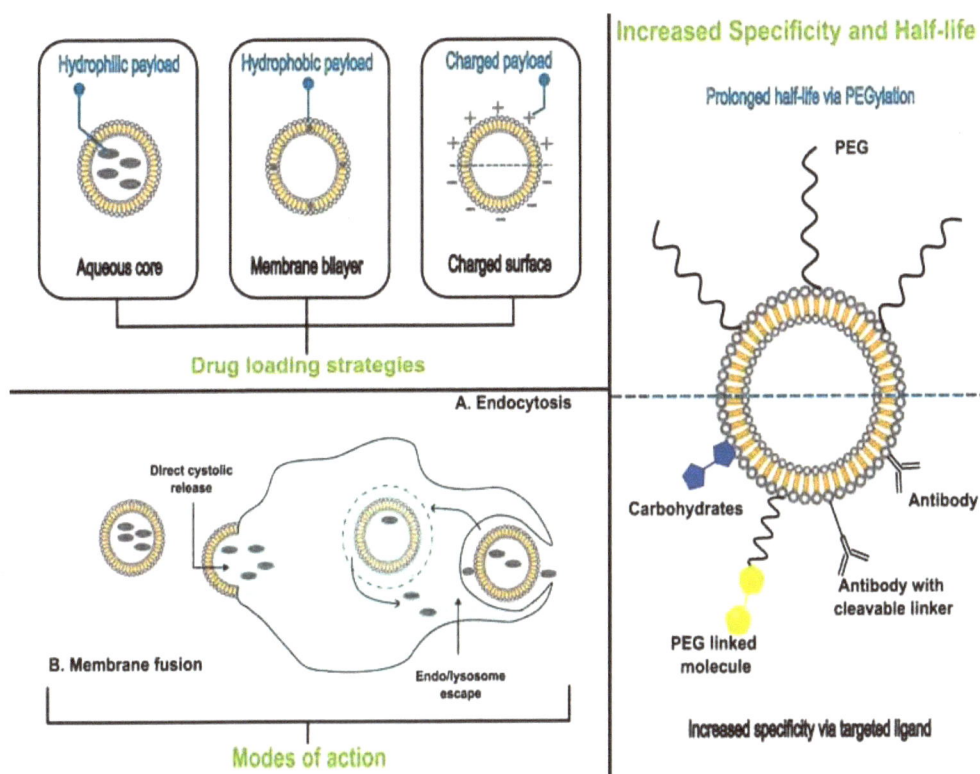

Fig. (2). Drug loading activity and mode of action in a cell, with a proper strategic mode of action and representing specificity to target cells.

Since this is seen to be promising for designing and manufacturing effective nanocarriers, researchers are now working with polymer NPs to create self-organizable assemblies as well as other amphiphile phase structures [67]. As possible nanocarriers for the model solid lipid stearyl alcohol, Angayarkanny *et al.* created micelle assemblies using Lauryl Esters of Tyrosine (LET) covered with polymer nanoparticles [68 - 71]. A little encapsulation of amino acid surfactant in the micelles is suggested by the fact that lauryl esters of phenylalanine and amino acid surfactant dispersions in pure lauryl ester of tyrosine separated spontaneously [68 - 71]. This has produced enough evidence to support the idea that polymer-coated LET micelles serve as appropriate SA encapsulation matrices, the strength of which is ascribed to an H-bonding contact between the hydroxyl (OH) group existing in the SA and the phenolic group found in LET [68 - 71]. As a result, several researchers are eager to create NPs from natural silk for anticancer treatment [72]. To demonstrate the harmless nature of NPs on healthy cells with enhanced efficacy that circumvents the drug resistance mechanism, F. P. Seib *et al.* [72] carried out an *in-vitro* investigation of silk NPs loaded with doxorubicin, as shown in Table **1**.

Table 1. Biopolymers alter the surface of nanoparticles to improve drug delivery.

Nanoparticle	Biopolymers	Drugs	Applications	References
PEG	Lipid	Paclitaxel	Increased *in-vitro* uptake.	[65]
Hydroxylated ethyl starch	PEG layer and mannose	N/A	Dendritic cell targeting.	[66]
MMT nanoparticles	Starch/ D-L-lactic acid	DMSO	Sustained release by *in-vitro* studies.	[71]
Gold nanoparticles	Gelatin	Doxorubicin	Sustained release by *in-vitro* studies.	[72]
Gold nanorods	Lipids	N/A	Increased bioimaging.	[73]

POLYSACCHARIDE-BASED BIOPOLYMER

Due to their lower toxicity and biocompatibility, polysaccharides have been identified as important components for stimulus responsiveness in DDSs [73]. Chitosan (polysaccharides) and alginates are safe to employ in DDSs either by themselves or in combination with certain surface modifications [74 - 76]. Some intriguing surface modification and manufacturing methods were proposed by Wang *et al.* to create biopolymers such as polylactic acid and chitosan nanoparticles for use in nanomedicine [77]. The emulsion diffusion method was used in the PLA procedures, including the creation of single and double emulsion systems.

Using this technique, lipophilic (hydrophobic) compounds were created. Using (i) Hydrophilic chemical entrapment, the salting out method creates raw NPs by emulsifying aqueous and organic phases in O/W and then adding distilled water; (ii) Nano-precipitation, which uses a syringe pump in an oil bath with a magnetic stirrer; and (iii) The emulsion evaporation method, which dissolves a drug agent in the polymer in organic solvent, emulsion, and then solvent evaporation in a vacuum.

The chitosan NPs (CNPs) were produced using the ionic gelation procedure, which entails adding sodium tripolyphosphate to the chitosan solution and rapidly homogenizing it to produce nanoparticles. In the reverse micelle procedure, a surfactant mixed in an organic solvent is combined with chitosan containing a cross-linking agent, stirred overnight, and then purified to yield CNPs. The spray drying method, which includes sprinkling the chitosan solution with compressed air, was also used to create these CNPs. Furthermore, at an adjusted pH of greater than 6.5, chitosan is blow-dried in an alkali solution using a hot air chamber, a process known as coacervation. By enhancing hydrophilicity, chitosan functionalization, target functionalization, pH-sensitive coating, and plasma treatment under a pressure gauge in a gaseous environment, surface modification approaches of the nanopolymers enhanced the *in vivo* treatment outcomes. Subsequent studies revealed that chitosan was supplemented with trimethyl chloride, galactoside, polyethylene glycol, thiolation, and target-modifying agents.

Alvarez-Lorenzo *et al.* established that ionic polysaccharides are pH-sensitive [78]. Through cross-linking networks, this sensitivity can be transmitted chemically or physically. The only cationic polysaccharide found in nature, chitosan, helps to speed up the release of medications at acidic pH since it prefers to increase at acidic pH and shrink at neutral/alkaline pH. This property is the opposite of anionic polysaccharides. Ions have an osmotic effect that raises the cross-linking density and the release rate. The medication is released as a result of an affinity-controlled mechanism brought on by ionic interactions. These properties make polysaccharides an excellent choice for the oral delivery of medications that target certain locations, particularly the colon. Therefore, susceptibility to the enzymes in the large intestine and resistance to the enzymes in the upper gastrointestinal tract that break down the drug may combine to deliver efficient site-specific release. The study also pointed out that variations in pH between tumor and healthy cells may be the cause of the carriers' triggering effect. The electric field may also affect the medication release cycles. Many factors, such as temperature, light, and redox activities, can potentially affect the rate of release. NPs based on polysaccharides are widely used in theragnostics. These days, self-assembly and surface modifications with cross-linkers are receiving more attention. There have been attempts to industrialize polymeric nanocarriers. Kim *et al.* manufactured and built an uncoated coating of chitosan and a coating of starch to use magnetic nanoparticles as a hyperthermic thermoseed [79]. The chitosan-coated magnetic nanoparticles were produced at 23°C in the presence of an AC magnetic field, as opposed to the starch-coated magnetite. The particle capture rate was 96% when an external magnetic field of 0.4 T was present. The chitosan-coated magnetic NPs increased the rate of KB cell capture by around 10.8% as compared to the normal vitality of 93.7% of L929 cells. *In-vitro* experiments revealed that chitosan-coated magnetic nanoparticles were compatible, making them useful and very promising for use in magnetically targeted hyperthermia (Table **2**).

Table 2. Comparison of lipid-based and polysaccharide-based biopolymers.

Feature	Lipid-based biopolymer	Polysaccharide-based biopolymer
Drug Loading Capacity	High for lipophilic drugs, moderate for hydrophilic drugs	High for hydrophilic drugs, moderate for lipophilic drugs
Release Profile	Sustained release, but potential for burst release	Controlled release, the potential for burst release depending on crosslinking
Targeting Capacity	Good targeting *via* surface functionalization (*e.g.*, ligands)	Good targeting through modification (*e.g.*, immune targeting, ligand functionalization)

MECHANISM OF DRUG DELIVERY THROUGH BIOPOLYMER

Numerous research studies based on the use of polymeric materials for drug delivery have been published [80 - 83]. Using a syringe, biopolymers may be injected into the body, where they solidify and create a semisolid depot [84]. Injectable implant systems are classified into five categories depending on the solidification process: hydrophobic lipophilic fatty acid-based injectable pastes, cross-linked systems *in situ*, thermally induced gelling systems, thermoplastic pastes, and precipitates *in situ*. To analyze this, the solvent is extracted from the surrounding tissue, causing precipitation that shields the protein and amide-based polymers from heat and shear stress, degrading [84].

The creation of nanocarriers requires materials with desirable qualities, such as biocompatibility and biodegradability [85]. According to research, a variety of natural polymers, including gliadin, albumin, chitosan, cellulose, and starch, meet these criteria [86]. Starch has previously been converted into NPs and is frequently utilized in medication administration [87]. Doxorubicin can be more effectively administered to the DU145 cell *via* entrapping it into the hydroxyethyl starch, a synthetic polymer made from starch, as opposed to directly targeting it [88].

Gliadin particles were isolated from gluten by M. Gulfam *et al.* using the desolation approach, which was previously employed for drug administration [89]. Nevertheless, they led to a reduced capacity to extract the particles from the aqueous phase and a low drug loading efficiency [89]. As a result, scientists used the electrospinning method, which eliminates the need for a surfactant, to prepare the gliadin NPs. The target NPs in the drug delivery for this investigation were gliadin NPs and gliadin gelatin nanocomposites loaded with cyclophosphamide [89]. The study found that gliadin nanoparticles released cyclophosphamide gradually over 48 hours, but that gliadin-gelatin composite nanoparticles increased drug release. Furthermore, 7% gliadin nanoparticles loaded with cyclophosphamide caused 24-hour-cultured breast cancer cells to undergo apoptosis [89]. Additionally, Western blotting verified the downregulation of the Bcl-2 protein. This gliadin nanoparticle coated with an anticancer medication can therefore be a potent tool for cancer treatment [89].

Yi Tian *et al.* [90] conducted another study with silk fibroin. By magnetizing the cocoon fiber of Bombyx mori, they created silk fibroin nanoparticles [90]. In the past decade, silk fibroin nanoparticles have been employed as a DDS for drug release and trapping. When loaded with doxorubicin, it has also demonstrated effectiveness as a lysosomotropic anticancer nano-drug carrier and has been able to overcome multidrug resistance [90]. On the other hand, a buildup of therapeutic drugs may result from the nonspecificity of tumor targeting. Consequently, specific concentrations of hydrophilic magnetic iron oxide nanoparticles (MNPs) in phosphate solution were included in a one-step potassium phosphate salting-out process [90]. As a result, the doxorubicin-loaded magnetic silk fibroin nanoparticles (NPs) showed a 100% survival rate and prevented tumor growth even on day 30. For future clinical research, it could thus be more advantageous to assess this in an *in-vivo* model [90]. To detect medication release by magnetic stimulation, Mohapatra *et al.* investigated a novel formulation that included magnetic nanoparticles in chitosan in addition to the antibiotic vancomycin [91]. Compared to the nonstimulated MNP, the one embedded in chitosan demonstrated more efficient and regulated drug release [91]. Having control over medication distribution, dosing schedules, and local concentration that can be adjusted to meet the patient's therapeutic demands is one of these distinctive qualities that would be extremely helpful to physicians [91].

Chitosan, a highly versatile substance, is a polymer extracted from the exoskeleton of crustaceans or other shell animals, and it is widely utilized in DDSs [92]. According to a study by Uchegbu *et al.*, chitosan drug delivery began in 1990 [93]. Chitosan is deacetylated chitin (the exoskeleton of many animals). It has shown promising results as a film-forming material required for drug delivery [93]. The amphiphilic character of chitosan provides the ability to form NPs in aqueous solution instead of using any cross-linking agents or ionic gelation agents [93]. The CNP less than 1 μm in dimension is stable for six months after drug loading [93]. Chitosan NPs showed oral bioavailability of hydrophobic drugs up to 6-fold, and when bound to anticancer agents, effectively delivered the drug to the tumor site by exhibiting tumoricidal activity without being toxic to normal cells [93]. When 200 nm dimension NPs were also administered to rabbit retinal cells, they were nontoxic, showing benefit as an ocular agent. It is also able to deliver peptides to the brain through the intravenous and oral routes [93].

Sun *et al.* critically reviewed the rational design and analyzed nanocarrier properties in DDSs [94]. After intravenous administration, he concluded that nanomedicine functions through a five-step CAPIR mechanism. This includes (1) Circulation through the compartments of blood, (2) Accumulation and aggregation in the tumor tissue through its leaky, hastily built vasculature, (3) Greater penetration into the tissues of tumors, (4) Internalization, and (5) Intracellular drug release [94]. Furthermore, researchers observed that nanomedicines exhibit variations in surface charge, dimensions, and stability challenges across different developmental or operational stages. To address these issues, efforts are directed at consolidating and optimizing these properties into a single nanomedicine, termed '3S transitions' for short, thereby overcoming such dilemmas [94]. Concerning the previous context and through an understanding of CAPIR and 3S transitions of nanocarriers, several modifications on the surface, stability, and size are being conducted [94]. Moreover, many researchers have synthesized amphiphilic self-assembled glycyrrhizic acid biotin-starch NPs (GaBS NPs) through a single-step esterification reaction, and observed the cellular uptake of free doxorubicin as well as GaBS NP loaded doxorubicin [94]. The cellular internalization of drug-loaded NPs was confirmed to be more rapid than that of the pure drug. A surface modification study was reported by Simon *et al.*, which synthesized a completely carbohydrate-based nanocarrier through surface-based modification using various sugar derivatives (hydroxyethyl starch, dextran, or glucose) through copper-free click chemistry. This study revealed a strong interaction between sugar moieties and plasma proteins [95]. The cellular uptake showed no nonspecific interaction between the carbohydrate NPs and phagocytic cells [95]. The advancement of peptide-based nanocarriers was also discussed by Wei *et al* [96]. According to them, peptides have multiple biomedical applications, like selfassembling them into nanocarriers in drug delivery, as a targeted ligand for tumor epithelial cells, for enhancing the penetrating ability of drugs loaded in them at the tumor site, or a triggering enzyme-responsive group to create a microenvironment for the tumor to resolve the problem of nanoparticle degradation [96]. This study also revealed that sometimes there are limitations of these NPs as drug agents due to their inability to accumulate at the tumor site and penetration issues, which peptides or surface-modified peptides can resolve [96]. Sanna *et al.* formulated some quinoxalinediones and entrapped them into pectin-derived peptides. This approach allowed them to target pancreatic cancer cells and observe the anticancer efficacy of the biomolecules [97]. Antiproliferative studies were performed where the targeted NP proved to be more potent than the nontargeted one [97].

Nanopolymeric materials exhibited better and more effective drug delivery than chemotherapeutic agents [98]. However, there were certain limitations to the penetration of nanocarriers into solid tumors of large diameters. Traditional polymers can only penetrate up to 50 μm; however, nanopolymers associated with anticancer agents can penetrate tumors of 100 μm diameter [98]. Dextran nanopolymers embedded in aldehyde-functionalized doxorubicin NPs (aldehyde dextran doxorubicin conjugates) showed such properties. 2D nanocarriers' monolayer studies also provided huge discrepancies when administered into *in-vivo* models [98]. As the efficacy was insignificant, the developed 3D cell structure was used to understand drug (doxorubicin) release with anticancer therapy for use in the childhood stage. The eradication of extracranial tumors (neuroblastoma) using drugs faces several challenges, including the reliance on expensive and time-consuming *in vivo* models, which can present discrepancies in results [98]. The role of the glycocalyx in the uptake of the Aldex-dox NPs was also studied. The study involved understanding the interaction between the dextran-based nanocarrier and the highly glycosylated tumor cell surface and their penetration ability [98]. Wang *et al.* extensively reviewed drug-loading strategies for nanocarriers [99]. They stated that major nanocarriers are developed to overcome problems of hydrophobic drugs, and researchers mainly focus on the environment or specific components without paying much attention to drug loading capacities [99]. There are three types of loading capacities of nanocarriers, wherein each system possesses its own release mechanism. In the case of the loading system on the surface, the loading of the drug is dependent upon absorption, while its release depends on desorption [100]. The matrix loading system is determined by the degradation of the carriers or diffusion since molecules of the drug are embedded in the nanocarriers [101]. The cavity loading system is dependent upon diffusion controlled by the shells, which provides the release of the drug. They conclude that further research and understanding can assist in the rational designing of nanocarriers for DDSs [102]. The primary function of nanocarriers is to encapsulate the drug within themselves until it reaches the tumor site following controlled drug release. The method used to understand this process of drug release is through dialysis [103]. This technique, however, cannot be applied to hydrophobic nanocarrier drug-loaded substances. Thus, to overcome such a problem, Bouchaala *et al.* developed a novel technique for the quantification of release of fluorescent moieties from LNCs using Fluorescence Correlation Spectroscopy (FCS) [104]. For this study, LNCs that were nanoemulsion droplets were encapsulated by the hydrophobic Nile red derivative NR668 [104]. The study showed that sharp contrast classical FCS parameters are more effective in drug release compared to less bright light [104]. Therefore, the researchers made use of the standard deviation of fluorescence fluctuations for quantitative analysis of the release of the dye from the nanocarriers. The drug release was found to be temperature dependent, and the rate declined to 37 °C after a 6-h duration of 50% delivery [104].

Similarly, a focused effort on newer techniques like the production of nanocrystals in a confined environment can be achieved within microfluidics channels and was reported by Fontana *et al.* [105]. The chapter previously mentioned that modified peptide moieties can serve as good nanocarriers for DDSs. In the latest work by Medea Neek *et al.*, the researchers found that the peptide alone fails to act as a nanocarrier at the targeted tumor site [106]. They concluded that the peptides could only function effectively as tumoricidal to the target cells if that embedded in a caged protein that has virus-like structures [106]. Moreover, the interaction of these virus-like structures and cage protein with the immune system has concluded that they have better efficacy as a DDS [106].

SYNERGY MECHANISMS IN BIOPOLYMER-DRUG COMBINATIONS

- **Controlled Drug Release**: Biopolymers can act as carriers for anti-cancer drugs, ensuring a sustained and controlled release of the drug at the tumor site. This controlled release prevents the premature degradation of the drug, increasing its bioavailability at the target site. The synergy arises when the biopolymer is capable of responding to the tumor microenvironment, such as pH, temperature, or enzyme activity, thus enhancing the drug's therapeutic effect. For instance, chitosan-based nanocarriers can be designed to release encapsulated drugs in response to acidic conditions typical of the tumor environment [79].
- **Targeted Delivery**: One of the key mechanisms of synergy between biopolymers and cancer treatments is their ability to improve the specificity of drug delivery. Biopolymers like polyethylene glycol (PEG) or polysaccharides can be conjugated with targeting moieties such as antibodies or peptides. These conjugates enable the selective delivery of therapeutic agents to cancer cells, minimizing damage to healthy tissues and reducing side effects. The polymer's functional groups can be designed to bind to specific receptors overexpressed on cancer cells, thereby increasing the local concentration of the therapeutic agent at the tumor site [78].
- **Enhancing Immunotherapy**: Biopolymers have also shown promise in enhancing the efficacy of immunotherapies. They can be used to deliver immune checkpoint inhibitors or cytokines directly to tumor sites. For example, polylactic-co-glycolic acid (PLGA) nanoparticles can deliver immune stimulants in a sustained manner, amplifying the immune response against tumors. The polymers can also act as adjuvants, stimulating immune cells like dendritic cells to enhance the presentation of tumor antigens, leading to a more robust anti-tumor immune response [79].
- **Nanoparticle Formation**: Biopolymers like alginate, chitosan, and hyaluronic acid can form nanoparticles that encapsulate chemotherapy drugs. These nanoparticles can improve drug solubility, stability, and bioavailability, leading to enhanced therapeutic effects. The surface properties of the nanoparticles can also be modified to target specific tumors. For instance, paclitaxel-loaded

chitosan nanoparticles have demonstrated enhanced anti-cancer effects due to their ability to target and penetrate tumor cells more effectively than the free drug [65].

BIOPOLYMER-BASED NANOPARTICLES' INTRACELLULAR MOVEMENT INTO CANCER CELLS

The intracellular trafficking of biopolymer-based nanoparticles by cancer cells is a complex and dynamic phenomenon that is essential for the targeted delivery of imaging agents and therapeutic drugs to specific cancer cell locations. The complex process entails intricate interactions between several biological elements and nanoparticles, which are crucial in determining the effectiveness of cancer treatment. Biopolymer-based nanoparticles—made from polymers like polylactic-coglycolic acid (PLGA), chitosan, or polydopamine (PDA)—offer unique advantages including biocompatibility, adaptability, and functionalization potential. The aforementioned attributes make them highly desirable candidates for cancer therapy since they enhance the effectiveness of medication delivery and imaging techniques [107].

Within cancer cells, biopolymer-based nanoparticles are transported *via* a sequence of crucial phases: Targeted delivery of drugs and imaging agents to particular cancer cell sites depends on the intricate and dynamic intracellular trafficking of biopolymer-based nanoparticles by cancer cells. The complex procedure involves complex interactions between many biological components and nanoparticles, which are essential for figuring out how well cancer treatment works.

- **Cellular Uptake:** The main stage is concerned with how cancer cells absorb nanoparticles. Phagocytosis and endocytosis are components of the process under consideration. The precise method of absorption depends on several variables, including the size of the nanoparticles, their surface charge, and the functionalization process. The nanoparticles may be modified to enhance the cellular absorption process by including ligands that bind preferentially to receptors that are overexpressed on the surfaces of cancer cells [107].
- **Endosomal Escape:** Following internalization, nanoparticles frequently get stuck in endosomes or lysosomes, forming a small area. For nanoparticles to have the desired therapeutic or imaging effects, they must be able to navigate across these compartments and enter the cytoplasm. Some biopolymer-based nanoparticles are engineered to react with the acidic endosome environment, releasing their payload into the cytoplasm. This particular process is essential to prevent cargo from degrading inside lysosomes [107].
- **Intracellular Trafficking:** After escaping endosomes, nanoparticles move

through the cytoplasm, following a series of intracellular trafficking pathways. Nanoparticles made of biopolymers might be designed to take advantage of biological pathways, which will boost their mobility inside the cell [107].

- **Nuclear Localization:** Because nanoparticles are designed to target the cell nucleus specifically, they can be employed for gene delivery treatments. Other challenges, such as the nuclear envelope, must be overcome to do this. Enhancing the nuclear entry of nanoparticles by surface modifications and the employment of certain cargo molecules might result in precise and targeted functions [107].

- **Delivery and Release of Cargo:** It is essential that biopolymer-based nanoparticles efficiently convey their payload, which might comprise genetic material, imaging agents, or therapeutic medications, after they have arrived at their assigned subcellular sites. Characteristics like pH sensitivity, enzymatic breakdown, or external stimuli like light or heat may all be used to tailor controlled release strategies [107].

- **Interaction with Organelles Inside Cells:** Nanoparticles may need to interact with several subcellular organelles, including the endoplasmic reticulum or mitochondria, to be involved in biological activities. Therapeutic results are affected by these interactions because they have a major effect on cellular functions [107].

- **Biodegradation and Clearance:** Since they enhance the safety and biocompatibility of the nanoparticles, they are important facets of the cellular destiny of biopolymer-based nanoparticles. The nanoparticles gradually break down as a result of a natural process called biodegradation that takes place as they traverse the intricate environment inside the cell. One notable feature that sets biopolymer-based nanoparticles apart is their ability to degrade over time, which lowers the possible hazards connected to prolonged accumulation within the cellular environment. The significance of biodegradation stems from its significant effects on treatments' long-term safety as well as their effectiveness [107].

- **Cellular Responses:** A comprehensive understanding of these diverse and complex reactions is essential, as it forms the foundation for improving and optimizing the use of these nanoparticles in cancer treatment [107]. Several interrelated processes are started when nanoparticles enter the cell, and these activities are essential in determining the overall therapeutic efficacy and safety of these therapeutic agents. Important physiological reactions, such as intricate alterations in gene expression, modification of signaling pathways, and immunological responses, are triggered.

- **Immune Reactions and Immunomodulation:** In addition to directly destroying cancer cells, these nanoparticles also engage with the immune system and elicit immunological reactions [107].

BIOPOLYMER THERAGNOSTIC USE IN ANTICANCER TREATMENT

As Anticancer Agents

Due to the large number of fatalities caused due to cancer, it is necessary to identify and develop therapeutic agents that are efficacious and have fewer side effects [93]. Chemotherapeutic agent research exhibits numerous advances and improvements. They still have negative consequences, like disrupting healthy cells and having inadequate transport to the intended location [108 - 111]. Starch and other polymers have been used in nanodrug formulations containing anticancer medicines due to their numerous benefits. These include natural abundance, non-cytotoxicity, biocompatibility, biodegradability, non-immunogenicity, air stability, and compatibility with the majority of medications.

Using the solvent emulsification diffusion approach, Prajakta Dandekar *et al.* synthesized a hydrophobic starch derivative called propyl-starch to comprehend its use in NP formulations that contained the anticancer drug docetaxel [112]. Because of its "nano size," the researchers saw an increase in the drug's availability and efficacy in the cancer cell line. Additionally, perinuclear localization of the NPs supported its specific connection with the nucleus (peri/intra), and cytotoxicity was not seen in *in-vivo* investigations [112].

Researchers have also employed protein-based polymers as antitumor and anticancer delivery vehicles [113]. Taurin *et al.* also explored EPR, which has been shown in several animal models. Even after six hours of treatment, nanomedicines demonstrated efficacy and up to 27 times the amount of accumulation at the tumor site compared to a typical medication [114, 115]. Apart from their biocompatibility and biodegradability, protein nanoparticles (NPs) have an advantage in synthesis because they can be made using complex coacervation techniques, emulsion solvent extraction techniques, or coacervation/desolvation techniques in mild conditions without the use of hazardous chemicals [116].

Biodegradable polyacrylamide nanoparticles with an amine function as a cancer theragnostic were thoroughly studied by Shouyan Wang *et al.*, who included investigations based on active cancer cell targeting, fluorescence imaging, and photodynamic treatment [117]. To develop the nano drug, biodegradable cross-linkers and primary amino moieties were added during the polymerization of NPs. Additionally, photodynamic and fluorescent agents conjugated with PEG and tumor surface ligands were introduced in the NP matrix [117].

The study showed the advantages of NPs (polymer-based) as multifunctional, biodegradable NCs for tumor treatment. Proteins are far safer than anticancer drug nanocarriers, claim Lohcharoenkal *et al.* [118]. As an anticancer medication, an albumin-bound nanocarrier with a particle size distribution of 130 nm demonstrated efficacy [118]. An example of the clinical viability of this strategy is the FDA's approval of albumin-bound paclitaxel (Abraham, ABI008) for metastatic breast cancer, which was also disclosed in their paper [118].

The scientific community has shown a great interest in polydopamine (PDA) because of its properties, which include biocompatibility, ease of manufacturing, and efficient binding of metal ions. Because of their exceptional qualities, multifunctional nanosystems with programmable logic devices (PDA) have been thoroughly researched.

Additionally, it looks at how these NPs are used in various imaging modalities and how they may be used therapeutically to treat cancer [119]. When mussel adhesive proteins come into contact with solid surfaces, they display adhesive qualities that are the origin of the PDA (polydopamine) phenomenon. By auto-oxidatively polymerizing dopamine (DA) under slightly alkaline conditions, Messersmith *et al.* created the adaptable material known as polydopamine (PDA) for the first time [120]. Among the benefits of using PDA are its broad range of optical absorption, consistent particle size, and a mild and solvent-free manufacturing procedure. It has been noted that this nano platform is used endogenously for tumor imaging, especially when combined with photoacoustic and fluorescence imaging methods [121]. The limits of traditional tumor detection methods, such as biopsies, which are insensitive and lack specificity, make this problem significant. Furthermore, imaging methods are prohibitively costly and vulnerable to certain limitations. It has been discovered that PDA-based nanoplatforms offer a very successful and noninvasive method of tumor diagnosis [121].

In Photodynamic Theory (PDT)

Chemotherapeutic medications have been demonstrated to be detrimental to healthy and normal cells, and often struggle with releasing the medication at the tumor location [122]. A novel hypothesis called photodynamic theory (PDT) has recently surfaced to overcome these restrictions because radiation treatment has been demonstrated to be harmful to healthy cells [123]. The execution of this theory requires the usage of oxygen, a drug-activating light, and a photosensitizer. Energy is produced by a photosensitizer light, which leads to the growth and production of ROS, which kills the malignant cells [124].

Porphyrin, when combined with a chlorine-based nanoformulation, is a possible sensitizer that has been used thus far to enhance the EPR of nanomedicines at the tumor site. Porphyrin has been loaded with liposomes, silicon particles, polymeric NP, and magnetic NP [125]. Fluorescence nanoparticles have been routinely employed to target breast, lung, and liver tumors, according to Alyssa Master *et al* [126]. The study also demonstrates the good performance of chitosan NP and human serum albumin when added to chlorinated NPs. Other photosensitizers, including phycocyanin, indocyanine, hypericin, and methylene blue, have shown remarkable effectiveness in inducing apoptosis in cancer cells when loaded with multiple NPs of polymeric conjugates [126].

Through optimal photosensitizer encapsulation, plasma-induced drug inactivation protection, premature drug leakage, improved uptake within tumor tissue and cells, and targeted drug release in the tumor environment and distribution, nanoparticles have offered a variety of ways to improve photosensitizer delivery to the targeted tumor component [126].

The FDA has deemed the amphiphilic tricarbocyanine dye indocyanine green (ICG) safe for use. ICG has a good signal-to-background ratio and is ideal for bioimaging applications since its emission maxima are about 800 nm [127]. Even under laser radiation, ICG coupled with DoX loaded in nano agents demonstrated longer blood circulation and greater stability, according to Sheng *et al.* [128].

Pei Wei *et al.* claimed that the FDA approved the use of an amine that was converted into layered double dihydroxide (LDH) and indocyanine green as a contrast agent for optical imaging *in vivo* experiments. LDH demonstrated superior non-cytotoxicity and biocompatibility. Using LDH-NH_2–ICG-coated chitosan, the stability of the nanocomposites in the *in-vivo* tests was confirmed. Furthermore, in noninvasive *in-vitro* imaging investigations employing LDHs–NH_2–ICG nanoparticles in combination with varying intravenous doses of chitosan coatings, the liver and lungs of nude mice under anesthesia showed a noticeable glow. The combination of chitosan and LDHs-NH_2–ICG was shown to boost the possibility for the development of organ-specific DDSs and *in vivo* contrast agents for cancer diagnosis and therapy [129].

According to Jin *et al.* [130], surface toxicity is a limitation of Au NPs, despite their significant usefulness in cancer phototherapy. Au NPs must be coated with organic polymers to lessen this [130]. Further modification is necessary to coat gold nanorods with polymer since organic polymer coating has both benefits and drawbacks [130]. It may be possible to conjugate with target molecules to improve photothermal treatment (PTT). The DDS of gold nanorods might be improved by combination therapies such as PTT, gene therapy, chemotherapy, and organic polymeric coating [130].

In Cancer Imaging

Millions of people have died from cancer worldwide, particularly in underdeveloped nations, despite the groundbreaking research that has been done in this area. Researchers in the field of nanomedicines are presently mostly focused on early-stage cancer diagnosis [131]. Since quantum dots, a semiconductor nanomaterial, are highly effective at optically imaging necrotic cells, they have been discovered to be particularly powerful in the detection of cancer when attached to biological molecules [132].

For the *in-vitro* diagnosis of cancers like Non-Hodgkin Lymphoma (NHL), Mansur *et al.* created, synthesized, and characterized novel multifunctional immunoconjugates consisting of an organic shell made of antibody-modified polysaccharide chitosan and a fluorescent inorganic core containing Quantum Dots (QDs) [133]. Transmission Electron Microscopy (TEM) images and UV-vis absorption data showed the creation of ultrasmall nanocrystals with average sizes between 2.5 and 3.0 nm. Furthermore, photoluminescence experiments showed that when the immunoconjugates were stimulated by UV light, they produced a "green" glow. Furthermore, the laser light scattering immunoassay confirmed that the antigen bound to the quantum dots, indicating that this technique may be useful for cancer early detection [133].

As a Nanocarrier

PTT, organic nanosystems and polymeric materials have also been produced [134]. The potential of melanin nanoparticles as effective drug-delivery vehicles for image-guided chemotherapy was investigated by Zhang *et al.* [135]. Melanin is a biopolymer that may be used to make an endogenous nano-DDS that is effective for image-guided chemotherapy due to its intrinsic photoacoustic, biocompatible, and recyclable properties as well as its capacity to bind to drugs. Sorafenib was included in the melanin nanoparticle formulation to increase the hydrophilicity of melanin. Melanin and sorafenib share π (pi) bond interactions. The SRFMNPs showed an anticancer impact comparable to that of the polymeric NPs, despite the MNPs system using a lower final loading dosage (SRF 4 mg kg^{-1}, once every two days for MNPs *vs.* SRF 3 mg kg^{-1}, three times every four days for polymeric NPs). This suggests that the MNP-based drug delivery system was superior and could at least have the same anticancer impact as conventional polymeric-NP-based DDSs. Additionally, the traditional nano platform employed in imaging-guided therapy requires functionalization, and the insertion of contrast agents is difficult and perhaps hazardous. Because MNP formulation requires less preparation, it is more suited for safe imaging-guided therapy [135].

It has been demonstrated that bovine serum albumin functions as an effective nanocarrier for drug delivery. Because of its affordability, sensitivity, and potential for noninvasive optical imaging in biological applications, recent research highlights its importance in fluorescence investigations. Bovine serum albumin (BSA) is negligible and lacks photoluminescence. Therefore, as fluorescent agents, Pan *et al.* [136] investigated a nano-DDS made of BSA doped with gold nanoclusters, iron NPs, and gold nanorods. Because of their emission and two-photon excitation, which fall within the near-infrared (NIR) "biological window" between 650 and 900 nm, gold nanorods and clusters are nontoxic, inert, and better functioning as fluorescent probes than photobleachable agents. They may also exhibit potential for imaging *in-vivo*. Magnetic Resonance Imaging (MRI) is one method of imaging for disease detection. Apoptosis was markedly induced in hepatocarcinoma cells by the combination loaded with the drug DOX, while fluorescence microscopy was used to visualize the delivery thanks to the photoluminescence of the nanocarriers [136].

The use of chitosan combined with fucoidan-functionalized gold nanorods as a nanocarrier for Photothermal Therapy (PTT) represents a promising area of research in cancer therapy [137]. The efficiency of the nanocarrier was examined by Manivasagan *et al.*, both *in-vitro* and *in-vivo* [137]. After being administered intravenously to the Balb/c mice model of breast cancer, fucoidan gold nanorods encased in chitosan were seen in an NIR absorbance. They were then exposed to radiation for five minutes after injection and six hours after injection at 54.4 °C. The study found that after 20 days of laser therapy, total tumor cell ablation was achieved [137].

CASE STUDIES AND EXPERIMENTAL DATA

1. **Chitosan and Paclitaxel Nanoparticles**: A study by El-Say *et al.* (2016) investigated the use of chitosan nanoparticles for the delivery of paclitaxel in breast cancer treatment. The results demonstrated that paclitaxel-loaded chitosan nanoparticles enhanced tumor accumulation, prolonged drug release, and improved cytotoxicity against cancer cells. The combination approach led to a significant reduction in tumor size in animal models compared to free paclitaxel. These findings indicate that the synergy between chitosan and paclitaxel could provide an effective means of overcoming the limitations of paclitaxel, such as poor solubility and systemic toxicity [138].

2. **PLGA and Doxorubicin**: In a study by Ranganathan *et al.* (2019), PLGA nanoparticles were used to encapsulate doxorubicin, a commonly used chemotherapy drug, and its therapeutic efficacy was evaluated in a mouse model of liver cancer. The study found that doxorubicin-loaded PLGA nanoparticles significantly increased the drug's tumor retention and decreased

systemic toxicity. The controlled release of doxorubicin from the PLGA nanoparticles resulted in prolonged therapeutic effects and enhanced tumor cell apoptosis. This combined approach showed a significant reduction in tumor volume compared to conventional chemotherapy [139].

3. **Hyaluronic Acid and Immunotherapy**: A promising case study involves the use of hyaluronic acid nanoparticles to enhance immune checkpoint blockade therapies. In a mouse model of melanoma, hyaluronic acid-based nanoparticles were used to deliver anti-PD-1 antibodies. The combination approach resulted in a more substantial anti-tumor immune response, leading to complete tumor regression in some cases. The hyaluronic acid nanoparticles facilitated better accumulation of the immune checkpoint inhibitors at the tumor site, enhancing the therapeutic effects of the immunotherapy [62].

Challenges in Optimizing Biopolymer-drug Combinations

While the combination of biopolymers and cancer therapies shows great promise, several challenges must be addressed to optimize these therapies:

1. **Formulation Stability**: One of the primary challenges in using biopolymers in combination therapies is the stability of the drug formulations. Many biopolymers are prone to degradation under physiological conditions, leading to premature drug release and loss of therapeutic efficacy. To overcome this, researchers are exploring the use of cross-linking agents, encapsulation techniques, and the development of novel biopolymer formulations that are more stable and can withstand the harsh conditions of the body.

2. **Compatibility of Biopolymers with Drugs**: The compatibility between biopolymers and therapeutic agents is another critical factor in the success of combination therapies. Some drugs may not effectively interact with certain biopolymers, leading to issues with loading efficiency, release profiles, and bioavailability. For instance, hydrophobic drugs may have poor encapsulation efficiency when using hydrophilic biopolymers, and *vice versa*. Researchers are continually working on optimizing biopolymer-drug interactions to improve drug loading capacity and enhance therapeutic effects.

3. **Toxicity and Immune Reactions**: Despite their biocompatibility, certain biopolymers may provoke immune responses, leading to inflammation or allergic reactions. For example, chitosan and other polysaccharides have been known to cause mild immune responses in some individuals. Ensuring the safety of biopolymer-based drug delivery systems is crucial, as the use of biopolymers in combination therapies may introduce new risks that need to be carefully evaluated through preclinical and clinical studies.

4. **Scalability and Cost**: The production of biopolymer-based drug delivery systems can be costly and time-consuming, particularly when scaling up from

laboratory research to clinical application. The complexity of formulating stable, effective, and safe biopolymer-drug combinations requires significant investment in research and development, which may limit the widespread adoption of these therapies in clinical settings.

CHALLENGES AND OUTLOOK

Despite recent advancements in drug delivery methods and the utilization of a variety of nanomaterials with appropriate properties, there are still significant problems with cancer care and therapy. Since drug delivery has advanced so dramatically in recent years, many believe that nanoparticles will fundamentally alter the healthcare system. However, only a small number of nanoformulations have progressed to clinical trials due to the difficulty in developing effective cancer nanotherapeutics. The toxicity and biocompatibility of nanomaterials in biological systems are greatly influenced by their physicochemical properties. To lower the possibility of unintentional toxicity to healthy cells, care must be taken throughout the production and characterization of the nano-biomaterials utilized to deliver medication. Nanocarriers may tend to create protein aggregates as a result of their interactions with biomolecules. This might interfere with the regular process of creating nanomedicine and make the nanodrugs ineffective in stopping the growth of cancer cells. Additionally, how nanomaterials are kept may have an impact on their pharmacological effectiveness. Drug delivery is further complicated by concerns about side effects associated with nanomaterials, which may or may not be noticeable or have an instant effect. When used to treat cancer, nanocarriers may interact with cells in ways that reduce the effectiveness of the treatment, perhaps leading to unexpected harm. Despite recent advancements in drug delivery methods and the utilization of a variety of nanomaterials with appropriate properties, there are still significant problems with cancer therapy. Since drug delivery has advanced so dramatically in recent years, many believe that nanoparticles will fundamentally alter the healthcare system. However, just a few nanoformulations have progressed to clinical trials due to the difficulty in developing effective cancer nanotherapeutics. The toxicity and biocompatibility of nanomaterials in biological systems are greatly influenced by their physicochemical properties. The capacity of chemotherapeutic medications to penetrate the blood-brain barrier is a growing concern among many others, since many of the chemotherapeutic medications are ineffective at doing so, which prevents them from binding to their respective target receptors. Researchers have created a number of nanocarrier molecules that can integrate or transport both hydrophilic and hydrophobic drug molecules, allowing them to deliver the medicine to a specified target by bridging the blood-brain barrier. These stable substances, which are shielded from enzymatic degradation before they reach their intended site, have low toxicity and an immunogenic impact on human bodies.

Large-scale production of biopolymers remains challenging despite their unique advantages. To produce biopolymer-based nanoparticles on a commercial scale with consistent quality and reproducibility, robust manufacturing processes are required. Innovations in scalable manufacturing methods, process optimization, and quality control will be crucial to bridging the gap between lab research and commercial applications. Finally, one of the most significant challenges is obtaining FDA approval for the *in-vivo* usage of bio-nanocarriers. Since the FDA has not issued any standards for products using nanomaterials, gaining FDA permission for the use of these drugs is another major challenge.

At the moment, the characteristics that are used are obtained directly from the specifications of bulk materials. Because regulatory decisions on nano-formulated medicines rely on individual assessments of benefits and risks, evaluations take a lot of time and delay commercialization. As multifunctional nanoplatforms proliferate, gaining regulatory approval is likely to grow more challenging.

CONCLUSION

Despite being established in an animal model, the latest design principles approach has not been translated or implemented. Understanding the characteristics of EP and R and their dynamics is essential in therapeutic settings. While employed in several attempts to enhance EP and R, the artificial rat tumor model does not accurately reflect human physiology or pathology in terms of transport or distribution. The *in-vitro* and *in-vivo* studies of nanocarriers as DDSs differed, according to clinical trials. Therefore, employing nanoparticles should be done with caution. To enhance the possibility of more effective anticancer nanomedicines, researchers must put removing existing nanomedicine side effects ahead of developing new or enhanced DDS target site drugs. In the last few decades, scientists have tried to bridge the gap between *in-vitro* and *in-vivo* studies by creating substitute methods, such as 3D cancer models. Our group intends to use a rodent tumor model to develop a clinical trial model that closely mimics human physiology and enhances the drug delivery capabilities of existing nanomedicines based on this extensive review of the literature.

Disclosure

Part of this chapter has previously been published in Advancement in Biopolymer Assisted Cancer Theranostics, Vol 6, Issue 10, 2023, pp. 3959–3983.

REFERENCES

[1]   Sung H, Ferlay J, Siegel RL, *et al.* Global cancer statistics 2020: GLOBOCAN estimates of incidence and mortality worldwide for 36 cancers in 185 countries. CA Cancer J Clin 2021; 71(3): 209-49.

[2]   Dy GW, Gore JL, Forouzanfar MH, Naghavi M, Fitzmaurice C. Global burden of urologic cancers, 1990–2013. Eur Urol 2017; 71(3): 437-46.
[http://dx.doi.org/10.1016/j.eururo.2016.10.008] [PMID: 28029399]

[3]   Zamorano JL, Lancellotti P, Rodriguez Muñoz D, *et al.* 2016 ESC Position Paper on cancer treatments and cardiovascular toxicity developed under the auspices of the ESC Committee for Practice Guidelines. Eur Heart J 2016; 37(36): 2768-801.
[http://dx.doi.org/10.1093/eurheartj/ehw211] [PMID: 27567406]

[4]   Miller KD, Nogueira L, Mariotto AB, *et al.* Cancer treatment and survivorship statistics, 2019. CA Cancer J Clin 2019; 69(5): 363-85.
[http://dx.doi.org/10.3322/caac.21565] [PMID: 31184787]

[5]   Mukerjee N, Maitra S, Ghosh A, Sharma R. Impact of CAR-T cell therapy on treating viral infections: unlocking the door to recovery. Hum Cell 2023; 36(5): 1839-42.
[http://dx.doi.org/10.1007/s13577-023-00942-2] [PMID: 37338785]

[6]   Fuentes AC, Szwed E, Spears CD, Thaper S, Dang LH, Dang NH. Denileukin diftitox (ontak) as maintenance therapy for peripheral t-cell lymphomas: three cases with sustained remission. Case Rep Oncol Med 2015; 2015(1): 1-5.
[http://dx.doi.org/10.1155/2015/123756] [PMID: 26240767]

[7]   Peer D, Karp JM, Hong S, Farokhzad OC, Margalit R, Langer R. Nanocarriers as an emerging platform for cancer therapy. Nano-enabled medical applications 2020; 61-91.
[http://dx.doi.org/10.1201/9780429399039-2]

[8]   Blanco E, Shen H, Ferrari M. Principles of nanoparticle design for overcoming biological barriers to drug delivery. Nat Biotechnol 2015; 33(9): 941-51.
[http://dx.doi.org/10.1038/nbt.3330] [PMID: 26348965]

[9]   Vaid P, Raizada P, Saini AK, Saini RV. Biogenic silver, gold and copper nanoparticles - A sustainable green chemistry approach for cancer therapy. Sustain Chem Pharm 2020; 16: 100247.
[http://dx.doi.org/10.1016/j.scp.2020.100247]

[10]   Barabadi H, Ovais M, Shinwari ZK, Saravanan M. Anti-cancer green bionanomaterials: present status and future prospects. Green Chem Lett Rev 2017; 10(4): 285-314.
[http://dx.doi.org/10.1080/17518253.2017.1385856]

[11]   Manikandan R, Manikandan B, Raman T, *et al.* Biosynthesis of silver nanoparticles using ethanolic petals extract of Rosa indica and characterization of its antibacterial, anticancer and anti-inflammatory activities. Spectrochim Acta A Mol Biomol Spectrosc 2015; 138: 120-9.
[http://dx.doi.org/10.1016/j.saa.2014.10.043] [PMID: 25481491]

[12]   Mishra RK, Ha SK, Verma K, Tiwari SK. Recent progress in selected bio-nanomaterials and their engineering applications: An overview. J Sci Adv Mater Devices 2018; 3(3): 263-88.
[http://dx.doi.org/10.1016/j.jsamd.2018.05.003]

[13]   Faridi Esfanjani A, Jafari SM. Biopolymer nano-particles and natural nano-carriers for nano-encapsulation of phenolic compounds. Colloids Surf B Biointerfaces 2016; 146: 532-43.
[http://dx.doi.org/10.1016/j.colsurfb.2016.06.053] [PMID: 27419648]

[14]   Utreja P, Jain S, Tiwary AK. Novel drug delivery systems for sustained and targeted delivery of anti-cancer drugs: current status and future prospects. Curr Drug Deliv 2010; 7(2): 152-61.
[http://dx.doi.org/10.2174/156720110791011783] [PMID: 20158482]

[15]   Siddique S, Chow JCL. Application of nanomaterials in biomedical imaging and cancer therapy. Nanomaterials (Basel) 2020; 10(9): 1700.
[http://dx.doi.org/10.3390/nano10091700] [PMID: 32872399]

[16]   Ovais M, Raza A, Naz S, *et al.* Current state and prospects of the phytosynthesized colloidal gold nanoparticles and their applications in cancer theranostics. Appl Microbiol Biotechnol 2017; 101(9): 3551-65.
[http://dx.doi.org/10.1007/s00253-017-8250-4] [PMID: 28382454]

[17]   Liu J, Cui L, Losic D. Graphene and graphene oxide as new nanocarriers for drug delivery applications. Acta Biomater 2013; 9(12): 9243-57.
[http://dx.doi.org/10.1016/j.actbio.2013.08.016] [PMID: 23958782]

[18]   Rigg A, Champagne P, Cunningham MF. Polysaccharide-based nanoparticles as pickering emulsifiers in emulsion formulations and heterogenous polymerization systems. Macromol Rapid Commun 2022; 43(3): 2100493.
[http://dx.doi.org/10.1002/marc.202100493] [PMID: 34841604]

[19]   Gopi S, Amalraj A, Sukumaran NP, Haponiuk JT, Thomas S. Biopolymers and their composites for drug delivery: a brief review. Macromol Symp 2018; 380(1): 1800114.
[http://dx.doi.org/10.1002/masy.201800114]

[20]   Song H, Liu X, Jiang L, Li F, Zhang R, Wang P. Current status and prospects of camrelizumab, a humanized antibody against programmed cell death receptor 1. Recent Patents Anticancer Drug Discov 2021; 16(3): 312-32.
[http://dx.doi.org/10.2174/22123970MTE09MDYg0] [PMID: 33563158]

[21]   Calzoni E, Cesaretti A, Polchi A, Di Michele A, Tancini B, Emiliani C. Biocompatible polymer nanoparticles for drug delivery applications in cancer and neurodegenerative disorder therapies. J Funct Biomater 2019; 10(1): 4.
[http://dx.doi.org/10.3390/jfb10010004] [PMID: 30626094]

[22]   Shen S, Wu Y, Liu Y, Wu D. High drug-loading nanomedicines: progress, current status, and prospects. Int J Nanomedicine 2017; 12: 4085-109.
[http://dx.doi.org/10.2147/IJN.S132780] [PMID: 28615938]

[23]   Ahmad MZ, Akhter S, Rahman Z, *et al.* Nanometric gold in cancer nanotechnology: current status and future prospect. J Pharm Pharmacol 2013; 65(5): 634-51.
[http://dx.doi.org/10.1111/jphp.12017] [PMID: 23600380]

[24]   Kim CS, Tonga GY, Solfiell D, Rotello VM. Inorganic nanosystems for therapeutic delivery: Status and prospects. Adv Drug Deliv Rev 2013; 65(1): 93-9.
[http://dx.doi.org/10.1016/j.addr.2012.08.011] [PMID: 22981754]

[25]   Abdel-Fattah WI, W Ali G. On the anti-cancer activities of silver nanoparticles. J Appl Biotechnol Bioeng 2018; 5(1): 43-6.
[http://dx.doi.org/10.15406/jabb.2018.05.00116]

[26]   Renn O, Goodwin DA, Studer M, Moran JK, Jacques V, Meares CF. New approaches to delivering metal-labeled antibodies to tumors: synthesis and characterization of new biotinyl chelate conjugates for pre-targeted diagnosis and therapy. J Control Release 1996; 39(2-3): 239-49.
[http://dx.doi.org/10.1016/0168-3659(95)00157-3]

[27]   Luo SH, Wu YC, Cao L, *et al.* Direct metal-free preparation of functionalizable polylactic acid-ethisterone conjugates in a one-pot approach. Macromol Chem Phys 2019; 220(5): 1800475.
[http://dx.doi.org/10.1002/macp.201800475]

[28]   Barua S, Mitragotri S. Challenges associated with penetration of nanoparticles across cell and tissue barriers: A review of current status and future prospects. Nano Today 2014; 9(2): 223-43.
[http://dx.doi.org/10.1016/j.nantod.2014.04.008] [PMID: 25132862]

[29]   Preetam S, Nahak BK, Patra S, *et al.* Emergence of microfluidics for next generation biomedical devices. Biosens Bioelectron X 2022; 10: 100106.
[http://dx.doi.org/10.1016/j.biosx.2022.100106]

[30]   Otto T, Sicinski P. Cell cycle proteins as promising targets in cancer therapy. Nat Rev Cancer 2017;

17(2): 93-115.
[http://dx.doi.org/10.1038/nrc.2016.138] [PMID: 28127048]

[31]   Leal-Esteban LC, Fajas L. Cell cycle regulators in cancer cell metabolism. Biochim Biophys Acta Mol Basis Dis 2020; 1866(5): 165715.
[http://dx.doi.org/10.1016/j.bbadis.2020.165715] [PMID: 32035102]

[32]   Kontomanolis EN, Koutras A, Syllaios A, *et al.* Role of oncogenes and tumor-suppressor genes in carcinogenesis: a review. Anticancer Res 2020; 40(11): 6009-15.
[http://dx.doi.org/10.21873/anticanres.14622] [PMID: 33109539]

[33]   Lipsick J. A history of cancer research: tumor suppressor genes. Cold Spring Harb Perspect Biol 2020; 12(2): a035907.
[http://dx.doi.org/10.1101/cshperspect.a035907] [PMID: 32015099]

[34]   Kaptain S, Tan LK, Chen B. Her-2/neu and breast cancer. Diagn Mol Pathol 2001; 10(3): 139-52.
[http://dx.doi.org/10.1097/00019606-200109000-00001] [PMID: 11552716]

[35]   Chen L, Liu S, Tao Y. Regulating tumor suppressor genes: post-translational modifications. Signal Transduct Target Ther 2020; 5(1): 90.
[http://dx.doi.org/10.1038/s41392-020-0196-9] [PMID: 32532965]

[36]   Nahak BK, Mishra A, Preetam S, Tiwari A. Advances in organ-on-a-chip materials and devices. ACS Appl Bio Mater 2022; 5(8): 3576-607.
[http://dx.doi.org/10.1021/acsabm.2c00041] [PMID: 35839513]

[37]   Preetam S, Dash L, Sarangi SS, Sahoo MM, Pradhan AK. Application of nanobiosensor in health care sector. Bio-Nano Interface: Applications in Food. Healthcare and Sustainability. 2022; pp. 251-70.
[http://dx.doi.org/10.1007/978-981-16-2516-9_14]

[38]   Ahmad A, Akhtar J, Ahmad M, *et al.* Bedaquiline: an insight into its clinical use in multidrug-resistant pulmonary tuberculosis. Drug Res 2024; 74(06): 269-79.

[39]   Singh A, Ansari VA, Ahsan F, Akhtar J, Khushwaha P, Maheshwari S. Viridescent concoction of genstein tendentious silver nanoparticles for breast cancer. Res J Pharm Technol 2021; 14(5): 2867-72.

[40]   Sharma A, Goyal A, Kumari S, *et al.* A comprehensive review on synthesis of silver nano-particles: an update. Nanosc Nanotechnol Asia 2024; 14(2): 3-23.

[41]   Behan FM, Iorio F, Picco G, *et al.* Prioritization of cancer therapeutic targets using CRISPR–Cas9 screens. Nature 2019; 568(7753): 511-6.

[42]   Slavin YN, Asnis J, Häfeli UO, Bach H. Metal nanoparticles: understanding the mechanisms behind antibacterial activity. J Nanobiotechnology 2017; 15(1): 65.
[http://dx.doi.org/10.1186/s12951-017-0308-z] [PMID: 28974225]

[43]   Singh AK. Comparative therapeutic effects of plant-extract synthesized and traditionally synthesized gold nanoparticles on alcohol-induced inflammatory activity in SH-SY5Y cells *in vitro*. Biomedicines 2017; 5(4): 70.
[http://dx.doi.org/10.3390/biomedicines5040070] [PMID: 29244731]

[44]   Barabadi H, Alizadeh A, Ovais M, Ahmadi A, Shinwari ZK, Saravanan M. Efficacy of green nanoparticles against cancerous and normal cell lines: a systematic review and meta-analysis. IET Nanobiotechnol 2018; 12(4): 377-91.
[http://dx.doi.org/10.1049/iet-nbt.2017.0120]

[45]   Rosenblum D, Joshi N, Tao W, Karp JM, Peer D. Progress and challenges towards targeted delivery of cancer therapeutics. Nat Commun 2018; 9(1): 1410.
[http://dx.doi.org/10.1038/s41467-018-03705-y] [PMID: 29650952]

[46]   Song R, Murphy M, Li C, Ting K, Soo C, Zheng Z. Current development of biodegradable polymeric materials for biomedical applications. Drug Des Devel Ther 2018; 12: 3117-45.
[http://dx.doi.org/10.2147/DDDT.S165440] [PMID: 30288019]

[47] Preetam S, Jonnalagadda S, Kumar L, *et al.* Therapeutic potential of lipid nanosystems for the treatment of Parkinson's disease. Ageing Res Rev 2023; 89: 101965.
[http://dx.doi.org/10.1016/j.arr.2023.101965] [PMID: 37268112]

[48] Pradhan AK, Sahoo SR, Satapathy S, Pradhan S, Prusty JS, Preetam S. Green synthesis of silver nanoparticles from various plant extracts and its applications: A mini review. World J Biol Pharm Health Sci 2022; 11(1): 50-61.

[49] Bhattacharjee R, Kumar L, Mukerjee N, *et al.* The emergence of metal oxide nanoparticles (NPs) as a phytomedicine: A two-facet role in plant growth, nano-toxicity and anti-phyto-microbial activity. Biomed Pharmacother 2022; 155: 113658.
[http://dx.doi.org/10.1016/j.biopha.2022.113658] [PMID: 36162370]

[50] Sharma K, Porat Z, Gedanken A. Designing natural polymer-based capsules and spheres for biomedical applications—a review. Polymers (Basel) 2021; 13(24): 4307.
[http://dx.doi.org/10.3390/polym13244307] [PMID: 34960858]

[51] Chahal A, Saini AK, Chhillar AK, Saini RV. Natural antioxidants as defense system against cancer. Asian J Pharm Clin Res 2018; 11(5): 38-44.
[http://dx.doi.org/10.22159/ajpcr.2018.v11i5.24119]

[52] Abdalla AME, Xiao L, Ullah MW, Yu M, Ouyang C, Yang G. Current challenges of cancer anti-angiogenic therapy and the promise of nanotherapeutics. Theranostics 2018; 8(2): 533-48.
[http://dx.doi.org/10.7150/thno.21674] [PMID: 29290825]

[53] Subbiah R, Veerapandian M, Yun KS. Nanoparticles: functionalization and multifunctional applications in biomedical sciences. Curr Med Chem 2010; 17(36): 4559-77.
[http://dx.doi.org/10.2174/092986710794183024] [PMID: 21062250]

[54] Ulbricht M. Design and synthesis of organic polymers for molecular separation membranes. Curr Opin Chem Eng 2020; 28: 60-5.
[http://dx.doi.org/10.1016/j.coche.2020.02.002]

[55] Patra JK, Das G, Fraceto LF, *et al.* Nano based drug delivery systems: recent developments and future prospects. J Nanobiotechnology 2018; 16(1): 71.
[http://dx.doi.org/10.1186/s12951-018-0392-8] [PMID: 30231877]

[56] Samal SK, Preetam S. Synthetic biology: refining human health. In: Suar M, Misra N, Dash C, editors Microbial engineering for therapeutics. Singapore: Springer Nature Singapore 2022; pp. 57-70.
[http://dx.doi.org/10.1007/978-981-19-3979-2_3]

[57] Salehi S, Shandiz SA, Ghanbar F, *et al.* Phytosynthesis of silver nanoparticles using *Artemisia marschalliana* Sprengel aerial part extract and assessment of their antioxidant, anticancer, and antibacterial properties. Int J Nanomedicine 2016; 11: 1835-46.
[PMID: 27199558]

[58] Khanra K, Panja S, Choudhuri I, Chakraborty A, Bhattacharyya N. Evaluation of antibacterial activity and cytotoxicity of green synthesized silver nanoparticles using *Scoparia dulcis*. Nano Biomed Eng 2015; 7(3): 128-33.
[http://dx.doi.org/10.5101/nbe.v7i3.p128-133]

[59] Venugopal K, Rather HA, Rajagopal K, *et al.* Synthesis of silver nanoparticles (Ag NPs) for anticancer activities (MCF 7 breast and A549 lung cell lines) of the crude extract of *Syzygium aromaticum*. J Photochem Photobiol B 2017; 167: 282-9.
[http://dx.doi.org/10.1016/j.jphotobiol.2016.12.013] [PMID: 28110253]

[60] Kanipandian N, Thirumurugan R. A feasible approach to phyto-mediated synthesis of silver nanoparticles using industrial crop *Gossypium hirsutum* (cotton) extract as stabilizing agent and assessment of its *in vitro* biomedical potential. Ind Crops Prod 2014; 55: 1-10.
[http://dx.doi.org/10.1016/j.indcrop.2014.01.042]

[61] Lee MK. Liposomes for enhanced bioavailability of water-insoluble drugs: *In vivo* evidence and recent

approaches. Pharmaceutics 2020; 12(3): 264.
[http://dx.doi.org/10.3390/pharmaceutics12030264] [PMID: 32183185]

[62]   Weiss VM, Lucas H, Mueller T, *et al.* Intended and unintended targeting of polymeric nanocarriers: the case of modified poly (glycerol adipate) nanoparticles. Macromol Biosci 2018; 18(1): 1700240.
[http://dx.doi.org/10.1002/mabi.201700240] [PMID: 29218838]

[63]   Dehsorkhi A, Castelletto V, Hamley IW. Self-assembling amphiphilic peptides. J Pept Sci 2014; 20(7): 453-67.
[http://dx.doi.org/10.1002/psc.2633] [PMID: 24729276]

[64]   Kamaly N, Yameen B, Wu J, Farokhzad OC. Degradable controlled-release polymers and polymeric nanoparticles: mechanisms of controlling drug release. Chem Rev 2016; 116(4): 2602-63.
[http://dx.doi.org/10.1021/acs.chemrev.5b00346] [PMID: 26854975]

[65]   Stolzoff M, Ekladious I, Colby AH, Colson YL, Porter TM, Grinstaff MW. Synthesis and characterization of hybrid polymer/lipid expansile nanoparticles: imparting surface functionality for targeting and stability. Biomacromolecules 2015; 16(7): 1958-66.
[http://dx.doi.org/10.1021/acs.biomac.5b00336] [PMID: 26053219]

[66]   Kang B, Okwieka P, Schöttler S, *et al.* Carbohydrate-based nanocarriers exhibiting specific cell targeting with minimum influence from the protein corona. Angew Chem Int Ed 2015; 54(25): 7436-40.
[http://dx.doi.org/10.1002/anie.201502398] [PMID: 25940402]

[67]   Begines B, Ortiz T, Pérez-Aranda M, *et al.* Polymeric nanoparticles for drug delivery: Recent developments and future prospects. Nanomaterials (Basel) 2020; 10(7): 1403.
[http://dx.doi.org/10.3390/nano10071403] [PMID: 32707641]

[68]   Angayarkanny S, Baskar G, Mandal AB. Nanocarriers of solid lipid from micelles of amino acids surfactants coated with polymer nanoparticles. Langmuir 2013; 29(23): 6805-14.
[http://dx.doi.org/10.1021/la400605v] [PMID: 23718941]

[69]   Telrandhe R. Anti-cancer potential of green synthesized silver nanoparticles-a review. Asian J Pharm Technol 2019; 9(4): 260-6.
[http://dx.doi.org/10.5958/2231-5713.2019.00043.6]

[70]   Jurj A, Braicu C, Pop LA, Tomuleasa C, Gherman C, Berindan-Neagoe I. The new era of nanotechnology, an alternative to change cancer treatment. Drug Des Devel Ther 2017; 11: 2871-90.
[http://dx.doi.org/10.2147/DDDT.S142337] [PMID: 29033548]

[71]   Hira I, Kumar A, Kumari R, Saini AK, Saini RV. Pectin-guar gum-zinc oxide nanocomposite enhances human lymphocytes cytotoxicity towards lung and breast carcinomas. Mater Sci Eng C 2018; 90: 494-503.
[http://dx.doi.org/10.1016/j.msec.2018.04.085] [PMID: 29853118]

[72]   Seib FP, Jones GT, Rnjak-Kovacina J, Lin Y, Kaplan DL. pH-dependent anticancer drug release from silk nanoparticles. Adv Healthc Mater 2013; 2(12): 1606-11.
[http://dx.doi.org/10.1002/adhm.201300034] [PMID: 23625825]

[73]   Pushpamalar J, Meganathan P, Tan HL, *et al.* Development of a polysaccharide-based hydrogel drug delivery system (DDS): An update. Gels 2021; 7(4): 153.
[http://dx.doi.org/10.3390/gels7040153] [PMID: 34698125]

[74]   Qi SS, Sun JH, Yu HH, Yu SQ. Co-delivery nanoparticles of anti-cancer drugs for improving chemotherapy efficacy. Drug Deliv 2017; 24(1): 1909-26.
[http://dx.doi.org/10.1080/10717544.2017.1410256] [PMID: 29191057]

[75]   Zhang Y, Tang L. The application of lncRNAs in cancer treatment and diagnosis. Recent Patents Anticancer Drug Discov 2018; 13(3): 292-301.
[http://dx.doi.org/10.2174/1574892813666180226121819] [PMID: 29485010]

[76]   da Silva Luz GV, Barros KV, de Araújo FV, *et al.* Nanorobotics in drug delivery systems for treatment

of cancer: a review. J Mater Sci Eng A 2016; 6: 167-80.
[http://dx.doi.org/10.17265/2161-6213/2016.5-6.005]

[77] Wang Y, Li P, Truong-Dinh Tran T, Zhang J, Kong L. Manufacturing techniques and surface engineering of polymer based nanoparticles for targeted drug delivery to cancer. Nanomaterials (Basel) 2016; 6(2): 26.
[http://dx.doi.org/10.3390/nano6020026] [PMID: 28344283]

[78] Alvarez-Lorenzo C, Blanco-Fernandez B, Puga AM, Concheiro A. Crosslinked ionic polysaccharides for stimuli-sensitive drug delivery. Adv Drug Deliv Rev 2013; 65(9): 1148-71.
[http://dx.doi.org/10.1016/j.addr.2013.04.016] [PMID: 23639519]

[79] Kim DH, Kim KN, Kim KM, Lee YK. Targeting to carcinoma cells with chitosan- and starch-coated magnetic nanoparticles for magnetic hyperthermia. J Biomed Mater Res A 2009; 88A(1): 1-11.
[http://dx.doi.org/10.1002/jbm.a.31775] [PMID: 18257079]

[80] Zhang Z, Li X, Han Y, *et al.* RAD54B potentiates tumor growth and predicts poor prognosis of patients with luminal A breast cancer. Biomed Pharmacother 2019; 118: 109341.
[http://dx.doi.org/10.1016/j.biopha.2019.109341] [PMID: 31545289]

[81] Yasuhara T, Suzuki T, Katsura M, Miyagawa K. Rad54B serves as a scaffold in the DNA damage response that limits checkpoint strength. Nat Commun 2014; 5(1): 5426.
[http://dx.doi.org/10.1038/ncomms6426] [PMID: 25384516]

[82] Sharma D, Kanchi S, Bisetty K. Biogenic synthesis of nanoparticles: A review. Arab J Chem 2019; 12(8): 3576-600.
[http://dx.doi.org/10.1016/j.arabjc.2015.11.002]

[83] Aziz N, Faraz M, Pandey R, *et al.* Facile algae-derived route to biogenic silver nanoparticles: synthesis, antibacterial, and photocatalytic properties. Langmuir 2015; 31(42): 11605-12.
[http://dx.doi.org/10.1021/acs.langmuir.5b03081] [PMID: 26447769]

[84] Chitkara D, Shikanov A, Kumar N, Domb AJ. Biodegradable injectable *in situ* depot-forming drug delivery systems. Macromol Biosci 2006; 6(12): 977-90.
[http://dx.doi.org/10.1002/mabi.200600129] [PMID: 17128422]

[85] Witika BA, Makoni PA, Matafwali SK, *et al.* Biocompatibility of biomaterials for nanoencapsulation: Current approaches. Nanomaterials (Basel) 2020; 10(9): 1649.
[http://dx.doi.org/10.3390/nano10091649] [PMID: 32842562]

[86] George A, Shah PA, Shrivastav PS. Natural biodegradable polymers based nano-formulations for drug delivery: A review. Int J Pharm 2019; 561: 244-64.
[http://dx.doi.org/10.1016/j.ijpharm.2019.03.011] [PMID: 30851391]

[87] Alp E, Damkaci F, Guven E, Tenniswood M. Starch nanoparticles for delivery of the histone deacetylase inhibitor CG-1521 in breast cancer treatment. Int J Nanomedicine 2019; 14: 1335-46.
[http://dx.doi.org/10.2147/IJN.S191837] [PMID: 30863064]

[88] Paleos CM, Sideratou Z, Theodossiou TA, Tsiourvas D. Carboxylated hydroxyethyl starch: a novel polysaccharide for the delivery of doxorubicin. Chem Biol Drug Des 2015; 85(5): 653-8.
[http://dx.doi.org/10.1111/cbdd.12447] [PMID: 25303215]

[89] Muhammad G, Ji-eun K, Min LJ, Boram K, Hyun CB, Geun CB. Anticancer drug-loaded gliadin nanoparticles induce apoptosis in breast cancer cells.
[http://dx.doi.org/ 10.1021/la300691n]

[90] Tian Y, Jiang X, Chen X, Shao Z, Yang W. Doxorubicin-loaded magnetic silk fibroin nanoparticles for targeted therapy of multidrug-resistant cancer. Adv Mater 2014; 26(43): 7393-8.
[http://dx.doi.org/10.1002/adma.201403562] [PMID: 25238148]

[91] Mohapatra A, Harris MA, LeVine D, *et al.* Magnetic stimulus responsive vancomycin drug delivery system based on chitosan microbeads embedded with magnetic nanoparticles. J Biomed Mater Res B Appl Biomater 2018; 106(6): 2169-76.

[http://dx.doi.org/10.1002/jbm.b.34015] [PMID: 29052337]

[92]   Baharlouei P, Rahman A. Chitin and chitosan: prospective biomedical applications in drug delivery, cancer treatment, and wound healing. Mar Drugs 2022; 20(7): 460.
[http://dx.doi.org/10.3390/md20070460] [PMID: 35877753]

[93]   Uchegbu IF, Carlos M, McKay C, Hou X, Schätzlein AG. Chitosan amphiphiles provide new drug delivery opportunities. Polym Int 2014; 63(7): 1145-53.
[http://dx.doi.org/10.1002/pi.4721]

[94]   Sun Q, Zhou Z, Qiu N, Shen Y. Rational design of cancer nanomedicine: nanoproperty integration and synchronization. Adv Mater 2017; 29(14): 1606628.
[http://dx.doi.org/10.1002/adma.201606628] [PMID: 28234430]

[95]   Simon J, Christmann S, Mailänder V, Wurm FR, Landfester K. Protein Corona mediated stealth properties of biocompatible carbohydrate-based nanocarriers. Isr J Chem 2018; 58(12): 1363-72.
[http://dx.doi.org/10.1002/ijch.201800166]

[96]   Wei G, Wang Y, Huang X, Hou H, Zhou S. Peptide-based nanocarriers for cancer therapy. Small Methods 2018; 2(9): 1700358.
[http://dx.doi.org/10.1002/smtd.201700358]

[97]   Sanna V, Nurra S, Pala N, *et al.* Targeted nanoparticles for the delivery of novel bioactive molecules to pancreatic cancer cells. J Med Chem 2016; 59(11): 5209-20.
[http://dx.doi.org/10.1021/acs.jmedchem.5b01571] [PMID: 27139920]

[98]   Yan L, Shen J, Wang J, Yang X, Dong S, Lu S. Nanoparticle-based drug delivery system: a patient-friendly chemotherapy for oncology. Dose Response 2020; 18(3): 1559325820936161.
[http://dx.doi.org/10.1177/1559325820936161] [PMID: 32699536]

[99]   Wang N, Cheng X, Li N, Wang H, Chen H. Nanocarriers and their loading strategies. Adv Healthc Mater 2019; 8(6): 1801002.
[http://dx.doi.org/10.1002/adhm.201801002] [PMID: 30450761]

[100] Ahuja G, Pathak K. Porous carriers for controlled/modulated drug delivery. Indian J Pharm Sci 2009; 71(6): 599-607.
[http://dx.doi.org/10.4103/0250-474X.59540] [PMID: 20376211]

[101] Perrigue PM, Murray RA, Mielcarek A, Henschke A, Moya SE. Degradation of drug delivery nanocarriers and payload release: A review of physical methods for tracing nanocarrier biological fate. Pharmaceutics 2021; 13(6): 770.
[http://dx.doi.org/10.3390/pharmaceutics13060770] [PMID: 34064155]

[102] Lee JH, Yeo Y. Controlled drug release from pharmaceutical nanocarriers. Chem Eng Sci 2015; 125: 75-84.
[http://dx.doi.org/10.1016/j.ces.2014.08.046] [PMID: 25684779]

[103] Din F, Aman W, Ullah I, *et al.* Effective use of nanocarriers as drug delivery systems for the treatment of selected tumors. Int J Nanomedicine 2017; 12: 7291-309.
[http://dx.doi.org/10.2147/IJN.S146315] [PMID: 29042776]

[104] Bouchaala R, Richert L, Anton N, *et al.* Quantifying release from lipid nanocarriers by fluorescence correlation spectroscopy. ACS Omega 2018; 3(10): 14333-40.
[http://dx.doi.org/10.1021/acsomega.8b01488] [PMID: 30411065]

[105] Fontana F, Figueiredo P, Zhang P, Hirvonen JT, Liu D, Santos HA. Production of pure drug nanocrystals and nano co-crystals by confinement methods. Adv Drug Deliv Rev 2018; 131: 3-21.
[http://dx.doi.org/10.1016/j.addr.2018.05.002] [PMID: 29738786]

[106] Neek M, Kim TI, Wang SW. Protein-based nanoparticles in cancer vaccine development. Nanomedicine 2019; 15(1): 164-74.
[http://dx.doi.org/10.1016/j.nano.2018.09.004] [PMID: 30291897]

[107] Bhattacharya T, Preetam S, Ghosh B, *et al.* Advancement in biopolymer assisted cancer theranostics. ACS Appl Bio Mater 2023; 6(10): 3959-83.
[http://dx.doi.org/10.1021/acsabm.3c00458] [PMID: 37699558]

[108] Bukowski K, Kciuk M, Kontek R. Mechanisms of multidrug resistance in cancer chemotherapy. Int J Mol Sci 2020; 21(9): 3233.
[http://dx.doi.org/10.3390/ijms21093233] [PMID: 32370233]

[109] Pećina-Šlaus N, Kafka A, Salamon I, Bukovac A. Mismatch repair pathway, genome stability and cancer. Front Mol Biosci 2020; 7: 122.
[http://dx.doi.org/10.3389/fmolb.2020.00122] [PMID: 32671096]

[110] Ruggiano A, Ramadan K. DNA–protein crosslink proteases in genome stability. Commun Biol 2021; 4(1): 11.
[http://dx.doi.org/10.1038/s42003-020-01539-3] [PMID: 33398053]

[111] Imre B, Pukánszky B. Compatibilization in bio-based and biodegradable polymer blends. Eur Polym J 2013; 49(6): 1215-33.
[http://dx.doi.org/10.1016/j.eurpolymj.2013.01.019]

[112] Dandekar P, Jain R, Stauner T, *et al.* A hydrophobic starch polymer for nanoparticle-mediated delivery of docetaxel. Macromol Biosci 2012; 12(2): 184-94.
[http://dx.doi.org/10.1002/mabi.201100244] [PMID: 22127828]

[113] Voci S, Gagliardi A, Fresta M, Cosco D. Antitumor features of vegetal protein-based nanotherapeutics. Pharmaceutics 2020; 12(1): 65.
[http://dx.doi.org/10.3390/pharmaceutics12010065] [PMID: 31952147]

[114] Taurin S, Nehoff H, Greish K. Anticancer nanomedicine and tumor vascular permeability; Where is the missing link?. J Control Release 2012; 164(3): 265-75.
[http://dx.doi.org/10.1016/j.jconrel.2012.07.013] [PMID: 22800576]

[115] Rasool M, Malik A, Waquar S, *et al.* New challenges in the use of nanomedicine in cancer therapy. Bioengineered 2022; 13(1): 759-73.
[http://dx.doi.org/10.1080/21655979.2021.2012907] [PMID: 34856849]

[116] Miao Y, Yang T, Yang S, Yang M, Mao C. Protein nanoparticles directed cancer imaging and therapy. Nano Converg 2022; 9(1): 2.
[http://dx.doi.org/10.1186/s40580-021-00293-4]

[117] Wang S, Kim G, Lee YEK, *et al.* Multifunctional biodegradable polyacrylamide nanocarriers for cancer theranostics-a "see and treat" strategy. ACS Nano 2012; 6(8): 6843-51.
[http://dx.doi.org/10.1021/nn301633m] [PMID: 22702416]

[118] Lohcharoenkal W, Wang L, Chen YC, Rojanasakul Y. Protein nanoparticles as drug delivery carriers for cancer therapy. BioMed Res Int 2014; 2014(1): 1-12.
[http://dx.doi.org/10.1155/2014/180549] [PMID: 24772414]

[119] Li M, Xuan Y, Zhang W, Zhang S, An J. Polydopamine-containing nano-systems for cancer multi-mode diagnoses and therapies: A review. Int J Biol Macromol 2023; 247: 125826.
[http://dx.doi.org/10.1016/j.ijbiomac.2023.125826] [PMID: 37455006]

[120] Lee H, Dellatore SM, Miller WM, Messersmith PB. Mussel-inspired surface chemistry for multifunctional coatings. Science 2007; 318(5849): 426-30.
[http://dx.doi.org/10.1126/science.1147241]

[121] Shao L, Li Y, Huang F, *et al.* Complementary autophagy inhibition and glucose metabolism with rattle-structured polydopamine@mesoporous silica nanoparticles for augmented low-temperature photothermal therapy and *in vivo* photoacoustic imaging. Theranostics 2020; 10(16): 7273-86.
[http://dx.doi.org/10.7150/thno.44668] [PMID: 32641992]

[122] Mansoori B, Mohammadi A, Davudian S, Shirjang S, Baradaran B. The different mechanisms of

cancer drug resistance: a brief review. Adv Pharm Bull 2017; 7(3): 339-48.
[http://dx.doi.org/10.15171/apb.2017.041] [PMID: 29071215]

[123] Gunaydin G, Gedik ME, Ayan S. Photodynamic therapy—current limitations and novel approaches. Front Chem 2021; 9: 691697.
[http://dx.doi.org/10.3389/fchem.2021.691697] [PMID: 34178948]

[124] Sai DL, Lee J, Nguyen DL, Kim YP. Tailoring photosensitive ROS for advanced photodynamic therapy. Exp Mol Med 2021; 53(4): 495-504.
[http://dx.doi.org/10.1038/s12276-021-00599-7] [PMID: 33833374]

[125] Rabiee N, Yaraki MT, Garakani SM, *et al.* Recent advances in porphyrin-based nanocomposites for effective targeted imaging and therapy. Biomaterials 2020; 232: 119707.
[http://dx.doi.org/10.1016/j.biomaterials.2019.119707] [PMID: 31874428]

[126] Master A, Livingston M, Sen Gupta A. Photodynamic nanomedicine in the treatment of solid tumors: Perspectives and challenges. J Control Release 2013; 168(1): 88-102.
[http://dx.doi.org/10.1016/j.jconrel.2013.02.020] [PMID: 23474028]

[127] Alander JT, Kaartinen I, Laakso A, *et al.* A review of indocyanine green fluorescent imaging in surgery. Int J Biomed Imaging 2012; 2012(1): 1-26.
[http://dx.doi.org/10.1155/2012/940585] [PMID: 22577366]

[128] Sheng Z, Hu D, Xue M, He M, Gong P, Cai L. Indocyanine green nanoparticles for theranostic applications. Nano-Micro Letters 2013; 5(3): 145-50.
[http://dx.doi.org/10.1007/BF03353743]

[129] Wei PR, Cheng SH, Liao WN, Kao KC, Weng CF, Lee CH. Synthesis of chitosan-coated near-infrared layered double hydroxide nanoparticles for *in vivo* optical imaging. J Mater Chem 2012; 22(12): 5503-13.
[http://dx.doi.org/10.1039/c2jm16447g]

[130] Jin N, Zhang Q, Yang M, Yang M. Detoxification and functionalization of gold nanorods with organic polymers and their applications in cancer photothermal therapy. Microsc Res Tech 2019; 82(6): 670-9.
[http://dx.doi.org/10.1002/jemt.23213] [PMID: 30767314]

[131] Bhatia SN, Chen X, Dobrovolskaia MA, Lammers T. Cancer nanomedicine. Nat Rev Cancer 2022; 22(10): 550-6.
[http://dx.doi.org/10.1038/s41568-022-00496-9] [PMID: 35941223]

[132] Liang Z, Khawar MB, Liang J, Sun H. Bio-conjugated quantum dots for cancer research: detection and imaging. Front Oncol 2021; 11: 749970.
[http://dx.doi.org/10.3389/fonc.2021.749970] [PMID: 34745974]

[133] Mansur AAP, Mansur HS, Soriano-Araújo A, Lobato ZIP. Fluorescent nanohybrids based on quantum dot-chitosan-antibody as potential cancer biomarkers. ACS Appl Mater Interfaces 2014; 6(14): 11403-12.
[http://dx.doi.org/10.1021/am5019989] [PMID: 24956063]

[134] Yang K, Yang G, Chen L, *et al.* FeS nanoplates as a multifunctional nano-theranostic for magnetic resonance imaging guided photothermal therapy. Biomaterials 2015; 38: 1-9.
[http://dx.doi.org/10.1016/j.biomaterials.2014.10.052] [PMID: 25457978]

[135] Zhang R, Fan Q, Yang M, *et al.* Engineering melanin nanoparticles as an efficient drug–delivery system for imaging-guided chemotherapy. Adv Mater 2015; 27(34): 5063-9.
[http://dx.doi.org/10.1002/adma.201502201] [PMID: 26222210]

[136] Pan UN, Khandelia R, Sanpui P, Das S, Paul A, Chattopadhyay A. Protein-based multifunctional nanocarriers for imaging, photothermal therapy, and anticancer drug delivery. ACS Appl Mater Interfaces 2017; 9(23): 19495-501.
[http://dx.doi.org/10.1021/acsami.6b06099] [PMID: 27476323]

[137] Manivasagan P, Hoang G, Santha Moorthy M, *et al.* Chitosan/fucoidan multilayer coating of gold

nanorods as highly efficient near-infrared photothermal agents for cancer therapy. Carbohydr Polym 2019; 211: 360-9.
[http://dx.doi.org/10.1016/j.carbpol.2019.01.010] [PMID: 30824100]

[138] El-Say KM. Maximizing the encapsulation efficiency and the bioavailability of controlled-release cetirizine microspheres using Draper–Lin small composite design. Drug Des Devel Ther 2016; 10: 825-39.
[http://dx.doi.org/10.2147/DDDT.S101900] [PMID: 26966353]

[139] Ranganathan S, Doucet M, Grassel CL, Delaine-Elias B, Zachos NC, Barry EM. Evaluating *Shigella flexneri* pathogenesis in the human enteroid model. Infect Immun 2019; 87(4): e00740-18.
[http://dx.doi.org/10.1128/IAI.00740-18] [PMID: 30642900]

Nano-based Biopolymer for Disease Targeting

Gaurish Narayan Singh[1], Nandani Jayaswal[2,*], Pooja Jaiswal[2] and Ganesh Lal[3]

[1] *Institute of Pharmacy, Deen Dayal Upadhyaya Gorakhpur University, 273009, Gorakhpur, Uttar Pradesh, India*

[2] *Faculty of pharmaceutical sciences, Mahayogi Gorakhnath University Gorakhpur, 273007, Gorakhpur, Uttar Pradesh, India*

[3] *KJ College of Pharmacy, Babatpur, 221006, Varanasi, Uttar Pradesh, India*

Abstract: Nano-based biopolymers are emerging as a powerful tool in targeted disease therapy, offering a promising combination of the precision of nanotechnology and the biocompatibility of natural polymers. In administering therapeutic chemicals specifically to certain disease areas, these nanoscale (1–100 nm) materials are intended to maximize therapeutic effectiveness while decreasing systemic adverse reactions. The integration of biopolymers, which are biodegradable and non-toxic, with nanoparticles allows for the creation of advanced drug delivery systems capable of responding to specific biological signals. This book chapter covers almost all current evolutions in nanomaterial-based medicament delivery techniques, with a focus on biopolymers such as polysaccharides (chitosan, alginate, cellulose, and starch) and proteins (albumin, collagen, gelatin, and silk fibroin). In cancer therapy, nano-based biopolymers can deliver chemotherapeutic agents precisely to tumour cells, minimizing damage to good tissues and overcoming the possible adverse effects linked with traditional radiation therapy. Additionally, these materials can be used for diagnostic purposes, enhancing imaging methods such as CT or MRI scans to better locate or characterize tumors. Regarding gene therapy, biopolymer-based nanoparticles can transfer genetic material, such as RNA or DNA, directly to specific cells, offering potential treatments for genetic disorders by correcting or silencing defective genes. Infectious diseases also benefit from this technology, with nano-based biopolymers delivering antimicrobial agents directly targeted to the site of infection, thereby increasing the drug's local concentration and efficacy. Furthermore, because these materials may carry anti-inflammatory medications straight to inflammatory tissues, they are being investigated for the treatment of inflammatory illnesses like rheumatic arthritis, which would lessen systemic protection against the negative effects. The possibilities of nano-based biopolymers in personalized medicine are examined in this chapter, along with issues including stability, scalability, and regulatory compliance.

* Corresponding author Nandani Jayaswal: Faculty of pharmaceutical sciences, Mahayogi Gorakhnath University Gorakhpur, 273007, India; E-mail: nandani.jaiswal123@gmail.com

Sudhanshu Mishra, Smriti Ojha, Shashi Kant Singh, Rishabha Malviya & Saurabh Kumar Gupta (Eds.)

Keywords: Biopolymers, Cancer therapy, Drug delivery systems, Gene therapy, Infectious diseases, Nano-based, Nanotechnology, Personalized medicine, Targeted disease therapy, Therapeutic agents.

INTRODUCTION

Drug administration often applies naturally occurring biodegradable polymers due to their accessibility, biodegradability, biocompatibility, and less toxic effects. In addition to polysaccharides, including chitosan, dextran, hyaluronic acid, agarose, carrageenan, cyclodextrin and alginate, it contains protein-based polymers such as gelatin, collagen, albumin, and soy. Because of their wide molecular weight ranges and batch variability, natural polymers present significant hurdles as drug carriers. Chitosan is the most often utilized natural polymer for drug administration because of its good cell and tissue integration, minimal cytotoxicity, capacity to change its surface, and lack of immunogenicity. It can also combine well with a range of different polymers. Although some biodegradable polymers have shown efficacy in various pre-clinical investigations, almost none of the actively targeted nanocarriers have progressed past clinical trials, and few passively targeted nanocarriers have been authorized for clinical usage. It is crucial to investigate novel candidates from this polymer family in order to achieve targeted and long-lasting medication release [1]. Protein and polysaccharide-based biopolymers are being researched because of their low levels of antibacterial activity, biodegradability, immunogenicity, and biocompatibility. Drugs, bioactive substances, and scaffolds for tissue engineering can all be transported by these nanoparticles. The DDS program utilizes nanotechnology to solve challenges facing the biomedical industry, such as low absorption, limited solubility, and adverse reaction. Nanoparticles' high surface area to volume ratio suggests that they might be useful in medicine. However, the mononuclear phagocyte system quickly clears non-modified nanoparticles. Synthetic approaches can manipulate nanoparticle characteristics and particle sizes, making them commonly used in the therapeutic sector. Covalent bonds between synthetic or biopolymers can modify nanoparticle surfaces [2]. Natural biopolymers are being explored for their biocompatibility, low toxicity, and immunogenicity, making them a valuable alternative to synthetic materials in nanomedicine. Their structural and chemical alterations are guaranteed by their biodegradability [3]. New developments in biopolymer technology are opening up the possibility for novel medication delivery methods that can enhance treatment results and patient compliance. These systems consider biopolymers' attributes like extraction technique, sustainable production, chemistry, surface characteristics, and biocompatibility [4]. The process of transporting an agent from the external environment to a designated target within the body is known as delivery. The possibility for specific cellular and molecular illness treatment is presented by developments in diagnostic agents, medications, and biological tools such as gene operators and nanotechnology. In order to transport to the target and shield them from deterioration, nanoparticles are usually integrated into vehicles. Current design paradigms focus on physical aspects, as nanoparticles interact with biological systems and can accumulate in non-target areas, affecting formulation effectiveness [5]. The first biopolymer nanoparticles were created using albumin [6]. Albumin and gelatin are the first naturally occurring proteins used to create nanoparticles, offering biodegradability, low immunogenicity, toxicity, stability, simplicity, and large-scale production. These nanoparticles offer surface modification and covalent drug attachment options [7]. Therapeutic chemicals may be delivered to the target tissue in a high-dose, targeted, and safe manner helpful to nanotechnology. It is possible to design biodegradable polymers into multilayered complex nanoparticles, which improves the therapeutic index and allows for drug targeting. Physical and chemical targeting strategies are used, with physical targeting influencing tissue accumulation and cell uptake. Chemical targeting uses molecular recognition units sensitive to external stimuli or pathological conditions. Advanced nanoparticles employ conjugation chemistry for tissue-specific chemical targeting and stimuli-responsive nanoparticles for chemical targeting [8]. This book chapter discusses the optimization of nanoparticle shape, size, and surface characteristics to improve physical targeting and chemical targeting. It also discusses the use of natural biopolymers to create biocompatible, low-immunogenic nanoparticles that offer improved drug encapsulation, controlled release, decreased toxicity, and biocompatibility. Understanding interactions between nanoparticles and biological barriers can improve medication delivery design and patient compliance.

FUNDAMENTALS OF NANOTECHNOLOGY AND BIOPOLYMERS

The twentieth century has been altered by advances in nanotechnology and its use in the fields of drugs and medicine. The field of nanotechnology studies incredibly microscopic structures. The Greek word "*nano*," which means "small," refers to minuscule size. Nanotechnology is a technique used to manufacture materials and devices with special properties by structuring small atoms, particles, or combinations. In nanotechnology, work is done either top-down, from immense structures to smaller ones, as in the case of optical photonics applications in semiconductor technology in this sector, or bottom-up, from individual atoms and molecules to nanostructures, which is more akin to chemistry and biology. Unique properties of materials include electrical conductivity, chemical reactivity, magnetism, visual effects, and nanotechnology, which focuses on materials with physical strength and sizes between 0.1 and 100 nm. Advances in nanotechnology and nanofabrication have significantly impacted drug delivery, enabling designs of nanoscale secondary structures and delivery systems made entirely of submicron components. As a result, controlled-release microchips with limitless modification potential have been created. Tissue engineering applications focus on nanoscale-sized parts, allowing for a deeper understanding of natural physiological processes. Nanoparticles overcome membrane barriers, making them advantageous over microparticles. Non-resorbable and biodegradable nanospheres are used for protein, small compounds, and therapies, improving bioavailability through absorption and decomposition [9]. Technological applications increasingly rely on material surface phenomena, requiring molecular control for polymers and bioactive materials. The layer-by-layer (LbL) technology has made it possible to create intricate, hierarchical structures for a different of applications, especially in the biological medical industry. This technique allows for the control of molecular architecture in multilayered structures. Materials with diameters between 0.1 and 100 nm are the focus of nanotechnology because of their special qualities that set them apart from bulk materials, such as electrical conductivity, chemical reactivity, magnetism, visual effects, and physical strength. The development of nanoparticles and electronics is only two of the many applications for nanotechnology, which works on a nanometer-scale length (1-100 nm) [10]. Drug delivery has been significantly impacted by the significant advancements in nanotechnology and nanofabrication over the past few years. Techniques have evolved from microelectronic fabrication and micromachining to designs ranging from delivery systems composed entirely of submicron components to secondary structures at the nanoscale. Transportation strategies for proteins, DNA, and small molecules have been greatly influenced by nanotechnology. This has resulted in the emergence of completely new and unexpected sectors. A novel bridge between formulation science and computer technology has enabled the development of a controlled-release microchip with infinite modification capabilities. Applications of tissue engineering have also tended to focus on creating and using nanoscale-sized parts. By producing artificial cells with the right physiologic characteristics, obstacles are being removed that lead to a deeper knowledge of natural physiological processes. Additionally, transfection systems on the nanoscales are being specially designed employing polymers for various uses in the era of genetic modification and gene delivery. By enhancing the formulation's retention through bio-adhesion, microsphere formulations increase the length of a drug's exposure and safeguard agents that are vulnerable to denaturation or degradation in hostile pH environments. Particle compositions smaller than one micron are connected with more challenges in terms of both administration and manufacturing (*e.g.*, aggregation). The capacity of nanoparticles to overcome membrane barriers, particularly in the small intestine's absorptive epithelium, is one of their main advantages over microparticles. Non-resorbable and biodegradable nanospheres have been employed by several organizations to deliver proteins, small compounds, and other therapies. The idea behind biodegradable nanospheres' attraction is improved bioavailability through absorption, which is followed by decomposition and the vehicle's elimination from the system [10]. Drug delivery techniques based on nanotechnology are being investigated for the therapy of fungal infections, gene therapy, cancer disease treatment, viral infection, and diabetes treatment. These systems, ranging from micelles to liposomes, interact with biomolecules within and on cell surfaces, allowing for targeted and improved safety. Surface-decorated systems can target specific tissues, enhancing drug delivery's specificity and allowing for targeted treatment [11].

Types of Biopolymers

As seen in Fig. (**1**), biopolymers are categorized according to their place of origin. Table **1** lists the different types of biopolymers and how they are used in disease targeting using nano-based systems.

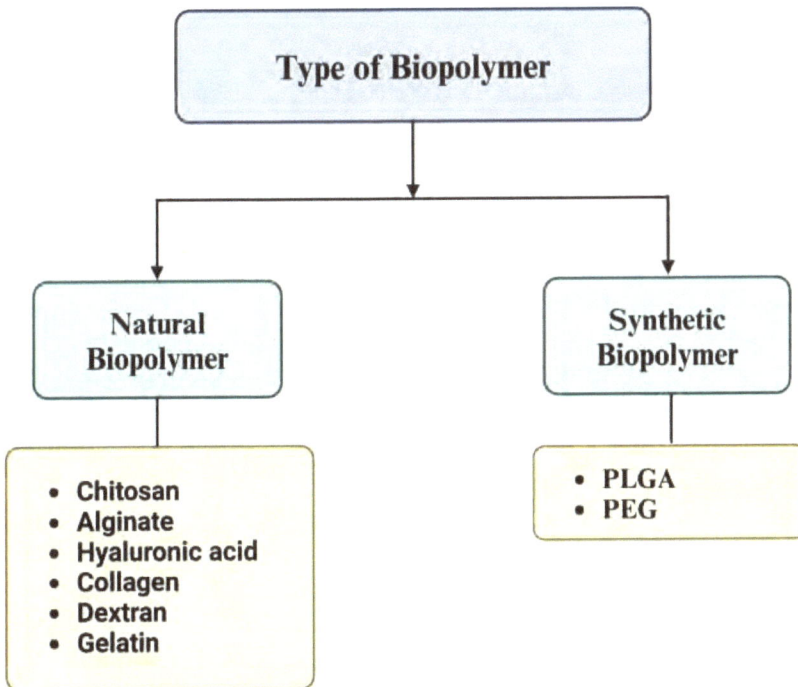

Fig. (1). Types of biopolymers for nano-based disease targeting.

Natural Biopolymers

Natural biopolymers have benefits over synthetic ones, such as being consumable, renewable, and biodegradable. However, their defective mechanical and water vapor boundary properties restrict their industrial use. The goal of studies is to improve these qualities in order to improve the physical and fluid boundary qualities of films made of biopolymers [12].

Table 1. A detailed overview of various biopolymer types and their applications in nano-based systems for disease targeting.

Sr. No.	Type of Biopolymer	Type of Nanoparticle	Application Area	Reference
1.	Chitosan	Nano capsules	Buccal treatments	[35]
2.	Polysaccharides	Biodegradable nanosheets, *in-situ* gels, gel formulation, IPN hydrogel, ocular films, ocular inserts, nanoparticles, and microparticles.	Improve drug delivery systems, and improve ocular bioavailability	[36]
3.	Hyaluronic acid	Nanoparticles, nanogels, nanoconjugates, nanoliposomes, nanocomposites and gold nanorods	Asthma, cancer, wound healing, dry eyes, and arthritis	[37]
4.	Poly (lactic-co-glycolic acid) and poly (lactic acid)	Microparticle and nanoparticle	Vaccination and cancer immunotherapy	[38]
5.	Polycaprolactone/Gelatin	Nanofiber	Wound Healing	[39]

Cellulose

A lengthy chain of sugar molecules makes up cellulose, a linear natural polymer that is sensitive to strong alkaline conditions but readily degraded by acid. As it is degradable in response to chemical and solution treatments, it is comparatively resistant to oxidizing agents [13]. Cellulose nanofibers (CNF), often referred to as nanofibrillated cellulose, are cross-linked, flexible, and elongated nanocellulose that is mechanically separated from cellulose fibers. They are produced by mechanical delamination after pre-treatment, reducing energy consumption and treatment time. CNF was first isolated in 1983 using bleached softwood fiber. Pretreatments like enzymatic hydrolysis, partial carboxymethylation, catalytic oxidation, and mechanical refining have been employed to reduce energy consumption [14]. Cellulose is a potential source for the creation of bio-based polymeric goods and provides environmentally benign substitutes for polymers based on petroleum. Challenges and research trends in cellulose-derived monomers are discussed, with potential applications in electrochemical, energy-storage, and biomedical fields [15]. Grafted Rosin (CNC-R) hybrids were effectively produced by the study using PHBV and CNCs. These hybrids showed enhanced mechanical strength, toughness, and thermal stability, and they may find use in bioactive food packaging materials [16].

Starch

Animal feed, industry, and nutrition ultimately depend on starch, a naturally occurring storage carbohydrate found in plants and algae. However, because of bulk trials, there is little specific information available on its structure and turnover [17]. Plants generate starch, a complex carbohydrate polymer that is vital for both animal and human energy and nourishment. Its metabolism is complex and consists of various enzymes/proteins [18]. Because of their intricate structure, bio-based polymers made from renewable resources can be difficult to produce, even though they are widely employed in many different industries. Reactive extrusion (REX) and enzymatic activity are ecologically beneficial technologies that have been employed in food, medicine, and energy for efficient conversion and usage. Recent advancements include the development of enzymatic reactive extrusion (eREX) under optimized conditions, the synthesis of food-related biopolymers using combined REX and enzymatic approaches, and the identification of novel REX–enzyme synergies that enhance processing efficiency and product functionality [19].

Gelatin

Gelatin is a non-immunogenic, biodegradable, and biocompatible substance that is frequently utilized in biomedical applications because of its capacity to stabilize surfaces, increase the steric barrier, and function as capping mediators [20]. Recent studies have incorporated gelatin into a new elastomeric aliphatic polyester (PGS) to modify its properties, despite its limitations in synthesis and tissue engineering applications [21]. The study investigates biopolymer films from pectin with varying degrees of esterification and gelatin, analyzing their thickness and spectral and refractometric characteristics. A study provides a comparative analysis of pectin films with gelatin films and refractometric data [22]. Another study evaluated the osteogenic properties of loading parathyroid hormone-related protein into nanocrystalline hydroxyapatite scaffolds for tissue engineering applications, finding increased bone volume and gene expression while decreasing Wnt inhibitors and SOST [23].

Chitosan

Chitosan, the second abundant Biopolymer, is extracted from shrimp shells through deacetylation of chitin. Its mechanical, chemical, and biological properties can be enhanced through chemical structure modification. The chitosan derivatives: substituted, crosslinked, carboxylic acid, ionic, and bound to specific molecules. The advantages and applications of these derivatives are highlighted, along with the synthetic routes of biopolymer modification [24]. A highly biocompatible biopolymer, chitosan, is a potential replacement for chitosan due to its biodegradability, bioadhesivity, and bioactivity. It can be dissolved under acidic conditions, adopt various conformations, and be functionalized for specific applications. Chitosan is also used in biomedical fields, trapping organic compounds and dyes, selectively separating binary mixtures, and acting as a catalyst or starting molecule for high-value products [25]. Chitosan is utilized for treating wounds because of its antimicrobial and healing properties. It focuses attention on chitosan's permeability to oxygen, antifungal, antibacterial, and bonding characteristics. Chitosan films are increasingly being explored as potential skin substitutes for the treatment of full-thickness wounds [26]. A sustainable agricultural resource, chitosan biopolymer is present in the shells of shrimp, shellfish, lobster, and crabs. It is non-toxic to the environment, biocompatible, and breaks down organically. It stimulates the development of plants and functions as an antibacterial, changing defense mechanisms and physiological processes [27]. Because of the strength and simplicity of synthesis, chitosan-based nanocarrier systems are widely used for drug administration in lung cancer and chemotherapeutic drugs, increasing their efficacy and cost [28].

Synthetic Biopolymer

A study discusses synthetic biopolymers, technological developments, lifespan analysis, biodegradability, and the most recent developments, obstacles, and prospects for recycling bioplastics [29]. Synthetic polymers offer advantages like tunable properties, endless forms, and established structures, making them easier to synthesize than naturally occurring polymers. They can restore tissue structure and function in damaged or diseased tissues [30].

Polycaprolactone

The Carothers group created the hydrophobic, semi-crystalline, biocompatible polymer known as polycaprolactone (PCL) in the early 1930s. It can be treated at low temperatures because of its high solubility, low melting point, and mix compatibility. Applications of PCL as degradable biomaterials are impacted by its degradation, which is ascribed to random hydrolytic chain scission [31]. Because of their superior biocompatibility and biodegradability, polycaprolactones (PCL) are employed extensively in biomedical applications, especially in tissue engineering and medical implants. As they are simple to produce and process, green biocomposites—which combine the advantageous qualities of individual constituents—are becoming more and more popular. Natural fiber-reinforced biopolymers improve their mechanical and thermal characteristics [32].

Poly(vinylalcohol)

Poly (vinyl acetate) is hydrolyzed to produce poly (vinyl alcohol) (PVA), a thermoplastic polymer. Inert, nontoxic, biocompatible, and soluble in water, acids, and highly polar solvents, it is the most polar synthetic polymer. PVA is utilized in the manufacturing of fibers, such as fake leather, transportation belts, gaskets, tubing, emulsifiers, adhesives, bonding surfaces such as paper, leather goods, textiles, and ceramics, as well as in surgical procedures. PVA water is treated with sodium sulfate, which contains formaldehyde and sulfuric acid, to render it insoluble [33]. In activated sludge (AD), PVOH, a polymer with a molecular weight between 14,000 and 2000, is primarily biodegradable. The hydroxyl group is oxidized and the C-C bond is cleaved as part of its catabolic route. The rate at which PVOH biodegrades is dependent on the inoculum and the molecular weight of the polymer. Although sludge only partially dissolves, low molecular weight polymers typically undergo rapid biodegradation [34].

MECHANISMS OF DISEASE TARGETING WITH NANO-BASED BIOPOLYMERS

Nano-based biopolymers offer a novel method for targeted disease treatment, utilizing nanoscale carriers made from biocompatible and biodegradable polymers. As biopolymers are biocompatible, low-toxicity, and modifiable, they are employed as drug delivery vehicles in tumor treatment. Research focuses on optimizing preparation processes and chemotherapy combination [40]. Fig. (**2**) describes the target of the cancer cell and active drug and nano-based Biopolymer.

Passive Targeting Strategies

The traditional cancer therapies that the passive targeted techniques are based on frequently fail to treat cancer very well. To increase the effectiveness of cancer treatment, pH-responsive nanocarriers have been created to deliver medications to the acidic extracellular fluids of tumor tissue or to endosomes or lysosomes found in cancer cells (Fig. **2**). This approach lowers systemic adverse effects, enables targeted drug distribution, and offers a great chance of increasing the effectiveness of cancer treatment [41]. We have included some of the studies that developed a kidney-targeting construct using Elastin-like polypeptides (ELPs) and a kidney-targeting peptide (KTP). The modified ELP showed longer plasma half-life and higher renal levels, suggesting potential for targeted renal disease treatment [42]. Elastin-like polypeptide (ELP), a thermoresponsive carrier, may passively gather and form in solid tumors when exposed to high temperatures. ELP was combined with three cell-penetrating peptides (CPP) and the anticancer drug DOXO to stop tumor development in mice. SynB1-ELP-DOXO was shown to have the lowest intrinsic cytotoxicity by fluorescence microscopy, which makes it a viable option for improving drug polymer conjugates that are thermally sensitive [43]. Elastin-like polypeptide, a thermally sensitive paclitaxel delivery vehicle that enhances water solubility and inhibits cell proliferation in MCF-7 cells, has demonstrated *in vitro* proof of concept [44].

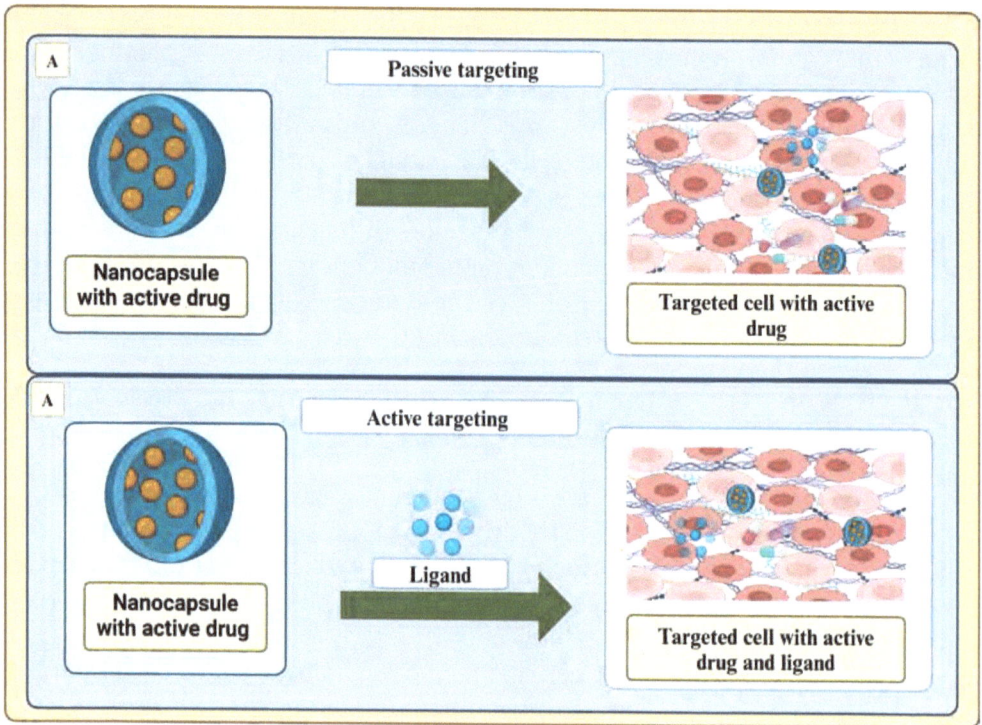

Fig. (2). Illustrations describe the mechanism of disease targeting by using the nano-based formulation and drug. (**A**) Passive target of Nano capsule and active drug for disease, and (**B**) Active target of Nano capsule, active drug with ligand for disease.

Active Targeting Strategies

Active targeting, as seen in Fig. (**2**), is the process of altering the surface of nano-based biopolymers by employing certain ligands that identify and attach to overexpressed receptors on sick cells. Diseased cells undergo pathological transformation, causing changes in shape, enzymatic profile, and microenvironment. Overexpressed markers require specific targeting approaches, making them more vulnerable to drug toxicity [45]. Bioactive substances are preserved by encapsulation, enhancing their anti-disease and health properties. The food and pharmaceutical industries depend heavily on micro- and nano-encapsulation techniques. Proteins, gums, and maltodextrin are examples of possible nanovehicles [46]. The present study describes the development of *in vivo* and *in vitro* investigations of magnetic resonance imaging (MRI)-active nanoparticles designed to target tumor cells with elevated folate receptor expression. The nanoparticles exhibit stability, non-toxicity, and biocompatibility. Based on *in vitro* tests, the nanoparticles were selectively absorbed in a way that was dependent on the folate receptor. The nanoparticle-Gd complexes significantly changed tumor cell signal intensity [47]. Since cancer is a major worldwide problem, advancements in nanomedicine are essential for efficient cancer therapy. In site-specific delivery systems enabled by nanotechnology, alginate, pullulan, cellulose, polylactic acid, chitosan, and other naturally occurring bioactive compounds can be employed. Biopolymeric nanoparticles (BNPs) are being developed for their biodegradability and cost-effectiveness. These systems aim to improve treatment efficiency, increase selectivity, and reduce toxicity [48]. The creation of reacetylated chitosan microspheres for regulated medication administration in the stomach cavity is presented in this work. The hydrophilic and hydrophobic substances can be effectively encapsulated using two microencapsulation techniques. Drug release, encapsulation effectiveness, and antimicrobial action are all strongly impacted by the reactivation time. Short reacetylation periods maintain the effectiveness of hydrophilic and lipophilic antibiotics against a variety of bacteria while allowing for effective control of their release [49]. Because of their durability, non-toxicity, biodegradability, and biocompatibility, biopolymeric nanoparticles are being utilized more and more as nanocarriers in biomedical applications. Animals, plants, algae, fungi, and bacteria can all produce these nanoparticles. Therapeutic agents such as bioactive chemicals, medications, antibiotics, antibacterial agents, extracts, and essential oils can be delivered through these potential methods [50].

Stimuli-responsive Targeting

The stimuli-responsive nanocarriers are programmable, controllable nanomaterials that are employed for solid tumor imaging, targeted treatment, and diagnostics. They can also be developed into stimuli-responsive nanomedicines, releasing drugs in response to physiological changes [51]. Stimuli-responsive bio-based polymeric systems are becoming more well-known as clever, adaptable instruments with enormous promise across a range of industries [52]. The work uses N-carboxyanhydride ring-opening polymerization, racemic amino acid glyco NCAs to regulate secondary conformation, and enthalpy-driven binding to create high molecular weight soluble in water O glycocopolypeptide polymers [53]. The different characteristics of nanogels (NGs), such as their tiny particle size, high encapsulation effectiveness, and ability to shield active substances from deterioration, are being researched. These cross-linked nanoparticles can swell and collapse in response to physical, chemical, or biochemical stimuli. Through their porous shape, NGs may encapsulate tiny molecules, including genes and medications, which are then released when external stimuli are provided through volume changes [54]. Poly (ethylene glycol)-b-poly (L-lysine) block copolymers embellished with lipoic acid and cis-1,2-cyclohexanedicarboxylic acid (CCA) serve as the structural building blocks of DOX, a powerful anticancer medication. These micelles are stable, non-toxic, and efficiently deliver DOX into HeLa cells, doubling its release under endosomal pH [55].

Cellular Uptake and Intracellular Targeting

Through cellular uptake and intracellular targeting, nano-based biopolymers ensure that therapeutic chemicals are released within the cell compartments and reach particular cells within the body. In this study, eight cell lines are used to examine how surface charge affects the cellular uptake and intracellular trafficking characteristics of chitosan-based nanoparticles (NPs). Studies show that the amount and rate of cellular absorption are positively associated with surface charge in hybridoma cells. While negatively and neutrally charged NPs often colocalize with lysosomes, positively charged NPs have the ability to escape from lysosomes and exhibit perinuclear relocation [56]. A specific medication delivery system was developed by functionalizing mesoporous silicon nanoparticles (MSNs) with hyaluronic acid and amino groups (MSN-NH$_2$). Three hyaluronic acid samples were used, and their cellular uptake in HeLa cells was studied. HA chain length significantly affected uptake, with MSN-NH$_2$ and MSN-HA having higher uptake than MSN-HAL and MSN-HAM. Cellular uptake experiments showed that MSN-HA samples inhibited internalization, suggesting two different mechanisms [57].

APPLICATIONS IN MEDICINE

Nano-based biopolymers offer several medicinal applications in addition to their biocompatibility, biodegradability, and task-specific flexibility. Fig. (**3**) describes how the nano-based biopolymer formulation is used to treat different kinds of disorders. Nano-based biopolymers are significantly impacting various fields such as gene therapy, cancer therapy, infectious disease and inflammatory disease, tissue engineering, imaging, and drug administration. Table **2** outlines the treatment of inflammatory diseases using nano-based biopolymers for disease targeting.

Cancer Therapy

Cancer therapy is crucial, with breast cancer being the top cancer, and prostate cancer being the most common. Black women have a 40% higher mortality rate despite a lower incidence. Early diagnosis, mammography screening, and treatment improvements are crucial [58]. Lung cancer's global importance necessitates ongoing diagnostics and therapeutic advancements, with recent research on proteomic and genetic biomarkers providing promising treatment guidance [59]. Studies on nanomedicine and biopolymers have shown that biopolymers are potentially useful for cancer therapy because of their biocompatibility, biodegradability, and tailored drug delivery capabilities. Because they are easily absorbed in the gastrointestinal system, nanopolymers have the potential to be cancer medication carriers. ChNPs can be loaded with cancer medications either directly or *via* encapsulation. Hydrophobic medications are stable and under control thanks to the solid core of chitosan nanospheres. ChNPs can be made more sensitive and selective by adding PEG, targeting ligands, and pH-sensitive polymer conjugates [60]. The poor efficacy of existing nanomedicine delivery technologies, which depend on passively targeting the EPR effect in malignancies, hinders clinical usage. Biopolymers, with excellent biocompatibility and inherent ligands on human cell surfaces, offer potential solutions for active targeting in tumor microenvironments [61].

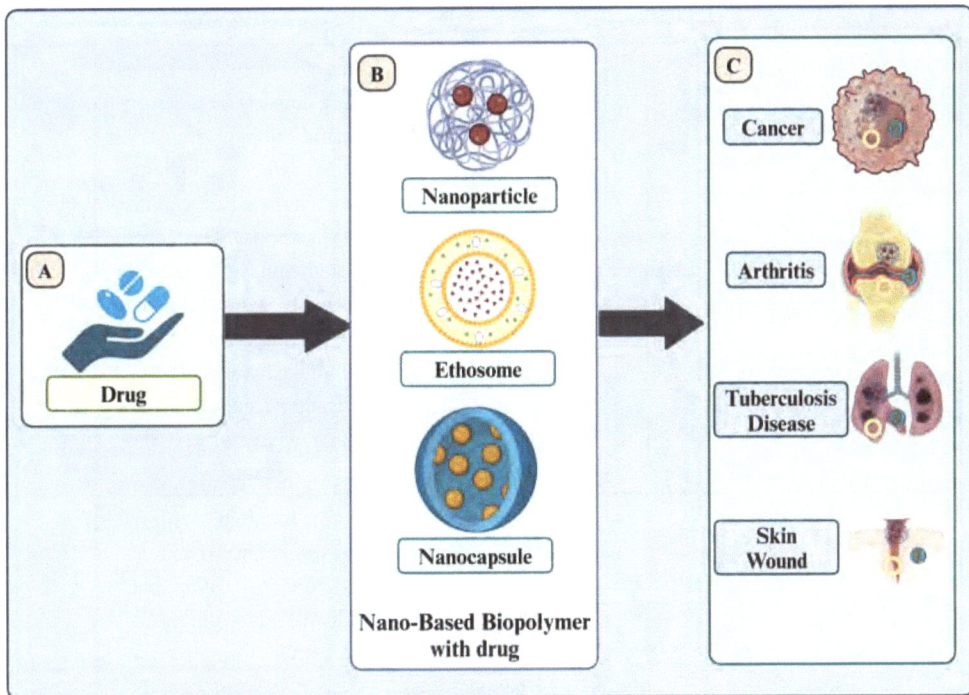

Fig. (3). A suitable drug and biopolymer can treat various types of diseases.

Table 2. Nano-based biopolymers for targeted distribution in the treatment of different types of disease.

Sr. No.	Disease	Biopolymer	Therapeutic Agent	Outcome	References
1	Diabetes, organ fibrosis, and arthritis	Polysaccharides	Nanoparticles	Polysaccharides, complex biopolymers, are utilized in nanoparticles (NPs) for therapeutics, diagnostics, and drug delivery. Using their superior pharmacokinetic qualities, these NPs enhance oral absorption, regulate drug release, boost *in vivo* retention, provide targeted administration, and have synergistic effects.	[99]
2.	Rheumatoid arthritis	Nanocarriers	Nanoparticles	Targeted therapy enhances drug delivery by targeting specific cell populations, but it has safety and toxicity issues. Surface PEGylation of nanomedicines can prevent resuscitation site recognition, extend bioavailability, and increase aggregation, but complex and expensive procedures are required for nanoparticles in RA treatment.	[100]
3.	Inflammatory bowel disease	Alginate and chitosan	Polymeric nanoparticles	Chronic inflammatory bowel disease (IBD) requires lifelong medication, but current therapies are ineffective. A new treatment paradigm includes early biologic agents, nanomedicine-based approaches, and nutritional interventions.	[101]

(Table 2) cont.....

Sr. No.	Disease	Biopolymer	Therapeutic Agent	Outcome	References
4.	Inflammatory-mediated diseases	Polyphenols	Liposomes, micelles, dendrimers	The rise in inflammatory diseases has led to the development of natural polyphenolic compounds with anti-inflammatory potential, which have been loaded into nanocarrier technologies to improve stability and bioavailability, but their properties and mechanisms of action are not fully understood.	[102]
5.	Inflammatory	Alginate	Nanoparticles	The study investigates the safe and efficient application of alginate nanoparticles aimed at macrophages as a noncondensing delivery system for plasmid DNA transfection.	[103]
6.	Acute kidney injury (AKI)	Hyaluronic acid	Liposome	The study utilized liposomes to improve curcumin solubility, creating hyaluronic acid-coated liposome nanocomplexes for targeted kidney injury treatment, demonstrating their biocompatibility.	[104]
7.	Pulmonary disease	Chitosan	Nanospheres	This chapter discusses the efficiency of fabricating chitosan nano- and microspheres, highlighting their potential for DNA delivery, but highlighting technological difficulties for homogeneous particles.	[105]

Bovine serum albumin's sensitivity, cost, and non-invasive optical imaging applications make it a viable nanocarrier for effective drug administration. Gold nanorods and clusters are perfect for *in vivo* imaging since they are nontoxic, inert, and excellent fluorescent probes, according to studies. Drug DOX with nanocarriers demonstrated effective delivery in hepatocarcinoma cells, resulting in apoptosis. The destiny of nanocarriers *in vivo* is analysed by their size, shape, and surface charge. While negatively charged particles reduce tumor formation, positively charged particles exhibit increased accumulation in cancers. Additionally, hydrophobicity is essential to the DDS nanocarrier process [62]. By enhancing medication delivery and treatment effectiveness, nanomedicine holds the potential to transform cancer therapy completely. Understanding nanoparticle behavior is crucial, and future studies should focus on developing sophisticated drug delivery systems. Interdisciplinary collaborations are essential [63]. Less research has been conducted on chitosan composite materials with integrated characteristics, and their derivatives have not yet reached their full potential. Chitosan-based drugs for cancer treatment are still in the laboratory stages. Although chitosan recovery, composites, and applications have advanced thanks to nanotechnology, the utilization of chitosan nanoparticles is still restricted. Chitosan's antioxidant potential is not being used, and its mechanism is not well understood. Although chitosan derivatives have shown higher antioxidant activity than chitosan, there is low enthusiasm to extend their applications [64, 65].

Reactive oxygen species (ROS) produced by DOX have the capacity to alter autophagic flux, damage DNA, and cause pathological SR/Ca^{2+} leakage. These mechanisms may ultimately lead to lipid peroxidation-dependent iron metabolism and several other forms of controlled necrosis [66]. It has been demonstrated that chitosan has anti-inflammatory qualities and lowers blood cholesterol, which can cause cardiotoxicity [67]. Chitosan increases targeted activities, circulation time, and safe medication administration to reduce the possibility of chemotherapy side effects. 5-FU can be made into controlled-release formulations with chitosan, allowing the medication to be released gradually over time. This could reduce the chance of cardiotoxicity by lowering peak concentrations of the medicine in the bloodstream while maintaining therapeutic drug levels in the targeted location. Colorectal cancer (CRC) accounts for approximately 600,000 deaths annually worldwide, making it one of the leading causes of cancer-related mortality. Five-fluorouracil (5FU) has been used for CRC therapy for thirty years. It is an antimetabolite and fluorinated pyrimidine [68]. Diarrhoeal stomatitis and damage to the mucosa of the gastrointestinal tract and the second most used chemotherapy medication linked to cardiotoxicity is 5-FU after anthracyclines [69]. Rats' liver, lungs, and renal become toxic when free cisplatin is administered intratracheally or intravenously. On the other hand, Maher Shaimaa *et al.* experimentally proved that Alginate functionalized AgNPs as an active drug nanoplatform, both with and without cisplatin used in chemotherapy in breast cancer, in contrast to reduced toxicity [70]. Fucoidan is a naturally occurring biopolymer with anticancer potential, present in several species of brown algae as well as several mammals.

The PI3K/AKT, MAPK, and caspase pathways are the primary pathways that fucoidan affects. The fucoidan-mediated activity of the AKT pathway seems to be significantly influenced by PTEN. Fucoidan as an adjuvant also appears to have positive effects, as evidenced by its increased efficacy in cancer treatment and decreased toxicity in healthy cells [71]. A variety of biological activities are displayed by piperlongumine (PL), an amide alkaloid that was isolated from *Piper longum* L. (long piper) and other piper plants [72]. Piperlongumine (PL), a new pro-oxidant agent, has garnered a lot of attention since it may increase oxidative stress levels in cancer cells, resulting in cancer-specific death. Due to PL's hydrophobic properties, its application in the clinic is restricted. Thus, PL was effectively introduced intracellularly into cancer cells with the use of nanoparticle encapsulation on chitosan and fucoidan. Ionic gelation turned fucoidan and chitosan into nanoparticles. In producing intracellular reactive oxygen species (ROS) from PL, the chitosan and fucoidan-based nanoparticles (CS–F NPs) effectively encapsulated PL, improved its water solubility and bioavailability, and effectively killed human prostate cancer cells [72]. The US FDA has approved four siRNA-based drugs: parisiran, givosiran, lumasiran, and inclisiran. Scientists are now paying more attention to siRNA medicines as a result. Biopolymers may be used as platforms for the delivery of siRNA medications and as therapeutic agents for effective cancer therapy [73]. One of the most popular and successful medications for the treatment of autoimmune and dermatological conditions is methotrexate (MTX). When using MTX for a long time, patients with rheumatoid arthritis and psoriasis are at high risk of liver damage. When the MTX metabolite MTX-polyglutamate (MTX-PG) accumulates inside cells, hepatocytes experience oxidative stress, inflammation, fibrosis, steatosis, and ultimately, death [74]. Chitosan protects against methotrixate-induced liver and kidney damage *via* influencing metalloproteinases and caspases [75].

Gene Therapy

Drug and gene delivery techniques with targeting capabilities, such as ligand-receptor interactions or surface modifications, can deliver therapeutic medications to specific cells or tissues with optimal efficacy and minimal off-target damage [76]. Through the use of exogenous nucleic acids for gene deletion, gene substitution, and gene suppression, gene therapy can alter the gene associated with the illness [77]. Malignant tumors such as gastrointestinal stromal tumors, breast, gastric, colorectal, lymphoma, and chronic myelogenous leukemia have all responded well to targeted gene therapy [78]. The 21–23 nucleotide siRNA has demonstrated potential as a gene therapy option. A sequence found in siRNA is complementary to a particular mRNA that codes for a protein [79]. This sequence has the ability to cause site-specific cleavage, which inhibits the production of proteins and suppresses their expression. Furthermore, siRNA synthesis is straightforward to perform without the need for cell expression systems. As a result, siRNA has significant uses in tumor-targeting treatments [80]. Due to their high bloodstream instability and ease of ribonuclease degradation, renal elimination, or mononuclear phagocyte phagocytosis, siRNAs have limited therapeutic utility. Additionally, the widespread usage of siRNAs has been restricted due to their poor endosomal escape upon endocytosis, reduced rate of absorption into tumor cells, and off-target effects [81, 82]. Because of their low immunogenicity, nontoxicity, great biocompatibility and biodegradability, and low manufacturing costs, chitosan (CS) and its substitution are therefore perfect vectors for siRNA dissemination. With cationic charges, the main amine groups on CS can readily form complexes or nanoparticles (NPs) by electrostatic interactions with negatively charged materials or siRNAs in slightly acidic environments [83]. Numerous chitosan derivatives, such as carboxymethylation of chitosan (CMC), O-carboxymethyl-chitosan, arginine-chitosan, and trimethyl chitosan (TMC), have been investigated in enhanced gene transfection. A cationic polymer called chitosan may form positively charged polyplexes by binding with negatively charged nucleic acids [76]. Chitosan contains amino groups, it can also encourage polyplex endosome escape through the "proton sponge effect" in addition to DNA condensation [84]. This improves the polymer's interactions with cellular membranes and promotes uptake by the cells. Several chitosan-based CRISPR/Cas9 delivery methods employ its pristine foundation, currently the transfection effectiveness of pristine chitosan is still rarely particularly high [76]. Taylor Catherine *et al.* claim that because the alginate molecule has numerous mucins contact sites and the mucin molecule has relatively few, the crosslinks in the mucin alginate gel network are caused by mucin–alginate interactions. The conditions in the lungs of people with cystic fibrosis who secrete *Pseudomonas aeruginosa* and the medicinal applications of alginate in oral dosage forms are likely connected to the creation of these mixed gels [85]. Cystic fibrosis (monogenic sickness) is caused by mutations in the CFTR gene on chromosome 7 and affects the kidney, liver, colon, pancreas, male genital system, lungs, and bones. Its clinical manifestations are diverse and intricate [86]. Alginate hydrogels have also been applied to the regeneration of skeletal muscle. Some significant drawbacks of using these hydrogels unaltered include their lack of cellular adhesion receptors, unpredictable disintegration, and diminished mechanical qualities. By modifying the alginate to ensure better cell adhesion and continuous distribution of bioactive chemicals, these shortcomings can be addressed. One such substance is the cross-linked hydrogel of oxidised alginate and gelatin (OA-GEL), which has gained interest lately as a possible material for tissue regeneration. The material's tunable qualities, which are dependent on the oxidation degree (OD) and the OA/GEL weight ratio, make it very helpful [87]. Platelet adhesion and aggregation are encouraged by chitosan. Chitosan increases the production of glycoprotein IIb / IIIa (GPIIb / IIIa) and intracellular free Ca^{2+} on platelet membranes, which in turn stimulates platelet adhesion and aggregation [88]. Chitosan is a good option for use as a vector for nonviral gene delivery because of its robust interactions with negatively charged nucleic acids and biomembranes, which promote effective cellular absorption [89]. On the other hand, ChNPs (Chitosan Nanoparticles) have the ability to react to particular internal and external stimuli that are unique to the tumour microenvironment. Since they are endogenous molecules like carbon dioxide and water, the breakdown products from the *in vivo* enzymatic breakdown of Ch are safe. ChNPs in a way that makes them react to particular internal and external stimuli that are typical of the tumour microenvironment. This minimises effects on healthy cells and permits regulated release of genes or medications only to cancer cells that are targeted. pH is an illustration of an internal stimulation. Normal tissues and blood have a pH of about 7.4, but tumour tissues have a pH of 6.5–7.2, which is slightly acidic. Lysosomes have a pH of 4.5–5, which is more acidic. Consequently, ChNPs may be altered to react to the tumor microenvironment's acidic pH, resulting in the release of certain drugs or genes [90]. CuVa polymers, which were modeled after the naturally occurring biopolymer curdlan, demonstrated enhanced biocompatibility and the capacity to deliver siRNA to solid tumor sites *in vitro* [91]. When antiPOL siRNA and guanidine-CO_2-functionalized chitosan were added *via* a pH-responsive nanoplatform at a neutral pH, POLR2 in TNBC tumors was downregulated with negligible to no damage. The siRNA nanodrug overcame intracellular trafficking at low pH by effectively internalizing into TNBC cells with hemizygous TP53 deletion [92]. Chitosan-silicon dioxide nanoparticles (NPs) were used to encapsulate a plasmid with a gene 1-siRNA specific to breast cancer (BCSG1-siRNA). MCF-7 cells showed higher apoptosis, decreased proliferation, and improved cytotoxicity under the gradual release of BCSG1-siRNA. Tumour growth was markedly reduced by BCSG1-siRNA, which downregulated BCSG1. Because of its strong siRNA delivery mechanism, chitosan-silicon dioxide NP-BCSG1-siRNA suppresses breast tumors [93].

Infectious Disease Treatment

Chitosan has mucoadhesive qualities, which enhance the potential for drug absorption, attributed to its positive charge. Additionally, its positive charge opens tight junctions, increasing uptake. Anti-TB medications such as isoniazid, prothionamide, and rifampicin are encapsulated in chitosan-based nanoparticles. By increasing the stability and bioavailability of medications and facilitating their targeted delivery to sick cells, these nanoparticles improve their therapeutic efficacy [94]. Chitosan nanoparticles were shown to exhibit antimycobacterial properties in tuberculosis H37Rv, with an MIC value of 1200 µg/mL and an MBC value of 2400 µg/mL [95]. Anti-fungal medication (itraconazole) shows limited solubility when taken orally. Jafarinejad *et al.* delivered itraconazole to the lungs using chitosan nanoparticles in a dry powder formulation. Increased intraconazole pulmonary deposition was the result of this [94]. The lipopeptide antibiotic daptomycin may be used in some circumstances to treat endophthalmitis caused by methicillin-resistant Staphylococcus aureus (MRSA). Specifically, mucoadhesive chitosan-coated alginate (CS-ALG) nanoparticles may enter the HCE and ARPE-19 cells and provide a possible way to administer daptomycin to treat bacterial endophthalmitis [95]. The starch-derived nano polyurethanes and native starch have been found in 60–97% anti-tuberculosis drug's loading efficiency. All drug-loaded carriers, with the exception of streptomycin, displayed a sustained release profile for starch-derived nano polyurethanes, according to the pH-dependent drug-releasing research conducted in this work [96]. Natural biopolymers are crucial to the creation and coating of ureteral stents because they provide a number of benefits, including improved medication delivery, biocompatibility, biodegradability and a decrease in problems. Alginate is a useful material for ureteral stents because of a number of important properties. Excellent biocompatibility is a well-known characteristic of alginate, which is vital for materials used in medical devices. Moreover, chitosan has strong antibacterial properties, it is commonly utilized to coat ureteral stents that are currently in use. The use of coatings can prevent biofilm growth and encrustation [97]. Recently, chitosan-based anti-coronavirus compounds were identified and published; *in vitro* and *ex vivo*, they effectively inhibited the infection of all low-pathogenic human coronaviruses. In fact, polymer-protein complexes that developed between this chitosan derivative and the recombinant S protein ectodomain were investigated to determine binding. They found that the polymer attaches itself to the spike protein of the coronavirus and prevents it from attaching itself to the cellular receptor [98].

Treatment of Inflammatory Diseases

Inflammation has been recognized since ancient times and is a complex process involving various mediators such as amines, peptides, lipids, enzymes, and complements. It contributes to the increase of chronic diseases, including heart disease, diabetes, intestinal problems, osteoporosis, and arthritis since it is necessary for the immune system to defend against infection and damage. Increased prostaglandins (PGs) are a major mediator of ocular inflammation, a frequent clinical issue. Cytokines, such as IL-6 and GM-CSF, also play a role in ocular inflammation. Neurokinins, such as SP, NKA, and NKB, play a role in modulating local ocular inflammation initiated by surgical perforation trauma [106]. Rheumatoid arthritis is treated with leflunomide (LEF), a DMARD that has low solubility and negative side effects after systemic exposure. An oral leflunomide-loaded nanoemulsion system was created using clove oil and chitosan's antirheumatic qualities to improve LEF's therapeutic impact and lessen its systemic adverse effects [107]. The triple-layered transdermal platform described in this study for the efficient treatment of RA consists of an electrospun PCL NF layer, a 3D-printed sodium alginate-based hydrogel layer with rosuvastatin-encapsulated core-shell lipid nano capsules, and a composite layer of PVA nanofibers/nanoparticles conjugated with diclofenac. The chemically related DIC is released by transdermal patches due to the esterases produced by the skin at inflammatory sites [108]. Hyaluronic acid (HA) has garnered attention to create target-specific and more efficient medication delivery solutions for Rheumatoid Arthritis. HA, a naturally occurring ligand for the CD44 receptors, is abundant in synovial fluid. The distribution of off-target medications can be decreased by employing the targeted delivery strategy with CD44 as the target. To address the prevalent problems with RA therapy, these novel approaches—HA-based surface-decorated nanocarriers, hydrogels, and MNs—offer greater patient adherence, fewer side effects, and personalized delivery [109]. In order to utilize the bioadhesive feature of the polysaccharide to modify the drug's unfavorable pharmacokinetic profile and enhance its anti-inflammatory effects when delivered rectally, mesalazine was microencapsulated in chitosan particles in an *in vivo* model of IBD [110]. Nanoparticle (NP)-based treatments have garnered much attention lately as viable substitutes for treating IBD. Because of their distinct physicochemical characteristics, biocompatibility, and mucosal adherence, chitosan nanoparticles are frequently used in the creation of novel drug delivery systems. The tissue's increased permeability and retention effect enable the epithelium's intracellular matrix to selectively absorb small chitosan NPs. By passively or actively directing these delivery systems towards inflammatory areas, medication concentration and bioavailability can be increased [111]. HA-DA, which may self-crosslink to form a hydrogel *in vivo*, was created by adding DA to HA. Without reducing immunological response, the hydrogel restores intestinal homeostasis by acting as an enteroprotective agent with a barrier function. Therefore, it avoids side effects, including infections and bacteremia, unlike the current first-line treatment for IBD (suppression of an overactive immune system) [112]. Qiu Lei *et al.* produced alginate hydrogel microspheres that contained bifidobacterium (Bac) and drug-modified nanoscale dietary fibers (NDFs). Drugs are delivered to the colorectum and shielded from acidic and multi-enzymatic conditions by the hydrogel microsphere. Gel microspheres played a protective role in the stomach, IL-1β monoclonal antibodies had a targeted effect on IBS, the regulated release of 5-ASA (5-aminosalicylic acid) in IBS was also supplied, and fermentation was used to remodel the gut microbiota. We believe that this makes it a viable approach to the creation of IBD site-specific therapy [113].

Combination of Drug and Nano-based Biopolymer

The integration of therapeutic drugs with nano-based biopolymers is a crucial process in the field of medicine. HereB, a combination of drugs and biopolymers, such as PLGA-polyelectrolyte NPs coated with chitosan and alginate, coated with LbL, offers a novel approach for controlled anticancer drug release, minimizing initial burst release and reducing toxicity [114]. The study suggests that chitosan nanoparticles, which contain the anticancer drug paclitaxel, have potential applications as carriers and an enhanced anticancer effect [115]. The study examined GemC18-loaded PEG-PLA micelles and self-assembled nanoparticles for their cytotoxicity effects on pancreatic cancer cell lines, finding enhanced cytotoxicity in both formulations[116]. Nano-delivery systems containing Dtx and Res show improved cancer treatment efficiency compared to free drugs. Co-delivery reduces multidrug resistance and fulfills anticancer therapy requirements [117]. Docetaxel-loaded dendritic copolymer NPs, co-delivery of autophagy inhibitors like chloroquine and chemotherapeutic drugs, significantly enhance cancer cell killing and decrease tumor volume and weight in severe immunodeficient mice [118].

PRECLINICAL STUDIES OF NANO-BASED BIOPOLYMER

The preclinical assessment of nano-based biopolymers offers crucial insights into their safety, biocompatibility, and efficacy in disease targeting. Table **3** describes the studies that involve both *in vitro* and *in vivo* experiments to evaluate the therapeutic potential of these materials for disease targeting. Preclinical studies are crucial for commercial use, so it is recommended to concentrate on plant and microbial polysaccharide-based nanoparticles[50].

TOXICITY AND SAFETY OF NANO-BASED BIOPOLYMER

The toxicity and safety, as well as the biocompatibility of nano-based biopolymers, are crucial prior to their clinical applications. Biopolymers are being used in tissue engineering due to their biocompatibility and non-toxic properties. Micro- and nanocellulose-based composites can effectively repair and regenerate damaged tissue. However, further research is needed on toxicity, *in vivo* degradation, and non-enzymatic degradation of cellulose. High-tech rapid prototyping is also needed for complex-shaped scaffolds [125]. Bio-based nanoparticles are being researched for their potential to reduce pollutants and reduce health risks, while also being used in precision farming to decrease reliance on artificial fertilizers and pesticides [126]. Here are some studies describing the CS/ZnO nanocomposites, synthesized using microbial methods with chitosan as stabilizer, significantly increased dye adsorption, achieved up to

95% degradation, and were safe for plant application and non-cytotoxic on human cell lines [127].

Table 3. The preclinical studies of nano-based Biopolymer for disease targeting.

Sr. No.	Type of Studies	Drug and Biopolymer	Target Disease	Outcomes	Referances
1	*In-vitro* and *in-vivo*	Curcumin-encapsulated Chitosan-Coated Poly (lactic-co-glycolic acid) Nanoparticles and Curcumin/Hydroxypropyl-β-Cyclodextrin	Alzheimer's Disease	CUR-CS-PLGA-NPs and CUR/HP-β-CD inclusion complexes exhibit stability, increased CUR release, higher cellular uptake, reduced CUR toxicity, antioxidant and anti-inflammatory activities *in vitro* studies. *In vivo* studies show CUR-CS-PLGA-NPs and CUR/HP-β-CD inclusion complexes enhance brain distribution of CUR, with higher bioavailability and distribution.	[119]
2.	*In-vivo* and *in-vitro*	Starch / Chitosan nanoparticles	Antimicrobial food packaging on the cherry tomatoes	The study found that CNP was more effective in inhibiting gram-positive bacteria, and 15% CNP was chosen for *in vivo* evaluation of starch/CNP films.	[120]
3.	*In vitro* models	Alginate-dextran sulfate-based nanoparticles (ADS-NPs) coated with chitosan-polyethylene glycol (PEG)-albumin.	Enzymatic degradations in the gastrointestinal (GI) tract	ALB-NPs enhance insulin permeability, preventing gastric release and sustaining it in intestinal pH, indicating potential for oral delivery.	[121]

(Table 3) cont.....

Sr. No.	Type of Studies	Drug and Biopolymer	Target Disease	Outcomes	Referances
4.	*In-Vivo, In-Vitro* and Cell cytotoxicity assay	Resveratrol-loaded zein/pectin nanoparticle	Anti-inflammatory activity	The study found that resveratrol, when encapsulated, showed significant anti-inflammatory properties, inhibiting the production of various inflammatory mediators and promoting the release of IL-10.	[122]
5.	*In-vitro* cytotoxicity and *in-vivo*	Quercetin loaded chitosan nanoparticles	Anticancer	The study showed a 67.28% release of QCT over 12 hours at pH 7.4, with a reduced IC50 value compared to free QCT. Treatment in tumor xenograft mice reduced tumor volume and increased serum superoxide dismutase levels.	[123]
6.	*In-vivo* biodistribution	Insulin-loaded alginate/dextran sulfate-based nanoparticles dual-coated with chitosan and technetium-99m-albumin.	Antihyperglycemic	The study shows that a biopolymer-based nanoparticulate coated with chitosan and albumin effectively reduces hyperglycemic effects in both T1D and T2D models, potentially enhancing oral bioavailability of insulin.	[124]

CHALLENGES AND CONSIDERATIONS

Nano-based biopolymers, despite their potential for disease targeting, face several challenges that hinder their widespread applications. Although there are challenges to overcome, including the requirement for accurate and steady drug delivery, structural integrity, and controlled release capabilities in complex *in vivo* situations, chitosan nanoparticles offer the potential to cure IBD [111]. There are challenges with biopolymer-based nanomedicines' toxicity, safety, efficacy, and uniformity in the treatment of cancer. Standardization is crucial for ensuring safety and facilitating industrial-scale production. To evaluate *in vivo* effects, thorough research on toxicity and effectiveness is required. A developing trend in nanomedicine is multimodal treatment, which combines many therapeutic techniques. A multifunctional approach should be used to manufacture biopolymer-based nanomedicines in order to combat multidrug resistance and enhance therapeutic effectiveness [61]. The nanomaterials offer high efficiency but pose health risks due to potential release into food products. Regulations are being established to minimize these risks. Biogenic nanoparticles, bio-renewable, reduce environmental contaminants and improve bio-packaging properties [128]. Biopolymeric nanoparticles are utilized in biomedical applications, such as drug delivery, imaging, tissue engineering, and sensor systems, due to their compatibility, degradability, and ability to target specific tumors. However, the dearth of standardized evaluation techniques and laws makes toxicity an issue [129]. Although they have practical uses, ASX nano-based drug delivery technologies have drawbacks. Nanoliposomes may be harmful to health, but nanoparticles have limited water solubility and excellent entrapment efficiency. There is little clinical research and no safety validation [130]. Nanotechnology in medical applications is improving disease treatment and diagnosis, but challenges remain, including safety and toxicity. To overcome these, understanding nanoparticle interactions, selecting appropriate properties, and addressing solid tumor entry are crucial. Current research focuses on active targeting and antibody conjugation, while controlled clinical studies are necessary for cytotoxicity [131]. Despite advancements in medical technologies, CNS disorders remain a global disease burden, and nanomedicine-based therapeutic strategies face challenges in complexity and AD neuropharmacology. Combining nanocarriers and drug targets could lead to more effective treatments [132].

FUTURE PERSPECTIVES

Nano-based biopolymers offer promising disease targeting due to their biocompatibility, precision, controlled release, and biodegradability, being increasingly used to tackle complex diseases like cancer, cardiovascular disorders, neurodegenerative disorders, and infectious diseases. Biopolymeric drug and gene delivery systems hold great promise for the future of medicine. They can enable targeted and personalized medicine, improving drug stability and bioavailability, and integrating smart materials and technologies. These advancements could lead to more effective and reliable therapies for various diseases. Furthermore, the development of combination therapies—in which several medications or genes are administered concurrently to address various facets of a disease or treat several coexisting conditions—may be made easier by biopolymeric delivery systems, leading to enhanced patient outcomes and synergistic effects [133]. The study presents a method for creating biopolymer films loaded with specific bioactive molecules, which have been successfully used as carriers for targeted antibiotics. By addressing the need for sustained medication efficacy, this research advances medication Delivery Systems (DDS). Through increasing medication efficacy, reducing adverse effects, and encouraging patient compliance, DDS's emphasis on biodegradable polymers has the potential to improve healthcare outcomes [134]. Drug delivery for illnesses like cancer has been transformed by nanotechnology, which incorporates medications into nanocarriers such as magnetic nanoparticles, synthetic polymers, and nanotubes. Biopolymeric nanocarriers show cost-effectiveness, biodegradability, and treatment efficacy [135]. Biopolymers and LbL technology are being researched for smart drug delivery systems, low immunological response, non-invasive devices, and theranostic systems. Further developments in pharmaceutical industry scaling are expected [136].

CONCLUSION

Nano-based biopolymers provide a substantial leap in targeted disease therapy, combining the accuracy of nanotechnology with the biocompatibility and biodegradability of biopolymers. These systems are designed to directly provide therapeutic chemicals, including medications or genetic material, to sick cells, resulting in increased effectiveness and less adverse effects. Because of their capacity to be functionalized with certain ligands, therapies may be delivered precisely to the afflicted areas—such as cancer cells—while preserving healthy tissue. Additionally, the controlled drug release capacity reduces the need for frequent dosing and improves patient compliance, enabling a prolonged therapeutic benefit. In addition to their great versatility, nano-based biopolymers are ideal for treating complex diseases, including cancer, infections, neurological conditions, and cardiovascular diseases.

Future research will probably concentrate on stimuli-responsive systems that release medications within sick tissues in reaction to external cues like pH or temperature changes. Furthermore, it is anticipated that the integration of medication delivery with thermonanostic (diagnostic) capabilities would improve real-time therapy efficacy monitoring. In conclusion, nano-based biopolymers have the potential to be very important in modern medicine as they provide safer, more precise, and more effective treatments for a range of diseases. Their adaptability and continued research make them an essential part of the healthcare breakthroughs of the future, especially in the areas of precision and customized medicine.

LIST OF ABBREVIATIONS

BNPs	Biopolymeric nanoparticles
CNF	Cellulose nanofibers
CS	Chitosan
CRC	Colorectal cancer
CMC	Carboxymethylation of chitosan
CCA	1,2-cyclohexanedicarboxylic acid
CPP	Cell-penetrating peptides
DOXO	Doxorubicin
DNA	Deoxyribonucleic Acid
DDS	Drug delivery system
ELPs	Elastin-like polypeptides
eREX	Enzymatic reactive extrusion
KTP	kidney-targeting peptide
MRI	Magnetic Resonance Imaging
MTX	Methotrexate
NGs	Nanogels
PCL	Polycaprolactone
PVA	Poly (vinyl alcohol)
PL	Piperlongumine
PTX	Paclitaxel
ROS	Reactive oxygen species
RNA	Ribonucleic Acid
REX	Reactive extrusion REX
5-FU	5-Fluorouracil
TMC	Trimethyl chitosan

REFERENCES

[1] George A, Shah PA, Shrivastav PS. Natural biodegradable polymers based nano-formulations for drug delivery: A review. Int J Pharm 2019; 561: 244-64.
[http://dx.doi.org/10.1016/j.ijpharm.2019.03.011] [PMID: 30851391]

[2] Jacob J, Haponiuk JT, Thomas S, Gopi S. Biopolymer based nanomaterials in drug delivery systems: A review. Mater Today Chem 2018; 9: 43-55.
[http://dx.doi.org/10.1016/j.mtchem.2018.05.002]

[3] Dumontel B, Conejo-Rodríguez V, Vallet-Regí M, Manzano M. Natural biopolymers as smart coating materials of mesoporous silica nanoparticles for drug delivery. Pharmaceutics 2023; 15(2): 447.
[http://dx.doi.org/10.3390/pharmaceutics15020447] [PMID: 36839771]

[4] Hasnain MS, Ahmed SA, Alkahtani S, *et al.* Biopolymers for drug delivery. In: Nayak AK, Hasnain MS, edss, Advanced Biopolymeric Systems for Drug Delivery 2020; 1-29.

[http://dx.doi.org/10.1007/978-3-030-46923-8_1]

[5]   Poon W, Kingston BR, Ouyang B, Ngo W, Chan WCW. A framework for designing delivery systems. Nat Nanotechnol 2020; 15(10): 819-29.
[http://dx.doi.org/10.1038/s41565-020-0759-5] [PMID: 32895522]

[6]   Scheffel U, Rhodes BA, Natarajan TK, Wagner HN Jr. Albumin microspheres for study of the reticuloendothelial system. J Nucl Med 1972; 13(7): 498-503.
[PMID: 5033902]

[7]   Sundar S, Kundu J, Kundu SC. Biopolymeric nanoparticles. Sci Technol Adv Mater 2010; 11(1): 014104.
[http://dx.doi.org/10.1088/1468-6996/11/1/014104] [PMID: 27877319]

[8]   Morachis JM, Mahmoud EA, Almutairi A. Physical and chemical strategies for therapeutic delivery by using polymeric nanoparticles. Pharmacol Rev 2012; 64(3): 505-19.
[http://dx.doi.org/10.1124/pr.111.005363] [PMID: 22544864]

[9]   Dobrzyńska□Mizera M, Dodda JM, Liu X, Knitter M, Oosterbeek RN, Salinas P, Pozo E, Ferreira AM, Sadiku ER. Engineering of bioresorbable polymers for tissue engineering and drug delivery applications. Advanced Healthcare Materials. 2024; 13(30): 2401674.
[http://dx.doi.org/10.1002/adhm.202401674]

[10]   Kubik T, Bogunia-Kubik K, Sugisaka M. Nanotechnology on duty in medical applications. Curr Pharm Biotechnol 2005; 6(1): 17-33.
[http://dx.doi.org/10.2174/1389201053167248] [PMID: 15727553]

[11]   Boisseau P, Loubaton B. Nanomedicine, nanotechnology in medicine. C R Phys 2011; 12(7): 620-36.
[http://dx.doi.org/10.1016/j.crhy.2011.06.001]

[12]   Rhim JW, Ng PKW. Natural biopolymer-based nanocomposite films for packaging applications. Crit Rev Food Sci Nutr 2007; 47(4): 411-33.
[http://dx.doi.org/10.1080/10408390600846366] [PMID: 17457725]

[13]   Motaung TE, Linganiso LZ. Critical review on agrowaste cellulose applications for biopolymers. Int J Plast Technol 2018; 22(2): 185-216.
[http://dx.doi.org/10.1007/s12588-018-9219-6]

[14]   Moohan J, Stewart SA, Espinosa E, *et al.* Cellulose nanofibers and other biopolymers for biomedical applications. A review. Appl Sci (Basel) 2019; 10(1): 65.
[http://dx.doi.org/10.3390/app10010065]

[15]   Shaghaleh H, Xu X, Wang S. Current progress in production of biopolymeric materials based on cellulose, cellulose nanofibers, and cellulose derivatives. RSC Advances 2018; 8(2): 825-42.
[http://dx.doi.org/10.1039/C7RA11157F] [PMID: 35538958]

[16]   Li F, Abdalkarim SYH, Yu HY, Zhu J, Zhou Y, Guan Y. Bifunctional reinforcement of green biopolymer packaging nanocomposites with natural cellulose nanocrystal–rosin hybrids. ACS Appl Bio Mater 2020; 3(4): 1944-54.
[http://dx.doi.org/10.1021/acsabm.9b01100] [PMID: 35025317]

[17]   Compart J, Li X, Fettke J. Starch-A complex and undeciphered biopolymer. J Plant Physiol 2021; 258-259: 153389.
[http://dx.doi.org/10.1016/j.jplph.2021.153389] [PMID: 33652172]

[18]   Apriyanto A, Compart J, Fettke J. A review of starch, a unique biopolymer – Structure, metabolism and in planta modifications. Plant Sci 2022; 318: 111223.
[http://dx.doi.org/10.1016/j.plantsci.2022.111223] [PMID: 35351303]

[19]   Xu E, Campanella OH, Ye X, Jin Z, Liu D, BeMiller JN. Advances in conversion of natural biopolymers: A reactive extrusion (REX)–enzyme-combined strategy for starch/protein-based food processing. Trends Food Sci Technol 2020; 99: 167-80.
[http://dx.doi.org/10.1016/j.tifs.2020.02.018]

[20] Divya M, Vaseeharan B, Abinaya M, *et al.* Biopolymer gelatin-coated zinc oxide nanoparticles showed high antibacterial, antibiofilm and anti-angiogenic activity. J Photochem Photobiol B 2018; 178: 211-8.
[http://dx.doi.org/10.1016/j.jphotobiol.2017.11.008] [PMID: 29156349]

[21] Aghajan MH, Panahi-Sarmad M, Alikarami N, *et al.* Using solvent-free approach for preparing innovative biopolymer nanocomposites based on PGS/gelatin. Eur Polym J 2020; 131: 109720.
[http://dx.doi.org/10.1016/j.eurpolymj.2020.109720]

[22] Su K, Wang C. Recent advances in the use of gelatin in biomedical research. Biotechnol Lett 2015; 37(11): 2139-45.
[http://dx.doi.org/10.1007/s10529-015-1907-0] [PMID: 26160110]

[23] Lozano D, Sánchez-Salcedo S, Portal-Núñez S, *et al.* Parathyroid hormone-related protein (107-111) improves the bone regeneration potential of gelatin–glutaraldehyde biopolymer-coated hydroxyapatite. Acta Biomater 2014; 10(7): 3307-16.
[http://dx.doi.org/10.1016/j.actbio.2014.03.025] [PMID: 24704694]

[24] Negm NA, Hefni HHH, Abd-Elaal AAA, Badr EA, Abou Kana MTH. Advancement on modification of chitosan biopolymer and its potential applications. Int J Biol Macromol 2020; 152: 681-702.
[http://dx.doi.org/10.1016/j.ijbiomac.2020.02.196] [PMID: 32084486]

[25] Jiménez-Gómez CP, Cecilia JA. Chitosan: A natural biopolymer with a wide and varied range of applications. Molecules 2020; 25(17): 3981.
[http://dx.doi.org/10.3390/molecules25173981] [PMID: 32882899]

[26] Bano I, Arshad M, Yasin T, Ghauri MA, Younus M. Chitosan: A potential biopolymer for wound management. Int J Biol Macromol 2017; 102: 380-3.
[http://dx.doi.org/10.1016/j.ijbiomac.2017.04.047] [PMID: 28412341]

[27] Chakraborty M, Hasanuzzaman M, Rahman M, *et al.* Mechanism of plant growth promotion and disease suppression by chitosan biopolymer. Agriculture 2020; 10(12): 624.
[http://dx.doi.org/10.3390/agriculture10120624]

[28] Kazmi I, Shaikh MAJ, Afzal O, *et al.* Chitosan-based nano drug delivery system for lung cancer. J Drug Deliv Sci Technol 2023; 81: 104196.
[http://dx.doi.org/10.1016/j.jddst.2023.104196]

[29] Mtibe A, Motloung MP, Bandyopadhyay J, Ray SS. Synthetic biopolymers and their composites: advantages and limitations—an overview. Macromol Rapid Commun 2021; 42(15): 2100130.
[http://dx.doi.org/10.1002/marc.202100130] [PMID: 34216411]

[30] Reddy MSB, Ponnamma D, Choudhary R, Sadasivuni KK. A comparative review of natural and synthetic biopolymer composite scaffolds. Polymers (Basel) 2021; 13(7): 1105.
[http://dx.doi.org/10.3390/polym13071105] [PMID: 33808492]

[31] Ali Akbari Ghavimi S, Ebrahimzadeh MH, Solati-Hashjin M, Abu Osman NA. Polycaprolactone/starch composite: Fabrication, structure, properties, and applications. J Biomed Mater Res A 2015; 103(7): 2482-98.
[http://dx.doi.org/10.1002/jbm.a.35371] [PMID: 25407786]

[32] Ilyas R, Zuhri M, Norrrahim M, *et al.* Natural fiber-reinforced polycaprolactone green and hybrid biocomposites for various advanced applications. Polymers (Basel) 2022; 14(1): 182.
[http://dx.doi.org/10.3390/polym14010182] [PMID: 35012203]

[33] Feldman D. Poly (vinyl alcohol) recent contributions to engineering and medicine. J Compos Sci 2020; 4(4): 175.
[http://dx.doi.org/10.3390/jcs4040175]

[34] Guo M, Trzcinski AP, Stuckey DC, Murphy RJ. Anaerobic digestion of starch–polyvinyl alcohol biopolymer packaging: Biodegradability and environmental impact assessment. Bioresour Technol 2011; 102(24): 11137-46.

[http://dx.doi.org/10.1016/j.biortech.2011.09.061] [PMID: 22001054]

[35]   Ortega A, da Silva AB, da Costa LM, *et al.* Thermosensitive and mucoadhesive hydrogel containing curcumin-loaded lipid-core nanocapsules coated with chitosan for the treatment of oral squamous cell carcinoma. Drug Deliv Transl Res 2023; 13(2): 642-57.
[http://dx.doi.org/10.1007/s13346-022-01227-1] [PMID: 36008703]

[36]   Karmakar S, Manna S, Kabiraj S, Jana S. Recent progress in alginate-based carriers for ocular targeting of therapeutics. Food Hydrocolloids for Health 2022; 2: 100071.
[http://dx.doi.org/10.1016/j.fhfh.2022.100071]

[37]   Vasvani S, Kulkarni P, Rawtani D. Hyaluronic acid: A review on its biology, aspects of drug delivery, route of administrations and a special emphasis on its approved marketed products and recent clinical studies. Int J Biol Macromol 2020; 151: 1012-29.
[http://dx.doi.org/10.1016/j.ijbiomac.2019.11.066] [PMID: 31715233]

[38]   Elmowafy EM, Tiboni M, Soliman ME. Biocompatibility, biodegradation and biomedical applications of poly(lactic acid)/poly(lactic-co-glycolic acid) micro and nanoparticles. J Pharm Investig 2019; 49(4): 347-80.
[http://dx.doi.org/10.1007/s40005-019-00439-x]

[39]   Lashkari M, Rahmani M, Yousefpoor Y, *et al.* Cell-based wound dressing: Bilayered PCL/gelatin nanofibers-alginate/collagen hydrogel scaffold loaded with mesenchymal stem cells. Int J Biol Macromol 2023; 239: 124099.
[http://dx.doi.org/10.1016/j.ijbiomac.2023.124099] [PMID: 36948335]

[40]   Wu X, Xin Y, Zhang H, Quan L, Ao Q. Biopolymer-based nanomedicine for cancer therapy: opportunities and challenges. Int J Nanomedicine 2024; 19: 7415-71.
[http://dx.doi.org/10.2147/IJN.S460047] [PMID: 39071502]

[41]   Manchun S, Dass CR, Sriamornsak P. Targeted therapy for cancer using pH-responsive nanocarrier systems. Life Sci 2012; 90(11-12): 381-7.
[http://dx.doi.org/10.1016/j.lfs.2012.01.008] [PMID: 22326503]

[42]   Bidwell GL III, Mahdi F, Shao Q, *et al.* A kidney-selective biopolymer for targeted drug delivery. Am J Physiol Renal Physiol 2017; 312(1): F54-64.
[http://dx.doi.org/10.1152/ajprenal.00143.2016] [PMID: 27784692]

[43]   Walker L, Perkins E, Kratz F, Raucher D. Cell penetrating peptides fused to a thermally targeted biopolymer drug carrier improve the delivery and antitumor efficacy of an acid-sensitive doxorubicin derivative. Int J Pharm 2012; 436(1-2): 825-32.
[http://dx.doi.org/10.1016/j.ijpharm.2012.07.043] [PMID: 22850291]

[44]   Moktan S, Ryppa C, Kratz F, Raucher D. A thermally responsive biopolymer conjugated to an acid-sensitive derivative of paclitaxel stabilizes microtubules, arrests cell cycle, and induces apoptosis. Invest New Drugs 2012; 30(1): 236-48.
[http://dx.doi.org/10.1007/s10637-010-9560-x] [PMID: 20938714]

[45]   Nag OK, Delehanty JB. Active cellular and subcellular targeting of nanoparticles for drug delivery. Pharmaceutics 2019; 11(10): 543.
[http://dx.doi.org/10.3390/pharmaceutics11100543] [PMID: 31635367]

[46]   Shishir MRI, Xie L, Sun C, Zheng X, Chen W. Advances in micro and nano-encapsulation of bioactive compounds using biopolymer and lipid-based transporters. Trends Food Sci Technol 2018; 78: 34-60.
[http://dx.doi.org/10.1016/j.tifs.2018.05.018]

[47]   Hajdu I, Bodnár M, Trencsényi G, *et al.* Cancer cell targeting and imaging with biopolymer-based nanodevices. Int J Pharm 2013; 441(1-2): 234-41.
[http://dx.doi.org/10.1016/j.ijpharm.2012.11.038] [PMID: 23246780]

[48]   Pathak N, Singh P, Singh PK, *et al.* Biopolymeric nanoparticles based effective delivery of bioactive

compounds toward the sustainable development of anticancerous therapeutics. Front Nutr 2022; 9: 963413.
[http://dx.doi.org/10.3389/fnut.2022.963413] [PMID: 35911098]

[49] Portero A, Remuñán-López C, Criado MT, Alonso MJ. Reacetylated chitosan microspheres for controlled delivery of anti-microbial agents to the gastric mucosa. J Microencapsul 2002; 19(6): 797-809.
[http://dx.doi.org/10.1080/0265204021000022761] [PMID: 12569028]

[50] Kučuk N, Primožič M, Knez Ž, Leitgeb M. Sustainable biodegradable biopolymer-based nanoparticles for healthcare applications. Int J Mol Sci 2023; 24(4): 3188.
[http://dx.doi.org/10.3390/ijms24043188] [PMID: 36834596]

[51] Fathi M, Abdolahinia ED, Barar J, Omidi Y. Smart stimuli-responsive biopolymeric nanomedicines for targeted therapy of solid tumors. Nanomedicine (Lond) 2020; 15(22): 2171-200.
[http://dx.doi.org/10.2217/nnm-2020-0146] [PMID: 32912045]

[52] Gao S, Tang G, Hua D, et al. Stimuli-responsive bio-based polymeric systems and their applications. J Mater Chem B Mater Biol Med 2019; 7(5): 709-29.
[http://dx.doi.org/10.1039/C8TB02491J] [PMID: 32254845]

[53] Pati D, Shaikh AY, Das S, et al. Controlled synthesis of O-glycopolypeptide polymers and their molecular recognition by lectins. Biomacromolecules 2012; 13(5): 1287-95.
[http://dx.doi.org/10.1021/bm201813s] [PMID: 22497456]

[54] Vicario-de-la-Torre M, Forcada J. The potential of stimuli-responsive nanogels in drug and active molecule delivery for targeted therapy. Gels 2017; 3(2): 16.
[http://dx.doi.org/10.3390/gels3020016] [PMID: 30920515]

[55] Wu L, Zou Y, Deng C, Cheng R, Meng F, Zhong Z. Intracellular release of doxorubicin from core-crosslinked polypeptide micelles triggered by both pH and reduction conditions. Biomaterials 2013; 34(21): 5262-72.
[http://dx.doi.org/10.1016/j.biomaterials.2013.03.035] [PMID: 23570719]

[56] Yue ZG, Wei W, Lv PP, et al. Surface charge affects cellular uptake and intracellular trafficking of chitosan-based nanoparticles. Biomacromolecules 2011; 12(7): 2440-6.
[http://dx.doi.org/10.1021/bm101482r] [PMID: 21657799]

[57] Nairi V, Magnolia S, Piludu M, et al. Mesoporous silica nanoparticles functionalized with hyaluronic acid. Effect of the biopolymer chain length on cell internalization. Colloids Surf B Biointerfaces 2018; 168: 50-9.
[http://dx.doi.org/10.1016/j.colsurfb.2018.02.019] [PMID: 29456044]

[58] Jayaswal N, Srivastava S, Kumar S, et al. Precision arrows: Navigating breast cancer with nanotechnology siRNA. Int J Pharm 2024; 662: 124403.
[http://dx.doi.org/10.1016/j.ijpharm.2024.124403] [PMID: 38944167]

[59] Srivastava S, Jayaswal N, Kumar S, et al. Unveiling the potential of proteomic and genetic signatures for precision therapeutics in lung cancer management. Cell Signal 2024; 113: 110932.
[http://dx.doi.org/10.1016/j.cellsig.2023.110932] [PMID: 37866667]

[60] Shanmuganathan R, Edison TNJI, LewisOscar F, Kumar P, Shanmugam S, Pugazhendhi A. Chitosan nanopolymers: An overview of drug delivery against cancer. Int J Biol Macromol 2019; 130: 727-36.
[http://dx.doi.org/10.1016/j.ijbiomac.2019.02.060] [PMID: 30771392]

[61] Jha A, Kumar M, Bharti K, Manjit M, Mishra B. Biopolymer-based tumor microenvironment-responsive nanomedicine for targeted cancer therapy. Nanomedicine 2024; 19(7): 633-51.
[http://dx.doi.org/10.2217/nnm-2023-0302]

[62] Bhattacharya T, Preetam S, Ghosh B, et al. Advancement in Biopolymer assisted cancer theranostics. ACS Appl Bio Mater 2023; 6(10): 3959-83.
[http://dx.doi.org/10.1021/acsabm.3c00458] [PMID: 37699558]

[63]   Alharbi H. Exploring the frontier of biopolymer-assisted drug delivery: advancements, clinical applications, and future perspectives in cancer nanomedicine. Drug Des Devel Ther 2024; 18: 2063-87.
[http://dx.doi.org/10.2147/DDDT.S441325] [PMID: 38882042]

[64]   Muthu M, Gopal J, Chun S, Devadoss AJP, Hasan N, Sivanesan I. Crustacean waste-derived chitosan: Antioxidant properties and future perspective. Antioxidants 2021; 10(2): 228.
[http://dx.doi.org/10.3390/antiox10020228] [PMID: 33546282]

[65]   da Cunha Menezes Souza L, Fernandes FH, Presti PT, Anjos Ferreira AL, Fávero Salvadori DM. Effect of doxorubicin on cardiac lipid metabolism-related transcriptome and the protective activity of Alda-1. Eur J Pharmacol 2021; 898: 173955.
[http://dx.doi.org/10.1016/j.ejphar.2021.173955] [PMID: 33617823]

[66]   Kong CY, Guo Z, Song P, *et al.* Underlying the mechanisms of doxorubicin-induced acute cardiotoxicity: oxidative stress and cell death. Int J Biol Sci 2022; 18(2): 760-70.
[http://dx.doi.org/10.7150/ijbs.65258] [PMID: 35002523]

[67]   Perdani MS, Juliansyah MD, Putri DN, *et al.* Immobilization of cholesterol oxidase in chitosan magnetite material for biosensor application. Int J Technol 2020; 11(4): 754-63.
[http://dx.doi.org/10.14716/ijtech.v11i4.3484]

[68]   Yusefi M, Chan HY, Teow SY, *et al.* 5-fluorouracil encapsulated chitosan-cellulose fiber bionanocomposites: synthesis, characterization and *in vitro* analysis towards colorectal cancer cells. Nanomaterials (Basel) 2021; 11(7): 1691.
[http://dx.doi.org/10.3390/nano11071691] [PMID: 34203241]

[69]   Sara JD, Kaur J, Khodadadi R, *et al.* 5-fluorouracil and cardiotoxicity: a review. Ther Adv Med Oncol 2018; 10: 1758835918780140.
[http://dx.doi.org/10.1177/1758835918780140] [PMID: 29977352]

[70]   Maher S, Kalil H, Liu G, Sossey-Alaoui K, Bayachou M. Alginate-based hydrogel platform embedding silver nanoparticles and cisplatin: characterization of the synergistic effect on a breast cancer cell line. Front Mol Biosci 2023; 10: 1242838.
[http://dx.doi.org/10.3389/fmolb.2023.1242838] [PMID: 37936720]

[71]   Van Weelden G, Bobiński M, Okła K, Van Weelden WJ, Romano A, Pijnenborg JMA. Fucoidan structure and activity in relation to anticancer mechanisms. Mar Drugs 2019; 17(1): 32.
[http://dx.doi.org/10.3390/md17010032] [PMID: 30621045]

[72]   Zhu P, Qian J, Xu Z, *et al.* Overview of piperlongumine analogues and their therapeutic potential. Eur J Med Chem 2021; 220: 113471.
[http://dx.doi.org/10.1016/j.ejmech.2021.113471] [PMID: 33930801]

[73]   Subhan MA, Torchilin VP. Biopolymer-based nanosystems for siRNA drug delivery to solid tumors including breast cancer. Pharmaceutics 2023; 15(1): 153.
[http://dx.doi.org/10.3390/pharmaceutics15010153] [PMID: 36678782]

[74]   Ezhilarasan D. Hepatotoxic potentials of methotrexate: Understanding the possible toxicological molecular mechanisms. Toxicology 2021; 458: 152840.
[http://dx.doi.org/10.1016/j.tox.2021.152840] [PMID: 34175381]

[75]   Şehirli AÖ, Sayıner S, Bilginaylar K, Özkayalar H, Aykaç A. Does chitosan introducc protection against methotrexate-induced hepatorenal injury in rats?. Clin Exp Health Sci 2024; 14(1): 39-44.
[http://dx.doi.org/10.33808/clinexphealthsci.1134320]

[76]   Lin M, Wang X. Natural biopolymer-based delivery of crispr/cas9 for cancer treatment. Pharmaceutics 2023; 16(1): 62.
[http://dx.doi.org/10.3390/pharmaceutics16010062] [PMID: 38258073]

[77]   Rupaimoole R, Slack FJ. MicroRNA therapeutics: towards a new era for the management of cancer and other diseases. Nat Rev Drug Discov 2017; 16(3): 203-22.

[http://dx.doi.org/10.1038/nrd.2016.246] [PMID: 28209991]

[78]   Zhang W, Xu W, Lan Y, He X, Liu K, Liang Y. Antitumor effect of hyaluronic-acid-modified chitosan nanoparticles loaded with siRNA for targeted therapy for non-small cell lung cancer. Int J Nanomedicine 2019; 14: 5287-301.
[http://dx.doi.org/10.2147/IJN.S203113] [PMID: 31406460]

[79]   Youngren-Ortiz SR, Gandhi NS, España-Serrano L, Chougule MB. Aerosol delivery of siRNA to the lungs. Part 2: nanocarrier-based delivery systems. Kona Powder Particle J 2017; 34(0): 44-69.
[http://dx.doi.org/10.14356/kona.2017005] [PMID: 28392618]

[80]   Majumder P, Bhunia S, Bhattacharyya J, Chaudhuri A. Inhibiting tumor growth by targeting liposomally encapsulated CDC20siRNA to tumor vasculature: Therapeutic RNA interference. J Control Release 2014; 180: 100-8.
[http://dx.doi.org/10.1016/j.jconrel.2014.02.012] [PMID: 24556418]

[81]   Borna H, Imani S, Iman M, Azimzadeh Jamalkandi S. Therapeutic face of RNAi: *in vivo* challenges. Expert Opin Biol Ther 2015; 15(2): 269-85.
[http://dx.doi.org/10.1517/14712598.2015.983070] [PMID: 25399911]

[82]   Haussecker D. Current issues of RNAi therapeutics delivery and development. J Control Release 2014; 195: 49-54.
[http://dx.doi.org/10.1016/j.jconrel.2014.07.056] [PMID: 25111131]

[83]   Sun P, Huang W, Jin M, *et al.* Chitosan-based nanoparticles for survivin targeted siRNA delivery in breast tumor therapy and preventing its metastasis. Int J Nanomedicine 2016; 11: 4931-45.
[http://dx.doi.org/10.2147/IJN.S105427] [PMID: 27729789]

[84]   Caprifico AE, Foot PJS, Polycarpou E, Calabrese G. Advances in chitosan-based CRISPR/Cas9 delivery systems. Pharmaceutics 2022; 14(9): 1840.
[http://dx.doi.org/10.3390/pharmaceutics14091840] [PMID: 36145588]

[85]   Taylor C, Pearson J, Draget K, Dettmar P, Smidsrod O. Rheological characterisation of mixed gels of mucin and alginate. Carbohydr Polym 2005; 59(2): 189-95.
[http://dx.doi.org/10.1016/j.carbpol.2004.09.009]

[86]   Castellani C, Assael BM. Cystic fibrosis: a clinical view. Cell Mol Life Sci 2017; 74(1): 129-40.
[http://dx.doi.org/10.1007/s00018-016-2393-9] [PMID: 27709245]

[87]   Lev R, Seliktar D. Hydrogel biomaterials and their therapeutic potential for muscle injuries and muscular dystrophies. J R Soc Interface 2018; 15(138): 20170380.
[http://dx.doi.org/10.1098/rsif.2017.0380] [PMID: 29343633]

[88]   Zadeh Mehrizi T, *et al.* A review study of the use of modified chitosan as a new approach to increase the preservation of blood products (erythrocytes, platelets, and plasma products): 2010-2022. Nanomed J 2023; 10(1): 16-32.
[http://dx.doi.org/10.22038/nmj.2022.65972.1693]

[89]   Gholap AD, Kapare HS, Pagar S, *et al.* Exploring modified chitosan-based gene delivery technologies for therapeutic advancements. Int J Biol Macromol 2024; 260(Pt 2): 129581.
[http://dx.doi.org/10.1016/j.ijbiomac.2024.129581] [PMID: 38266848]

[90]   Antoniou V, Mourelatou EA, Galatou E, Avgoustakis K, Hatziantoniou S. Gene therapy with chitosan nanoparticles: modern formulation strategies for enhancing cancer cell transfection. Pharmaceutics 2024; 16(7): 868.
[http://dx.doi.org/10.3390/pharmaceutics16070868] [PMID: 39065565]

[91]   Luan X, Sansanaphongpricha K, Myers I, Chen H, Yuan H, Sun D. Engineering exosomes as refined biological nanoplatforms for drug delivery. Acta Pharmacol Sin 2017; 38(6): 754-63.
[http://dx.doi.org/10.1038/aps.2017.12] [PMID: 28392567]

[92]   Xu J, Liu Y, Li Y, *et al.* Precise targeting of POLR2A as a therapeutic strategy for human triple negative breast cancer. Nat Nanotechnol 2019; 14(4): 388-97.

[http://dx.doi.org/10.1038/s41565-019-0381-6] [PMID: 30804480]

[93]   Cui L, Zheng R, Liu W, *et al.* Preparation of chitosan-silicon dioxide/BCSG1-siRNA nanoparticles to enhance therapeutic efficacy in breast cancer cells. Mol Med Rep 2018; 17(1): 436-41.
[http://dx.doi.org/10.3892/mmr.2017.7887] [PMID: 29115613]

[94]   Pachouri C, Patel B, Shroti S, Shukla S, Pandey A. Recent trends in nanoparticles based drug delivery for tuberculosis treatment. Int J Med Nano Res 2021; 8: 035.
[http://dx.doi.org/10.23937/2378-3664.1410035]

[95]   Wardani G, M M, Sudjarwo SA. *In vitro* antibacterial activity of chitosan nanoparticles against *Mycobacterium tuberculosis*. Pharmacogn J 2017; 10(1): 162-6.
[http://dx.doi.org/10.5530/pj.2018.1.27]

[96]   Desai SK, Mondal D, Bera S. Polyurethane-functionalized starch nanocrystals as anti-tuberculosis drug carrier. Sci Rep 2021; 11(1): 8331.
[http://dx.doi.org/10.1038/s41598-021-86767-1] [PMID: 33859215]

[97]   Khan SA, Rahman ZU, Javed A Natural biopolymers in the fabrication and coating of ureteral stent: An overview. Biomaterials Advances 2024; p. 214009.
[http://dx.doi.org/10.1016/j.bioadv.2024.214009]

[98]   Das A, Ghosh S, Pramanik N. Chitosan biopolymer and its composites: Processing, properties and applications-A comprehensive review. Hybrid Advances 2024; p. 100265.
[http://dx.doi.org/10.1016/j.hybadv.2024.100265]

[99]   Allawadhi P, Singh V, Govindaraj K, *et al.* Biomedical applications of polysaccharide nanoparticles for chronic inflammatory disorders: Focus on rheumatoid arthritis, diabetes and organ fibrosis. Carbohydr Polym 2022; 281: 118923.
[http://dx.doi.org/10.1016/j.carbpol.2021.118923] [PMID: 35074100]

[100]   Nasra S, Bhatia D, Kumar A. Recent advances in nanoparticle-based drug delivery systems for rheumatoid arthritis treatment. Nanoscale Adv 2022; 4(17): 3479-94.
[http://dx.doi.org/10.1039/D2NA00229A] [PMID: 36134349]

[101]   Antunes JC, Seabra CL, Domingues JM, *et al.* Drug targeting of inflammatory bowel diseases by biomolecules. Nanomaterials (Basel) 2021; 11(8): 2035.
[http://dx.doi.org/10.3390/nano11082035] [PMID: 34443866]

[102]   Rakotondrabe TF, Fan MX, Muema FW, Guo MQ. Modulating inflammation-mediated diseases *via* natural phenolic compounds loaded in nanocarrier systems. Pharmaceutics 2023; 15: 699.
[http://dx.doi.org/10.3390/pharmaceutics15020699]

[103]   Jain S, Amiji M. Tuftsin-modified alginate nanoparticles as a noncondensing macrophage-targeted DNA delivery system. Biomacromolecules 2012; 13(4): 1074-85.
[http://dx.doi.org/10.1021/bm2017993] [PMID: 22385328]

[104]   Huang J, Guo J, Dong Y, *et al.* Self-assembled hyaluronic acid-coated nanocomplexes for targeted delivery of curcumin alleviate acute kidney injury. Int J Biol Macromol 2023; 226: 1192-202.
[http://dx.doi.org/10.1016/j.ijbiomac.2022.11.233] [PMID: 36442556]

[105]   Masotti A, Ortaggi G. Chitosan micro- and nanospheres: fabrication and applications for drug and DNA delivery. Mini Rev Med Chem 2009; 9(4): 463-9.
[http://dx.doi.org/10.2174/138955709787847976] [PMID: 19356124]

[106]   Sharma AK, Arya A, Sahoo PK, Majumdar DK. Overview of biopolymers as carriers of antiphlogistic agents for treatment of diverse ocular inflammations. Mater Sci Eng C 2016; 67: 779-91.
[http://dx.doi.org/10.1016/j.msec.2016.05.060] [PMID: 27287177]

[107]   Zewail MB, El-Gizawy SA, Asaad GF, Shabana ME, El-Dakroury WA. Chitosan coated clove oil-based nanoemulsion: An attractive option for oral delivery of leflunomide in rheumatoid arthritis. Int J Pharm 2023; 643: 123224.
[http://dx.doi.org/10.1016/j.ijpharm.2023.123224] [PMID: 37451327]

[108] Elshabrawy HA, Abo Dena AS, El-Sherbiny IM. Triple-layered platform utilizing electrospun nanofibers and 3D-printed sodium alginate-based hydrogel for effective topical treatment of rheumatoid arthritis. Int J Biol Macromol 2024; 259(Pt 2): 129195.
[http://dx.doi.org/10.1016/j.ijbiomac.2023.129195] [PMID: 38184049]

[109] Priya S, Daryani J, Desai VM, Singhvi G. Bridging the gap in rheumatoid arthritis treatment with hyaluronic acid-based drug delivery approaches. Int J Biol Macromol 2024; 271(Pt 1): 132586.
[http://dx.doi.org/10.1016/j.ijbiomac.2024.132586] [PMID: 38795889]

[110] Palma E, Costa N, Molinaro R, *et al.* Improvement of the therapeutic treatment of inflammatory bowel diseases following rectal administration of mesalazine-loaded chitosan microparticles *vs* Asamax®. Carbohydr Polym 2019; 212: 430-8.
[http://dx.doi.org/10.1016/j.carbpol.2019.02.049] [PMID: 30832877]

[111] Liu X, Dong Y, Wang C, Guo Z. Application of chitosan as nano carrier in the treatment of inflammatory bowel disease. Int J Biol Macromol 2024; 278(Pt 4): 134899.
[http://dx.doi.org/10.1016/j.ijbiomac.2024.134899] [PMID: 39187100]

[112] Zhang G, Song D, Ma R, *et al.* Self-crosslinking hyaluronic acid hydrogel as an enteroprotective agent for the treatment of inflammatory bowel disease. Int J Biol Macromol 2024; 273(Pt 2): 132909.
[http://dx.doi.org/10.1016/j.ijbiomac.2024.132909] [PMID: 38848832]

[113] Qiu L, Shen R, Wei L, *et al.* Designing a microbial fermentation-functionalized alginate microsphere for targeted release of 5-ASA using nano dietary fiber carrier for inflammatory bowel disease treatment. J Nanobiotechnology 2023; 21(1): 344.
[http://dx.doi.org/10.1186/s12951-023-02097-6] [PMID: 37741962]

[114] Chai F, Sun L, He X, *et al.* Doxorubicin-loaded poly (lactic-co-glycolic acid) nanoparticles coated with chitosan/alginate by layer by layer technology for antitumor applications. Int J Nanomedicine 2017; 12: 1791-802.
[http://dx.doi.org/10.2147/IJN.S130404] [PMID: 28424550]

[115] Li F, Li J, Wen X, *et al.* Anti-tumor activity of paclitaxel-loaded chitosan nanoparticles: An *in vitro* study. Mater Sci Eng C 2009; 29(8): 2392-7.
[http://dx.doi.org/10.1016/j.msec.2009.07.001]

[116] Daman Z, Ostad S, Amini M, Gilani K. Preparation, optimization and *in vitro* characterization of stearoyl-gemcitabine polymeric micelles: A comparison with its self-assembled nanoparticles. Int J Pharm 2014; 468(1-2): 142-51.
[http://dx.doi.org/10.1016/j.ijpharm.2014.04.021] [PMID: 24731731]

[117] Jurczyk M, Kasperczyk J, Wrześniok D, Beberok A, Jelonek K. Nanoparticles loaded with docetaxel and resveratrol as an advanced tool for cancer therapy. Biomedicines 2022; 10(5): 1187.
[http://dx.doi.org/10.3390/biomedicines10051187] [PMID: 35625921]

[118] Zhang X, Yang Y, Liang X, *et al.* Enhancing therapeutic effects of docetaxel-loaded dendritic copolymer nanoparticles by co-treatment with autophagy inhibitor on breast cancer. Theranostics 2014; 4(11): 1085-95.
[http://dx.doi.org/10.7150/thno.9933] [PMID: 25285162]

[119] Zhang L, Yang S, Wong LR, Xie H, Ho PCL. *In vitro* and *in vivo* comparison of curcumin-encapsulated chitosan-coated poly(lactic- *co* -glycolic acid) nanoparticles and curcumin/hydroxypropyl-β-cyclodextrin inclusion complexes administered intranasally as therapeutic strategies for alzheimer's disease. Mol Pharm 2020; 17(11): 4256-69.
[http://dx.doi.org/10.1021/acs.molpharmaceut.0c00675] [PMID: 33084343]

[120] Shapi'i RA, Othman SH, Nordin N, Kadir Basha R, Nazli Naim M. Antimicrobial properties of starch films incorporated with chitosan nanoparticles: *In vitro* and *in vivo* evaluation. Carbohydr Polym 2020; 230: 115602.
[http://dx.doi.org/10.1016/j.carbpol.2019.115602] [PMID: 31887886]

[121] Lopes M, Derenne A, Pereira C, *et al.* Impact of the *in vitro* gastrointestinal passage of biopolymer-based nanoparticles on insulin absorption. RSC Advances 2016; 6(24): 20155-65.
[http://dx.doi.org/10.1039/C5RA26224K]

[122] Liu Y, Liang X, Zou Y, Peng Y, McClements DJ, Hu K. Resveratrol-loaded biopolymer core–shell nanoparticles: bioavailability and anti-inflammatory effects. Food Funct 2020; 11(5): 4014-25.
[http://dx.doi.org/10.1039/D0FO00195C] [PMID: 32322856]

[123] Baksi R, Singh DP, Borse SP, Rana R, Sharma V, Nivsarkar M. *In vitro* and *in vivo* anticancer efficacy potential of Quercetin loaded polymeric nanoparticles. Biomed Pharmacother 2018; 106: 1513-26.
[http://dx.doi.org/10.1016/j.biopha.2018.07.106] [PMID: 30119227]

[124] Lopes M, Aniceto D, Abrantes M, *et al.* *In vivo* biodistribution of antihyperglycemic biopolymer-based nanoparticles for the treatment of type 1 and type 2 diabetes. Eur J Pharm Biopharm 2017; 113: 88-96.
[http://dx.doi.org/10.1016/j.ejpb.2016.11.037] [PMID: 28007370]

[125] Khalil HPSA, Jummaat F, Yahya EB, *et al.* A review on micro- to nanocellulose biopolymer scaffold forming for tissue engineering applications. Polymers (Basel) 2020; 12(9): 2043.
[http://dx.doi.org/10.3390/polym12092043] [PMID: 32911705]

[126] Saraswat P, *et al.* Applications of bio-based nanomaterials in environment and agriculture: A review on recent progresses. Hybrid Advances 2023; 4: 100097.
[http://dx.doi.org//10.1016/j.hybadv.2023.100097]

[127] Khola Tazeen S, Khan Niazi MB, Abdulaziz Binobead M, Ahmed T, Shahid M. Characterization and toxicity evaluation of chitosan/ZnO nanocompoite as promising nano-biopolymer for treatment of synthetic wastewater. J King Saud Univ Sci 2024; 36(10): 103432.
[http://dx.doi.org/10.1016/j.jksus.2024.103432]

[128] Jafarzadeh S, Nooshkam M, Zargar M, *et al.* Green synthesis of nanomaterials for smart biopolymer packaging: challenges and outlooks. J Nanostructure Chem 2024; 14(2): 113-36.
[http://dx.doi.org/10.1007/s40097-023-00527-3]

[129] Sreena R, Nathanael AJ. Biodegradable biopolymeric nanoparticles for biomedical applications-challenges and future outlook. Materials (Basel) 2023; 16(6): 2364.
[http://dx.doi.org/10.3390/ma16062364] [PMID: 36984244]

[130] Chen S, Wang J, Feng J, Xuan R. Research progress of Astaxanthin nano-based drug delivery system: Applications, prospects and challenges?. Front Pharmacol 2023; 14: 1102888.
[http://dx.doi.org/10.3389/fphar.2023.1102888] [PMID: 36969867]

[131] Thakuria A, Kataria B, Gupta D. Nanoparticle-based methodologies for targeted drug delivery—an insight. J Nanopart Res 2021; 23(4): 87.
[http://dx.doi.org/10.1007/s11051-021-05190-9]

[132] Nguyen TT, Dung Nguyen TT, Vo TK, *et al.* Nanotechnology-based drug delivery for central nervous system disorders. Biomed Pharmacother 2021; 143: 112117.
[http://dx.doi.org/10.1016/j.biopha.2021.112117] [PMID: 34479020]

[133] Skorik YA. Biopolymers in drug and gene delivery systems 20. MDPI 2023; p. 17099.
[http://dx.doi.org/10.3390/books978-3-0365-8710-3]

[134] Tatullo M, Marrelli B, Facente A, Paduano F, Qutachi O. Novel biopolymer spray formulation for drug delivery in precision dentistry. Heliyon 2024; 10(16): e36038.
[http://dx.doi.org/10.1016/j.heliyon.2024.e36038] [PMID: 39224339]

[135] Rayhan MA, Hossen MS, Niloy MS, Bhuiyan MH, Paul S, Shakil MS. Biopolymer and biomaterial conjugated iron oxide nanomaterials as prostate cancer theranostic agents: a comprehensive review. Symmetry 2021; 13(6): 974.
[http://dx.doi.org/10.3390/sym13060974]

[136] Vilela C, Figueiredo ARP, Silvestre AJD, Freire CSR. Multilayered materials based on biopolymers as drug delivery systems. Expert Opin Drug Deliv 2017; 14(2): 189-200.
[http://dx.doi.org/10.1080/17425247.2016.1214568] [PMID: 27488175]

CHAPTER 7

AI-powered Biopolymers Engineering: Advancements in Drug Delivery and Efficacy

Anindita De[1,*], **Sonali Jayronia**[1], **Gowthamarajan Kuppusamy**[2] and **Young Joon Park**[3]

[1] *Department of Pharmaceutics, College of Pharmacy, JSS University, Noida 201301, India*

[2] *Department of Pharmaceutics, JSS College of Pharmacy, JSS Academy of Higher Education & Research, Ooty 643001, Tamil Nadu, India*

[3] *Department of Formulation and Drug Delivery , College of Pharmacy, Ajou University, 206 Worldcup-ro, Yeongtong-gu, Suwon-si 16499, Republic of Korea*

Abstract: The application of AI within biopolymer engineering is a monumental step in the advancement of drug delivery systems. With the aid of AI, it is possible for researchers to design and develop biopolymer materials to the highest degree, which means that drug effects, targeting accuracy, and treatment modulation for each patient can be improved to a large extent, which brings out the concept of personalized medicine. This chapter presents the state-of-the-art knowledge concerning the topic of biopolymer engineering utilizing AI, with a focus on Nanomedicine and Drug Delivery, respectively, in the context of the most fundamental and interesting areas–drug delivery and therapy and Machine Learning for Polymer Design. The developing frontiers of AI and biopolymer technology have the potential to alter drug discovery, delivery and administration processes with the patient in mind, leading to better compliance and control of the remedy for the patient.

Keywords: Advanced drug delivery systems, AI-driven biopolymer engineering, Machine learning, Personalized medicine, Precision drug targeting.

INTRODUCTION

In recent years, the focus has shifted towards the use of biopolymers as carriers of drugs. These substances are naturally occurring polymers obtained from renewable biological sources such as plants, animals, and microorganisms. The biocompatibility and biodegradability of these materials make them a good choice for building structures for drug delivery systems, providing safety and efficacy with respect to the environment [1]. In addition, the biopolymers also provide a

---

* **Corresponding author Anindita De:** Department of Pharmaceutics, College of Pharmacy, JSS University, Noida, India; E-mail: aninditanirupa@gmail.com

Sudhanshu Mishra, Smriti Ojha, Shashi Kant Singh, Rishabha Malviya & Saurabh Kumar Gupta (Eds.)

unique approach for controlled and sustained delivery of drugs, which makes them very useful for a number of therapeutics. Drug delivery is an example of biopolymers that include chitosan, alginate and gelatin and polylactic acid, PLA: each of which can be used based on its unique characteristics to increase drug action [2].

Nevertheless, the classic techniques for engineering these biodegradable materials for the purpose of drug delivery have posed considerable challenges. One of the main drawbacks is the ability to foresee and fine-tune the behavior of biopolymers within the subsets of biological systems. For instance, polymer degradation rate, interactions between the drug and the polymer, as well as the immune response of the body, are parameters that are hard to predict using normal methods. More often than not, these challenges lead to a semblance of an arduous process which is prone to wastage of resources and time and in certain instances, does not work at all [3].

The use of Artificial Intelligence (AI) in combination with machine learning (ML) in the synthesis of biopolymers is changing the perspective of designing these materials and optimizing them for drug delivery systems. Generally, the process of developing biopolymers has been overly reliant on experimentation, which is slow, expensive, and lengthy due to biological complexities. On the contrary, AI and ML make the whole process more accurate and efficient. With AI, it becomes possible to design new structures of biopolymers, knowing how they will behave biologically based on extensive polymer properties, biological interactions, and clinical data. For example, it is possible to employ complex learning algorithms to hyperparameterize the collected data to perform inference tasks. Specifically, key attributes such as polymers' degradation rates, drug release kinetics, and overall biocompatibility can be predicted accurately [4]. This predictive capability encourages scientists to engineer biopolymers for specific purposes and minimize the extent of trial analytics required. By way of illustration, one can use Generative Adversarial Networks (GANs), which have generative capabilities in creating new types of polymers with specific desired characteristics in mechanical strength and chemical properties. A study was conducted where they used a GAN to fabricate a new biopolymer with better stability and drug encapsulation efficiency for the delivery of tedious-to-formulate hydrophobic drugs [5].

Such an approach, driven by AI, not only streamlines the development process but also makes it possible to fabricate biopolymers with newer functionalities that the conventional methods could not realize. The use of AI in the engineering of biopolymers is a significant step in dealing with the deficiency mentioned earlier. In this respect, AI, especially those programs that employ ML and deep learning techniques, create an avenue for designing and developing biopolymers with better properties and functions for specific needs. For example, such AI mechanisms can comprehend and process large volumes of information about the existing polymers, their architecture, and even predict their behavior within biological systems. With this ability, they can design advanced drug delivery systems in the form of biopolymers that can target specific sites, reduce side effects and enhance the quality of life of the individuals [6]. A significant example of using AI in the field of biopolymer engineering is the development of custom drug delivery systems. AI can facilitate the design of such bio-polymeric drug delivery systems considering individual genetic information, disease state and it can help optimize a specific biopolymeric delivery system by controlling the release profile of the drug from the biopolymer in order to administer the drug at the right time and appropriate dosage for the maximum therapeutic efficacy [7].

This section of the book investigates the way AI is changing the landscape of biopolymer engineering, particularly in drug delivery systems. AI techniques allow researchers to design and fabricate biopolymer materials with great accuracy and speed, which seems likely to benefit the pharmaceutical R&D processes greatly. Such an AI revolution in this field is driven by the technology's ability to process large datasets, forecast outcomes, and restructure biopolymers to enhance effectiveness in pharmaceutical applications. In this chapter, we tried to provide different examples and case studies that highlight the role of AI in the development of biopolymers, among other advanced materials. These unique materials made *via* AI techniques and processes offer enhanced targeting and efficacy of drug delivery systems, but create a better environment for more effective treatments, making modern medicine more personalized.

AI IN BIOPOLYMER DESIGN AND SYNTHESIS FOR DRUG DELIVERY

AI has contributed to the development of biopolymer creation and design methodologies by allowing scientists to explore chemical space in a matter of hours. Forming biopolymers was very labour-intensive in the olden days; a lot of time was also wasted trying to experiment on the best possible material for drug delivery, which mostly depended on trial and error. However, the coming of AI, particularly the ML algorithms, has made it possible for scientists to assess and estimate the properties of numerous hypothetical polymer structures in a very short time (Fig. **1**). This aspect minimizes the time taken to search for suitable biopolymers that are required for a certain application in drug delivery, for instance, when controlled release, targeted delivery, and biocompatibility of the biopolymer are of concern. For instance, random polymer interactions can be modelled, and the efficiency of drug encapsulation, and it is how long it takes to degrade and how long it takes to release them, can also be modelled in this way using AI [5, 8]. AI significantly enhances the speed of the design of new polymers, which are tailored for better therapeutic benefits. Also, these AI-intermediated interfaces preserve the history of the design outcomes in the previous designs and keep updating themselves, which enhances the design definition in the later stages of the design process [9]. Thus, this fast search and improvement of the biopolymer designs leads not only to reduced time for development but also to the emergence of new opportunities for designing highly specialized materials that fit into the multifarious requirements of current drug delivery systems.

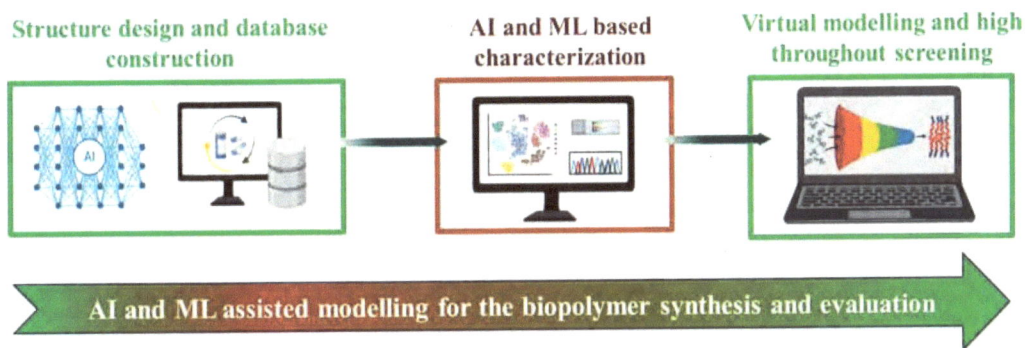

Structure design and database construction	AI and ML based characterization	Virtual modelling and high throughout screening

AI and ML assisted modelling for the biopolymer synthesis and evaluation

Fig. (1). AI and ML-assisted modelling design for the biopolymer synthesis and evaluation.

An interesting example in the context of controlled drug delivery, the drug release properties of Polyethylene glycol-polycaprolactone (PEG-PCL) copolymers are controlled, generally by modulating the molecular weight, composition, crystallinity, or any other structural property of the copolymer, and that can be further optimized through the use of ML operating methods. One of the goals of AI modelling is the throughput screening of synthesis conditions of PEG-PCL systems and obtained experimental results in order to determine the most feasible optimal conditions for polymer-drug systems to be formed. AI's predictive in PEG-PCL ratio and reaction parameters evaluation that improves the rates of biodegradation and drug loading, minimizing the laboratory experimentation costs and time in the designing process of the poly copolymer for a specific therapeutic application [10].

Machine Learning for Polymer Design

ML algorithms, especially the deep learning methods, have made groundbreaking contributions to the field of biopolymer engineering by providing sophisticated methods for the prediction, optimization and design of biopolymer properties such as solubility, degradation rates, mechanical strength, and biocompatibility. These advanced computational methods permit manipulation of large amounts of data containing numerous chemical structures, their physical properties, performance data of polymers and the performance of biopolymers in order to fashion new polymers of specific functionality. This, in turn, has greatly improved the designing and utilization of biopolymers, especially when it comes to applications like pharmacokinetics. For instance, in biopolymer-driven drug delivery mechanisms, deep learning models have been used to determine the self-assembly of block copolymers. Block copolymers can self-organize in a variety of shapes, such as micelles, which are critical for the delivery of drugs. For example, a neural network model can be provided with favorable data containing known polymer compositions, lengths of hydrophobic and hydrophilic blocks and stability of micelles, then this model can suggest the best configuration enhancing drug incorporation and release rates. One example is synthesizing PEG-b-PLA block copolymer and predicting its properties. ML models also help to fine-tune the PEG and PLA block lengths of the corresponding copolymer, allowing to obtain desired particle size, blood stability, and drug release profiles of resulting micelles [11]. The development of this technology enhances drug delivery systems by reinforcing the stability of micelles in circulation and drug release only in tumors, sparing the healthy tissues from the potential toxicity of the drugs. To engineer working micelle systems, one has to understand and predict the micellization patterns of different copolymer systems based on their architecture and environmental conditions, which is not only complicated but also takes a long time through regular laboratory procedures.

ML models have been created by researchers that take the data from the block copolymers that were synthesized before and estimate the parameters for new copolymer designs. These models consider the block length, polymer type, solvents used, temperature and predict the micelles' formation and stability as well as drug loading capacity. For example, BEPO®: bioresorbable diblock mPEG-PDLLA and triblock PDLLA-PEG-PDLLA based *in situ* forming depots with flexible drug delivery kinetics modulation is research in which the data on block copolymers was used in developing a neural network that could estimate the ability of new copolymer shapes to form stable self-assembling structures, known as micelles. This ML and AI designed a new block copolymer with quite stable micelles, which could encapsulate an anti-cancer drug with a low log p value, showing active targeting of the drug to the diseased tissue [11, 12]. This is a very useful predictive power which, as one could imagine, will mitigate the demerit of getting polymer development through a lot of experimenting with thin film fabrication. There is no need to synthesize and qualitatively assess rationally hundreds of polymers; instead, quantitative screens are applied, aided by mathematical modeling to pick a few that are likely to deliver good results. Thus, the period and expenditure necessary for the invention of new drug delivery systems are lessened, and the whole activity becomes more efficient.

Furthermore, it is possible in certain instances to implement an ML algorithm to analyze an additional polymer structure that was not previously available. Because of looking into large and diverse datasets and making connections that may not be overtly visible to human beings, such approaches can, in principle, invent polymers, which would be superior compared to those that are already available. Other examples of ML and AI include successful high temperature polymers, such as perfluoroalkoxy alkanes, polyether ether ketone (PEEK) and fluorinated ethylene propylene, whose synthesis was obtained through search in the polymer base using ML and AI. The unique thermal stability of these materials is conferred by the presence of heteroatoms in the chain of thermoplastic polymers [13]. In addition to forecasting micelle formation, deep learning systems are also capable of integrating both the vast amount of experimental and computational data for purposes such as improving the efficiency of biopolymer drug delivery systems. This entails fine-tuning the interactions between the polymers and the drugs, forecasting degradation processes, as well as controlling drug release rates and localization in the patient's body. For instance, ML models help to predict the effect on the drug release kinetics from a polymeric device of reconstructing the device in terms of the shape and/or arrangement of the polymers, which facilitates the design of drug delivery systems providing long-term controlled drug administration for days or even weeks [14]. Machine and deep learning are proving to be integral components in biopolymers engineering for delivery of drugs. These allow finding new polymers of particular properties, making the process of building these systems faster, and increasing the efficiency of drug delivery systems many times. As ML approaches are improving and data available is becoming more extensive, the possibilities of further development of biopolymer-based medical technologies are limitless, which implies that even more effective and customized treatment regimens will be available in the future [11, 14, 15].

Generative Models for Polymer Synthesis

Generative models, including Generative Adversarial Networks (GANs) and Variational Autoencoders (VAEs), have dramatically improved the process of creating new polymers by allowing the production of new materials with specific designs [16]. These models, which belong to the wider class of ML techniques, are capable of producing new polymer structures that achieve certain performance targets, thus facilitating the search for materials for a wide range of use cases, including drug delivery, tissue engineering, and medical devices (Fig. **2**). Hypothetically, the generative model can analyze previous semantic datasets of polymers and their characteristics before coming up with designs for new polymers that are manufactured and tested.

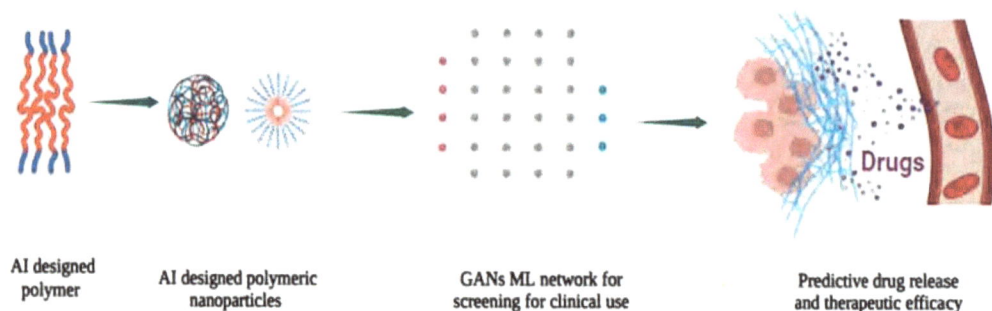

| AI designed polymer | AI designed polymeric nanoparticles | GANs ML network for screening for clinical use | Predictive drug release and therapeutic efficacy |

Fig. (2). Anticipation of the therapeutic efficacy of the drug-loaded bio-polymer by GANs modelling.

Specifically, GANs are composed of two neural networks, a generator that generates an imaginary polymer and a discriminator network that determines how well this imaginary polymer scores by comparing it with real polymer data. This process causes the generator to learn how to synthesize better quality structural polymers. In contrast, VAEs embed various polymer characteristics in a latent space and retrieve them as potential polymer structures tailored to specific properties, such as solubility, elasticity, or degradation time [17].

To develop new biopolymers for the construction of tissue engineering scaffolds with modified mechanical properties, a GANs approach was used. In tissue engineering, it is also important to have materials that can bear load while being flexible at the same time; these materials should also be able to grow cells and facilitate tissue formation [18]. Conventional materials in the targeted area are rather useful; however, they lack the ideal properties sought in this case application. In this instance, the GAN was successively fed with known biopolymers, their tensile strength, elasticity, biodegradability, and so on. The GAN model generator suggested several new polymer structures, which were analysed in terms of their predicted mechanical characteristics and considered superior to the existing materials. Among them, the modification of the chitosan with the other gums and polymers created a long list of GAN model-based polymers for tissue engineering and drug delivery [8, 19 - 22]. The discriminator network augmented these suggestions by discarding designs that were unlikely to comply with the requirements based on the previous experimental results. Following the creation of a number of promising biopolymer architectures, the new materials were synthesized and tested in the laboratory. They exhibit some interesting characteristics. One such polymer demonstrated some wonderful mechanical properties for which there has not been any application in tissue engineering before. One of the example is melt electrowriting (MEW) method has been used to fabricate a Polylactic acid / Polycarbonate (PLA/PCL) scaffold which exhibits anisotropic property during a uniaxial tensile loading where varying the Young's modulus, ultimate tensile stress, and strain to failure are possible depending on the direction of the tensile load applied. This anisotropy in the composite PLA/PCL scaffolds makes it an interesting material for medical applications such as heart tissue engineering. It displayed the perfect combination of strength and ease of bending, which makes it appropriate for use in tissue scaffolds [18, 23, 24]. When evaluated *in vivo*, these polymers supported cell growth and attachment but also exhibited a much higher biocompatibility with controlled biodegradability than the previous materials [25]. Similar efforts are also being investigated in the case of Drug Delivery System (DDS), where it is conceived that generative models will be utilized for a better design of polymers that either coat drugs, have desired degradation rates, or control the release of the drug. The ability to design and engineer new polymer backbones *in silico* prior to laboratory work has reduced the time and financial constraints associated with producing innovative biomaterials. It is also anticipated that nanobiomaterials such as these and algorithms that design such materials will continue to decrease costs for medical treatments and enhance the customization of medicine.

AI-polymer Enhanced Drug Delivery Systems

The emergence of AI has significantly contributed to the advancement of drug delivery systems, particularly in the design and use of biopolymeric carriers. Such carriers are often loaded with therapeutic agents constructed with a high level of accuracy in order to deliver results with minimal side effects. AI has enhanced the efficiency of this system for the on/off release profile of the drugs, all designed before actual testing, owing to its ML and modeling techniques.

Predictive Modelling for Drug-polymer Interactions

AI systems are quite useful in predicting the interactions that may exist between different drugs and biopolymers, which is important in creating suitable carriers for drug delivery. These models employ extensive databases and computational methods to attempt to forecast the behavior of various drugs toward a certain biopolymer system. AI assists drug delivery systems to become more efficient and effective with the addition of drug loading capacity, stability, and release profiles, and further enhances the systems [26, 27].

Expansion of the definition of AI algorithms to include, for example, ML techniques allows such algorithms to be trained with large amounts of data that incorporate various properties of drugs (structure, solubility, stability properties, *etc.*) and their relations to traits of biopolymers (chemical composition, molecular weight, structure, *etc.*). The ability of this predictive model allows experts to easily screen suitable polymers for the drug at hand in enhancing the encapsulation and release profiles of the said drug. Key Advantages of AI in drug-biopolymer interaction prediction are:

- **Scale-up Drug Loading:** The models can improve the polymer design in such a way that it would be possible to load more drugs into the carrier. This helps in achieving the desired therapeutic effect while reducing the bioavailability needed [28].
- **Enhancing Stability:** With the help of interaction prediction algorithms, polymers that enhance the stability of the drug are selected so that the drug does not get degraded during storage and transportation [29]. This is critical for biomolecules like proteins and nucleic acids that are very sensitive to changes in their environment.
- **Customizable Release Profiles:** The application of AI can assist in formulating carriers that manipulate the speed at which a drug is released for its extended use. Such an approach would optimize therapeutic management and minimize adverse effects by keeping the drug concentration within the therapeutic range [30]. For instance, in this field, AI was used to build a model where its primary task was to estimate how well various anticancer agents would 'stick' to a biomaterial made of, for example, Chitosan or Polyethylene glycol (PEG) [24, 31]. All this data was gathered in the form of a database, which contained, apart from drugs, the ways in which these drugs interacted with various biopolymers.
- **Development of the Model:** For this AI model, regression models as well as neural networks were employed to extract the relationship between molecular descriptors of molecules (for example, hydrophobicity, size, charge) and the drug-biomatrix binding interactions [32].
- **Selection of Optimal Polymer:** The simulations predicted that using certain polymers, in particular a PEGylated chitosan derivative, would best sustain the release of the doxorubicin drug, an anticancer agent. The high binding affinity of the said polymer anticipated its capability of drug encapsulation and drug-eluting ability in a prolonged duration [33].
- **Preclinical Studies:** In preclinical studies, formulations using the AI-selected polymer showed significantly improved therapeutic outcomes compared to standard formulations. For instance, the AI-designed drug delivery system resulted in a controlled release profile that maintained effective drug concentrations in the bloodstream over several days, leading to enhanced tumor targeting and reduced side effects [34].

Researchers are increasingly using AI as one of the tools in predicting drug-biopolymer interactions and thus assisting in the rational design and optimization of drug delivery systems. The capacity to predict the binding affinities and interactions associated with specific biopolymer drug carriers not only streamlines the entire process of development but also improves the performance and drug delivery capability of the biopolymer. The progression deepening the use of AI in drug delivery might bring a paradigm shift to the way patients receive treatment, enabling the use of more efficient and tailored therapies.

AI-driven Optimization of Drug Release Profiles

Artificial Intelligence algorithms, especially reinforcement learning, have been employed in devising drug delivery profiles of biopolymer matrices. By altering the drug release scenarios, the AI can highlight the conditions that would attain the highest therapeutic effect, with the least side effects [35]. In another case study around delivering insulin to diabetic patients, artificial intelligence was used to modify the release profile of insulin, which was embedded in a biodegradable implant [36, 37]. The engineered system, in turn, enabled constant implants of insulin, eliciting better glucose control in animal studies.

AI IN THE DEVELOPMENT OF BIOPOLYMER-BASED NANOPARTICLES

In recent years, especially in the area of drug delivery, the use of biopolymer-based nanoparticles has increased markedly. This is not surprising considering that such systems enable efficient encapsulation of drugs, protection while in circulation, and improvement of their bioavailability. Given the advances made, AI functionalities are being incorporated in the design and applications of these nanomaterials for improved drug-delivery systems. Data analytics is performed using algorithms like ML or deep learning. These may include, for example, drug delivery, biopolymers, nanoparticles or any other topics' performance or even their possible properties. Consequently, these profiles can be used to correlate different constants and enable the rational design of nanoparticles. Special enhancements optimised by AI comprise:

- **Particle Size:** The performance of nanoparticles highly depends on their size, as it affects the extent of their distribution within cells, the rate of uptake, and the release of the drug. AI can help determine the most suitable particle size to achieve the best therapeutic benefits without causing toxicity [29, 38].
- **Surface Charge:** The interaction of nanoparticles with biological systems, including the ability to be taken up by cells and how long they stay in the bloodstream, is influenced by the surface charge. AI models can assist in surface charge optimization for improved nanoparticle stability and targeting [39].
- **Polymer Composition:** The selection of biopolymers in use when fabricating nanoparticles is quite critical for the biocompatibility of the nanoparticles as well as for their drug-carrying capacity. AI can offer assistance in the evaluation of the different polymer compositions so that the one with the best drug loading and release profile is used.

The use of artificial intelligence (AI) in science is proving to be invaluable, particularly in the development of biopolymer-based nanoparticles. One notable example involves a group of scientists who synthesized chitosan nanoparticles to address the poor water solubility of indomethacin, a nonsteroidal anti-inflammatory drug (NSAID) [40, 41]. Among several options, indomethacin was selected for nanoparticle delivery. The researchers employed a machine learning algorithm trained on data from previously developed nanoparticles—such as particle size, surface charge, polymer volume, and drug loading efficiency. This AI-driven approach was used to identify the optimal conditions for producing effective chitosan nanoparticles.

- **Prediction of Optimal Parameters:** Based on the model predictions, the optimal nanoparticle size for cellular uptake enhancement and attenuation of possible immune system clearance would be about 150 nanometers [8, 21]. Additionally, the model predicted that a surface charge of +20 mV would enhance the stability of the nanoparticles and their interaction with the target cells, which in turn would facilitate drug delivery.
- **Polymer Composition Optimization:** The ratios of chitosan and drug were dovetailed in a way, according to the AI analysis, that would ensure efficient encapsulation without affecting the drug release characteristics [42]. This resulted in a formulation that entailed the use of a chitosan derivative that improved the solubility of the anti-inflammatory compound used.
- **Experimental Validation:** Based on the AI forecasts, the chitosan-made nanoparticles were produced by the researchers in accordance with the already suggested parameters. The *in vitro* studies showed that these nanoparticles were far more effective than conventional formulations in enhancing the solubility and bioavailability of the anti-inflammatory drug [43]. The formulated nanoparticles exhibited a drug release profile that was extended in nature, thus improving the therapeutic effectiveness of the drug in the cellular models.
- **Impact on Therapeutic Outcomes:** The implementation of AI as one of the tools during the design process made it possible not only to expedite the development of these nanoparticles but also to introduce a more efficacious formulation in preclinical animal studies for the reduction of inflammation. Owing to improved bioavailability and controlled release properties, lesser amounts were required for dosing, translating to lower side effects, thus showing the possibilities of such designs towards achieving a high patient therapeutic index.

The use of AI in the manufacture of biopolymer-based nanoparticles represents a paradigm shift in drug delivery systems. AI system significantly optimizes primary parameters, including particle size, surface charge, and polymer composition, thereby making it possible to formulate nanoparticles with improved drug solubility and bioavailability. It also quickens the designing phase, in return enabling strategies which are more efficient and tailored to the individual patient, as in the case of visual therapy, where treatment outcomes are always better than average for such patients.

AI in Stimuli-responsive Drug Delivery Systems

Stimuli-responsive biopolymers can release drugs in response to specific biological or environmental triggers (*e.g.*, pH, temperature) [44]. AI is used to design these systems, ensuring precise control over drug release. AI was employed to design a pH-responsive biopolymer for targeted drug delivery to cancer cells [45]. The AI model predicted the polymer's response to the acidic tumor microenvironment, enabling the selective release of the drug at the tumor site, thereby minimizing damage to healthy tissues.

CHALLENGES AND FUTURE DIRECTIONS FOR THE AI-POWERED BIOPOLYMERS ENGINEERING FOR DRUG DELIVERY

Despite the fact that engineering biopolymers with AI techniques holds promise, there are still some challenges (Fig. **3**). They are in relation to: large and high-quality data sets required for training models for AI, applying AI within the experimental work process, and finally, the application of materials designed by AI in the real world.

Data Availability and Quality

The success of AI models, above all in intricate domains like biopolymer engineering, depends on the presence of huge and well-structured datasets [38, 46]. These datasets help in training the AI models, which in turn help in making predictions and optimizing a few aspects of drug delivery systems, such as the polymers used, interactions of the drugs, and the expected clinical outcomes of the drug delivery system. Nevertheless, in biopolymer engineering, the greatest challenge is the availability of data, and even the existing data is usually too poor in quality to build effective AI models that can take into account the challenges of biopolymer-drug associations.

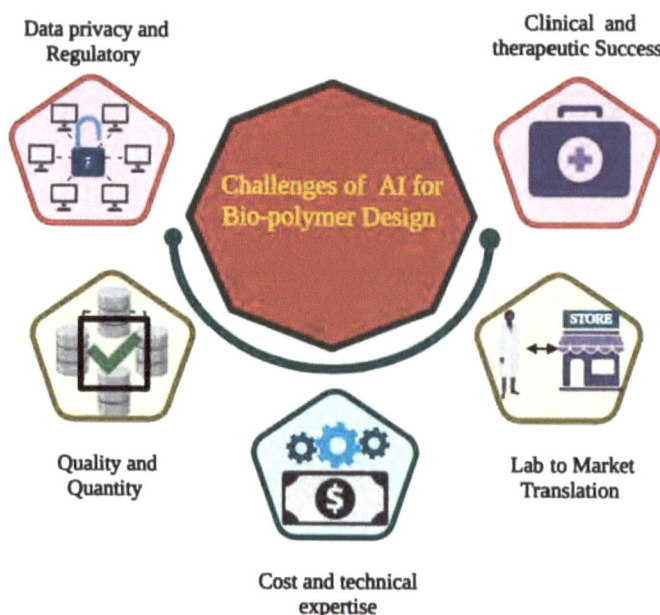

Fig. (3). Challenges for the AI-powered biopolymers engineering for drug delivery.

A key concern in this domain is the scarcity of large and well-organized databases. A lack of uniformity in data bases characterizes biopolymer engineering. This is different from more developed areas like genomics or imaging, where there is a rich supply of repositories. Most of the professionals involved in the field of polymer synthesis or drug delivery, and even nanoparticle fabrication, create databases in single instances, and such data sets remain hidden from the overall scientific society. There is therefore a compartmentalization of data, where polymer content or how fast it is degraded or how quickly it releases the drug are likely to be kept in different systems or internal archives. For instance, in the case of a research group working on developing nanoparticles to transport drugs, the group might be interested in the mechanical properties of one specific type of biopolymer or how well a drug can be loaded into a polymer at a particular condition, but such information is unlikely to appear in any database or be accessible for others. This problem of a lack of central and standardized data makes it difficult for any AI models to draw conclusions that can cut across various polymers and applications.

In conjunction with availability, the second factor that raises red flags is the quality of the data. In order for the AI models to give trustworthy and exact forecasts, the input data has to cover a large area and be correct [18, 47]. Nevertheless, the available datasets in biopolymer engineering are often either incomplete or, on the contrary, rather mess, with factors such as different experimental conditions, errors in measuring, and the physical impossibility of reproduction in different studies. For example, the values for drug release profiles or kinetic rates of polymer degradation may vary due to different laboratory practices or conditions under which testing was carried out. In turn, this 'noise' exists within the AI models and contributes to the degradation of the predictions made, which in turn affects the performance of the models adversely.

In order to address these issues, it is imperative that the sectors of research, industry, and regulatory processes work together. There is a need for a framework that will facilitate the development of versatile databases that will be compliant with every research institution and every pharmaceutical company. External partnerships, through the pharmaceutical industry, can have access to sensitive information on how biopolymers were formulated, the exact type of clinical trials conducted, and the outcomes recorded, and regulatory authorities like the FDA and EMA can help in formulating standards outlining how data should be collected and presented for uniformity and high-quality standards. For example, if a university has researchers who are working on biopolymer-based nanoparticles and their applications, and if that university collaborates with a company that develops such nanoparticles, both parties may be able to create a database with petabytes of polymer related data that includes above mentioned studies, use of polymers in drug delivery, drug loading and release, and efficacy assays. Such a partnership would allow researchers to build better AI technologies that would be useful in any drug delivery system.

It would facilities great advancements if accessible databases that have biopolymer characteristics as well as drug interaction information were created. Just as there are repositories such as the Protein Data Bank (PDB) or the Genomic Data Commons (GDC) in other sectors, establishing a bio polymer engineering repository may quicken research and creativity. This is because adequate data to train artificial intelligent systems can be collected where researchers are encouraged to upload and share data. In addition, this would promote transparency and reproducing of results as it would be mandatory for research to comply with a certain data collection procedure and methods.

One instance of data-sharing initiatives that has proven to be successful can be found in material science, where, for example, the Materials Project has provided free online databases of the properties of countless materials. Such endeavors have made it possible to incorporate AI models in predicting material properties and making the process of finding new materials easier. In the same way, AI models deployed for the design of efficient CAD systems for drug delivery devices using biopolymers can benefit tremendously from a networked available database of reasonable size that carries information on polymer properties, their degradation times, as well as interaction with drugs.

The success of innovations in biopolymer engineering and drug delivery that use AI significantly depends on data availability and data quality. Overcoming the current challenges in the two fields will necessitate partnerships, especially between academia, industry, and regulatory entities, to develop harmonised, elaborate data collections. This increased level of data quality and availability will facilitate more accurate predictions on the interaction of biopolymers with drug molecules, enabling the realization of more targeted and efficient drug delivery systems. The creation of publicly available repositories and networks of data will form an important foundation towards the advancement of AI capabilities in this area of work, thereby promoting creativity and speeding up the conversion of research to clinical practice.

Integration with Experimental Workflows

For AI models to be fully effective in biopolymer studies and in drug delivery as well, they need to be embedded within an experimental cycle. While the power of AI predictions has been proven already, these systems require experimental confirmation in order to be used effectively. The cycle of predicting, validating, and fine-tuning is necessary for the enhancement of AI capabilities and for science-based outputs of the technology. Nonetheless, there remain challenges in the efficient incorporation of AI into experimental systems that include usability to non-professionals and interfacing with experimental apparatus.

Bridging the AI Technology and the Practical Work

In order to use AI in experimental work, it is necessary to develop appropriate tools and platforms that are simple to use and that also do not disrupt any existing laboratory processes. This is because, in this area, a greater number of AI models are created by computer scientists and engineers who do not always bear in mind what happens in a real laboratory. This has made it hard for researchers with little knowledge of computational modeling to use these models in their work, which has made them hard to integrate into practice.

In order to solve this problem, AI tools must be built with their intended audience in mind so that biopolymer researchers without data science experience can comfortably make use of them. This can be accomplished by using simple graphical user interfaces (GUIs), making everything in the program 'drag and drop' or including detailed procedures where a user can load the data, run the algorithm and obtain results in a coded environment without writing any code at all. For instance, an AI tool within a biopolymer research site could include fixed boxes for inputs where users can add materials such as the polymer mixture's elements, how much drugs were incorporated in the system, or the rate of breakdown of the polymer and the AI would carry out calculations on what would be the behavior of the polymers, the release kinetics, or the biological response of the polymers.

As an example, drug discovery offers a clear-cut example of how AI systems can be incorporated into experimental work, as AI-based tools are more and more introduced to automated laboratory facilities. Specifically, in one such example, scientists derive the possible molecular skeletons of new drugs and use AI to predict the achieved molecular structure. These predictions are accomplished in high-throughput robotic systems, which find new compounds and validate their activity in real time, forming the cycle of prediction and validation in a fully automatic mode. In the same vein, it is not hard to imagine how biopolymer engineering would redefine itself by incorporating the AI models into automated systems for the synthesis and testing of polymers, allowing very rapid turnaround of polymer designs generated by the AI.

An AI system may, for example, foretell that a specific composition of biopolymers will lead to the greatest enhancement in the stability and drug release profile of an anticancer agent. The subsequent action to this AI output may be the direct feeding of the polymer synthesis command to the robotic machine or center that is capable of synthesizing polymers and performing unique investigations of the synthesized material. If the experimental results are consistent with what the AI assumed, the model's accuracy is affirmed in a real sense. In case there are differences, the actual experimental data can be recycled into the AI model so that its forecast is improved over time and becomes more precise through numerous times of such iterations. One of the critical measures in fast-tracking the use of AI in biopolymers research is the creation of systems able to connect and work with laboratory devices. In contemporary labs, various sophisticated devices such as High Performance Liquid Chromatography (HPLC), Scanning Electron Microscope (SEM), and particle size distribution devices are commonly used to evaluate the properties of polymers and the drug release rates [5, 8, 48]. There is a need for AI models to perform an AI data assimilation process before integration into these workflows; this entails assimilating data from these instruments in real time, processing the data, and returning the data into the experimental loop [49].

In point of fact, such systems can indeed be built with API, in other words, Application programming interfaces that connect the communicative roots of AI and laboratory devices, thereby enabling automatic collection of experimental data and limiting the extent of data entry done manually by researchers. For example, suppose the researcher formulated a biopolymer of a particular ant that he read precedes its molecules in order to make biopolymer nanoparticles. After synthesizing the nanomaterials, once again, the particle size analyzer with its adjoining measurements would send the data to the AI, whereby the AI would then use this information to compare its guessed size variation with the real size deviation. The constant interfacing of the AI algorithms with the relevant experimental apparatus would enhance the turnaround time for the validation process by enabling the modification of the predictions or the experimental conditions with real-time feedback from the users.

Another point, however, is that the successful merging of AI in the experimental processes is also affected by having scientists, biopolymer researchers, and machinery manufacturers work together. To this end, computational specialists should always collaborate with practitioners and developers to ensure that the design of AI applications aligns with laboratory needs. On the other hand, practitioners should also equip functional laboratories, and develop smart devices and software suitable for AI integration [8, 50]. For instance, transportation systems might be built where a computational group creates a self-assembled biopolymer structure prediction program and an experimental group provides the program with experimental data. At the same time, an equipment developer may create a new model of a particle size analyzer or a drug release tester with AI capabilities, enabling the use of the AI system by the experimental team as part of their routine.

The future of AI integration in biopolymer engineering may involve fully automated workflows where AI not only predicts polymer properties but also controls the experimental process. In such a system, an AI model might predict the optimal formulation for a biopolymer-based drug carrier, which an automated robotic system would then synthesize. This system could measure key properties like drug loading efficiency, particle size, or release rate, feeding the data back into the AI model in real-time. The AI would analyze the data in order to improve the accuracy of its forecasts and fine-tune the composition in several cycles without human assistance. Such AI technologies would be essential for testing theory and the improvement of processes aimed at the design of advanced systems for the controlled delivery of drugs [51]. The full potential of AI and improve research in the biopolymer domain can be achieved when developing AI tools for researchers who are not highly computationally skilled and also develop tools that can connect with laboratory devices appropriately. This will require the combined skills of computational scientists, experimental researchers and manufacturers of the tools, for effective integration to occur, which will in turn enhance drug delivery and other innovations.

Clinical Translation

The translation of AI-designed biopolymers into clinical practice is a multifaceted and highly challenging process, requiring not only scientific innovation but also rigorous regulatory compliance, extensive testing, and successful scaling for large-scale manufacturing. While AI models offer immense potential in designing biopolymers with optimized properties for drug delivery and therapeutic applications, the journey from concept to clinical application involves several key stages.

Firstly, AI-generated biopolymers must undergo rigorous preclinical testing to ensure their safety, efficacy, and biocompatibility. This includes *in vitro* studies to evaluate their interactions with drugs, cellular uptake, and degradation, followed by *in vivo* studies to assess pharmacokinetics, biodistribution, and toxicity in animal models. The data from these studies must demonstrate not only the therapeutic potential of the biopolymer but also its superiority over existing materials to justify moving into clinical trials.

Secondly, navigating the complex regulatory landscape is crucial. Regulatory bodies, such as the FDA or EMA, require detailed documentation on the safety, efficacy, and manufacturing processes of any new therapeutic material [30, 52]. For AI-designed biopolymers, this involves providing clear, reproducible evidence that AI-driven design methods lead to predictable and consistent outcomes. AI developers must work closely with regulatory agencies to ensure that these novel materials meet established guidelines for biocompatibility, toxicity, and manufacturing quality control.

Scaling up the production of AI-designed biopolymers for commercial use presents additional challenges. AI-optimized polymers, which may work effectively in the lab, must be manufactured consistently and in large quantities to meet clinical demands. This requires the development of robust manufacturing processes that maintain the integrity and performance of the biopolymer at scale. Ensuring that the AI-designed structures can be produced with the same precision and functionality on an industrial scale is essential for their successful integration into clinical practice.

Partnerships between AI developers, material scientists, pharmaceutical companies, and regulatory experts are essential to overcoming these challenges. Collaboration across these fields ensures that the biopolymers not only meet the necessary regulatory and safety standards but also offer viable solutions for large-scale production. By working together, stakeholders can streamline the path to market, making AI-designed biopolymers a reality in modern healthcare. An example of this collaborative effort is seen in the development of chitosan-based nanoparticles for drug delivery. A partnership between AI researchers, material scientists, and a pharmaceutical company successfully brought a novel nanoparticle formulation from AI prediction to clinical testing [40 - 42]. By optimizing the polymer composition, particle size, and drug loading, the AI model predicted a formulation with superior drug solubility and bioavailability. Through close collaboration, the team was able to scale up production and navigate regulatory hurdles, leading to the successful launch of a new drug delivery system in clinical trials.

Lack of AI-specific Regulatory Frameworks

The traditional biopolymer-based drug delivery systems have precise regulation methods under regulatory agencies such as the U.S. Food and Drug Administration (FDA), the European Medicines Agency (EMA), and other global regulating agencies.

But the use of AI in the design of the polymer introduces a complexity that is beyond the capacity of existing regulations [53].

One of the biggest challenges to AI-based polymer design is transparency and understandability, since these algorithms tend to be "black boxes" and the reasoning behind the produced polymer structures is hard to see. For instance, an AI model could suggest a new biodegradable polymer with better mechanical properties, but without reasoning how it is designed to this result, researchers may struggle to test or modify the design [54]. In addition, legal uncertainty and regulation guidelines is a major challenges, with the absence of AI regulations being a huge challenge in the adaptation of AI-generated content in the context of existing safety and compliance requirements. This issue is exacerbated by a lack of uniformity in scrutinizing the characteristics of biopolymers forecasted by AI, which constrains regulatory acceptability [55]. Incorporating AI-designed polymers into commercial products is complicated and unclear given the lack of an agreed-upon method for assessing critical properties like biodegradability, toxicity, and stability [56].

Complexity in Biopolymer Characterization and Validation

Artificial intelligence-based drug delivery systems (DDS) often introduce new biopolymers with complex molecular topologies, which sometimes do not align with current regulatory requirements, resulting in approval challenges. Physicochemical characterization is a significant challenge as AI-optimized biopolymers require detailed analysis of aspects such as mechanical strength, degradation rates, and polymer-drug interactions [57]. For instance, a nanoparticle-based polymer designed by AI can enhance drug solubility and controlled release, but without optimized and approved procedures, its long-term stability and degradation in physiological use and regulatory approval are uncertain [58]. Furthermore, ensuring Good Manufacturing Practice (GMP) compliance with AI-driven designs may give rise to concerns about batch-t--batch variations, potentially leading to performance variability [59]. Variability in synthesis processes or raw materials could lead to alterations in polymer behavior, requiring rigorous quality assurance. In addition, current regulatory frameworks must be revised to encompass AI-generated prediction models for assessing biopolymer stability and performance. For instance, if an AI model estimates a polymer's degradation profile in different storage conditions, specialists have to create guidelines for validating the predictions against empirical data to ensure safety and effectiveness in actual use [6].

Biocompatibility and Long-term Safety Considerations

AI-synthesized biopolymers need to undergo rigorous toxicological screening to ensure they meet safety standards, since even minor algorithmic modifications could significantly impact their biological interactions. Immunogenicity and biodegradability are of particular concern, since AI-induced structural modifications could lead to unwanted immune reactions or effects on how the polymer breaks down in the body. For instance, an AI-optimized biopolymer nanoparticle formulated for controlled release of a drug can be structurally modified to enhance solubility, but this modification could cause an immune reaction, necessitating further safety evaluation [29].

In addition, degradation products should be thoroughly assessed since AI-synthesized polymers can degrade into fragments with unseen or potentially harmful properties. For example, when a polymer degrades into acidic degradation products, it can cause local inflammation or tissue damage. Another major issue is long-term stability since AI-optimized DDS needs to maintain its integrity, efficacy, and safety in real storage and physiological conditions [35]. For a reference, a drug-releasing polymer that lasts for six months needs to be studied to ascertain whether temperature changes, humidity, or enzymatic exposure triggers premature degradation or loss of activity, thus ensuring its therapeutic reliability [60].

CHALLENGES IN LARGE-SCALE CLINICAL TRIALS

High Costs and Investment Risks

Transitioning from preclinical to clinical trials of AI-based biopolymer DDS poses significant economic hurdles due to the high cost of production, investor reluctance, and limited possibilities for financing [29]. AI-formulated biopolymers require complex and precise manufacturing processes at times, including multi-step polymerization or nanofabrication, increasing the cost of production. An AI-optimized machine learning-based nanoparticle carrier for cancer therapy may involve alien or specially designed monomers that increase large-scale production costs and render it economically unviable [61].

In addition, regulatory uncertainty regarding AI-driven biomaterials discourages investment since there are no well-established channels for clearance, thereby increasing concerns over prolong characterization and approval periods and delays. It is risky to invest in large-scale clinical trials, which could cost hundreds of millions of dollars, without the absence of clarity. Additionally, since conventional DDS systems present fewer perceived risks and established regulatory protocols, pharmaceutical firms and funding agencies might be inclined towards them as opposed to AI-driven counterparts. For instance, a drug company might rather invest in a well-established liposomal delivery technology than an AI-designed polymer whose long-term stability is uncertain, limiting funding for innovative AI-driven biopolymers [62]. These financial barriers must be overcome with improved legal regulations, incentives for AI-driven innovation, and intelligent collaborations among academics, businesses, and government agencies.

RECOMMENDATIONS FOR ADDRESSING REGULATORY AND CLINICAL TRIAL CHALLENGES

The successful clinical translation of AI-driven biopolymer-based DDS needs the development of AI-specific regulatory guidelines, collaboration, adaptive clinical trial designs, and improvements in AI transparency and validation.

Regulatory bodies need to offer specialized frameworks for biopolymer engineering with standardized evaluation criteria developed in association with AI developers [63].

Good Machine Learning Practices (GMLP) implementation will enhance the transparency of AI-driven DDS development [64]. Precompetitive partnerships and regulatory sandboxes can speed approvals by fostering collaborations among academia, industry, and regulatory agencies, while open-access AI databases and controlled testing environments will make review easier. Moreover, adaptive clinical trial designs incorporating AI for patient stratification, dosing adjustment, and *in silico* simulations based on digital twin models have the potential to lower costs and timelines substantially.

Real-world data integration will also facilitate faster regulatory approvals. AI transparency through explainable AI (XAI) methods, standardized validation routes, and blockchain implementation for secure and auditable records of drug formulation will establish regulatory confidence. It is possible to overcome the regulatory, technical, and ethical challenges of AI-based biopolymer DDS with a multi-faceted approach, but through strategic partnership and innovative validation strategies, AI can transform biopolymer engineering and the future of drug delivery.

POTENTIAL SOCIAL IMPLICATIONS OF THE WIDESPREAD USE OF AI-POWERED BIOPOLYMER ENGINEERING

The global implementation of AI-based biopolymer technology has profound societal consequences, covering ethical issues and economic instability, all the way up to environmental sustainability and medicine accessibility. If AI-optimized biodegradable implants are still unaffordable owing to proprietary algorithms and unique manufacturing processes, low-income nations will find it difficult to finance these innovative medical therapies. Furthermore, AI-driven biopolymer engineering also poses ethical issues related to job loss and workforce transformation. The computerization of material design, previously carried out through conventional trial-and-error laboratory methods, could decrease demand for certain research careers while increasing demand for AI and data science expertise.

On the brighter side, AI-driven biopolymer innovations have the potential to bring environmental sustainability, particularly reducing plastic waste in the form of eco-friendly, degradable materials. AI is able to fine-tune polymers with tailor-made breakdown rates to serve as substitutes for one-time plastics, for example, AI-created biodegradable food packaging that disintegrates under specific conditions without releasing harmful microplastics. But large-scale adoption of these alternatives relies on international regulations, consumer acceptability, and industrial preparedness to move away from traditional petroleum-based products. In addition, the ethical considerations of AI decision-making in medicine and biomaterial engineering need to be taken into account. AI-based biopolymer DDS can provide personalized treatment by adapting polymers to an individual's genetic makeup for the best drug delivery. Although technology offers incredible therapeutic precision, it does raise concerns of data privacy, algorithmic bias, and informed consent. Overall, while biopolymer engineering powered by AI holds the potential to revolutionize health, sustainability, and industry, it should be implemented with due consideration for its social, economic, and ethical consequences. Policymakers, scientists, and business leaders must work together to ensure equitable access, worker flexibility, and ethical AI stewardship to optimize benefits and limit unforeseen adverse consequences.

CONCLUSION

The use of AI in biopolymer engineering is a significant advancement in drug delivery systems, allowing for the development of tailored drug delivery polymers. In this way, polymorphic drug delivery systems pharmacies can not only be engineered for extended or controlled release but also be adjustable to patients' precise demands, thanks to AI modeling and analysis. Such personalization, in turn, increases the efficiency and safety of the therapies due to the ability to control a certain range of drug doses and their delivery site with much precision, thus minimizing adverse effects and improving the efficiency of the treatment. The future of biopolymer engineering with the use of AI is very enriching, but still there are many issues such as dealing with different regulatory requirements, producing high-quality products on a significant scale and clinical fruition. But as things develop, these limitations will be worked on with the unrelenting creativity and teamwork that researchers, clinicians and regulators have. There is a promising future in the use of AI-assisted biopolymer research based on the potential it has to enhance drug delivery systems, thereby increasing the efficiency of prescription drugs, improving the rates of patient recovery and enhancing health care across the world.

REFERENCES

[1] Arif ZU, Khalid MY, Sheikh MF, Zolfagharian A, Bodaghi M. Biopolymeric sustainable materials and their emerging applications. J Environ Chem Eng 2022; 10(4): 108159.
[http://dx.doi.org/10.1016/j.jece.2022.108159]

[2] Biswas A, Kumar S, Choudhury AD, *et al.* Polymers and their engineered analogues for ocular drug delivery: Enhancing therapeutic precision. Biopolymers 2024; 115(4): e23578.
[http://dx.doi.org/10.1002/bip.23578] [PMID: 38577865]

[3] Malheiro V, Duarte J, Veiga F, Mascarenhas-Melo F. Exploiting pharma 4.0 technologies in the non-biological complex drugs manufacturing: innovations and implications. Pharmaceutics 2023; 15(11): 2545.
[http://dx.doi.org/10.3390/pharmaceutics15112545] [PMID: 38004525]

[4] Gianti E, Percec S. Machine learning at the interface of polymer science and biology: how far can we go?. Biomacromolecules 2022; 23(3): 576-91.
[http://dx.doi.org/10.1021/acs.biomac.1c01436] [PMID: 35133143]

[5] Liang Y, Wei X, Peng Y, Wang X, Niu X. A review on recent applications of machine learning in mechanical properties of composites. Polym Compos 2024.
[http://dx.doi.org/10.1002/pc.29082]

[6] Tran H, Gurnani R, Kim C, *et al.* Design of functional and sustainable polymers assisted by artificial intelligence. Nat Rev Mater 2024; 9(12): 866-86.
[http://dx.doi.org/10.1038/s41578-024-00708-8]

[7] Chaput JC. Redesigning the genetic polymers of life. Acc Chem Res 2021; 54(4): 1056-65.
[http://dx.doi.org/10.1021/acs.accounts.0c00886] [PMID: 33533593]

[8] Wang W, Ye Z, Gao H, Ouyang D. Computational pharmaceutics - A new paradigm of drug delivery. J Control Release 2021; 338: 119-36.
[http://dx.doi.org/10.1016/j.jconrel.2021.08.030] [PMID: 34418520]

[9] Upadhya R, Kosuri S, Tamasi M, *et al.* Automation and data-driven design of polymer therapeutics. Adv Drug Deliv Rev 2021; 171: 1-28.
[http://dx.doi.org/10.1016/j.addr.2020.11.009] [PMID: 33242537]

[10] El Yousfi R, Brahmi M, Dalli M, *et al.* Recent advances in nanoparticle development for drug delivery: A comprehensive review of polycaprolactone-based multi-arm architectures. Polymers (Basel) 2023; 15(8): 1835.
[http://dx.doi.org/10.3390/polym15081835] [PMID: 37111982]

[11] Roberge C, Cros JM, Serindoux J, *et al.* BEPO®: Bioresorbable diblock mPEG-PDLLA and triblock PDLLA-PEG-PDLLA based *in situ* forming depots with flexible drug delivery kinetics modulation. J Control Release 2020; 319: 416-27.
[http://dx.doi.org/10.1016/j.jconrel.2020.01.022] [PMID: 31931049]

[12] Chen Y, Dai L, Shi K, Pan M, Yuan L, Qian Z. Cabazitaxel-loaded thermosensitive hydrogel system for suppressed orthotopic colorectal cancer and liver metastasis. Adv Sci (Weinh) 2024; 11(33): 2404800.
[http://dx.doi.org/10.1002/advs.202404800] [PMID: 38934894]

[13] Tao L, Chen G, Li Y. Machine learning discovery of high-temperature polymers. Patterns 2021; 2(4)
[http://dx.doi.org/10.1016/j.patter.2021.100225]

[14] Gupta AK, Choudhari A, Kumar A, *et al.* Composites for drug-eluting devices: emerging biomedical applications. In: Applications of Biotribology in Biomedical Systems 2024; 251-311.
[http://dx.doi.org/10.1007/978-3-031-58327-8_10]

[15] Sarker IH. Machine learning: algorithms, real-world applications and research directions. SN Comput Sci 2021; 2(3): 160.
[http://dx.doi.org/10.1007/s42979-021-00592-x]

[16] Anstine DM, Isayev O. Generative models as an emerging paradigm in the chemical sciences. J Am Chem Soc 2023; 145(16): 8736-50.
[http://dx.doi.org/10.1021/jacs.2c13467] [PMID: 37052978]

[17] Akkem Y, Biswas SK, Varanasi A. A comprehensive review of synthetic data generation in smart farming by using variational autoencoder and generative adversarial network. Eng Appl Artif Intell 2024; 131: 107881.
[http://dx.doi.org/10.1016/j.engappai.2024.107881]

[18] Khan MUA, Aslam MA, Bin Abdullah MF, Hasan A, Shah SA, Stojanović GM. Recent perspective of polymeric biomaterial in tissue engineering– a review. Mater Today Chem 2023; 34: 101818.
[http://dx.doi.org/10.1016/j.mtchem.2023.101818]

[19] Bao Z, Bufton J, Hickman RJ, Aspuru-Guzik A, Bannigan P, Allen C. Revolutionizing drug formulation development: The increasing impact of machine learning. Adv Drug Deliv Rev 2023; 202: 115108.
[http://dx.doi.org/10.1016/j.addr.2023.115108] [PMID: 37774977]

[20] Salma H, Melha YM, Sonia L, Hamza H, Salim N. Efficient prediction of *in vitro* piroxicam release and diffusion from topical films based on biopolymers using deep learning models and generative adversarial networks. J Pharm Sci 2021; 110(6): 2531-43.

[http://dx.doi.org/10.1016/j.xphs.2021.01.032] [PMID: 33548245]

[21]   Gao XJ, Ciura K, Ma Y, *et al.* Toward the integration of machine learning and molecular modeling for designing drug delivery nanocarriers. Adv Mater 2024; 36(45): 2407793.
[http://dx.doi.org/10.1002/adma.202407793] [PMID: 39252670]

[22]   De A, Ko YT. A tale of nucleic acid–ionizable lipid nanoparticles: Design and manufacturing technology and advancement. Expert Opin Drug Deliv 2023; 20(1): 75-91.
[http://dx.doi.org/10.1080/17425247.2023.2153832] [PMID: 36445261]

[23]   Shahverdi M, Seifi S, Akbari A, Mohammadi K, Shamloo A, Movahhedy MR. Melt electrowriting of PLA, PCL, and composite PLA/PCL scaffolds for tissue engineering application. Sci Rep 2022; 12(1): 19935.
[http://dx.doi.org/10.1038/s41598-022-24275-6] [PMID: 36402790]

[24]   Kumar R, Singh R, Kumar V, Ranjan N, Gupta J, Bhura N. On 3D printed thermoresponsive PCL-PLA nanofibers based architected smart nanoporous scaffolds for tissue reconstruction. J Manuf Process 2024; 119: 666-81.
[http://dx.doi.org/10.1016/j.jmapro.2024.04.008]

[25]   Bouakaz BS, Habi A, Grohens Y, Pillin I. Organomontmorillonite/graphene-PLA/PCL nanofilled blends: New strategy to enhance the functional properties of PLA/PCL blend. Appl Clay Sci 2017; 139: 81-91.
[http://dx.doi.org/10.1016/j.clay.2017.01.014]

[26]   Mathers A, Fulem M. Drug–polymer compatibility prediction *via* COSMO-RS. Int J Pharm 2024; 664: 124613.
[http://dx.doi.org/10.1016/j.ijpharm.2024.124613] [PMID: 39179010]

[27]   Pires FQ, Alves-Silva I, Pinho LAG, *et al.* Predictive models of FDM 3D printing using experimental design based on pharmaceutical requirements for tablet production. Int J Pharm 2020; 588: 119728.
[http://dx.doi.org/10.1016/j.ijpharm.2020.119728] [PMID: 32768526]

[28]   Aumklad P, Suriyaamporn P, Panomsuk S, Pamornpathomkul B, Opanasopit P. Artificial intelligence-aided rational design and prediction model for progesterone-loaded self-microemulsifying drug delivery system formulations. Sci Eng Health Stud 2024; pp. 24050002-.
[http://dx.doi.org/10.69598/sehs.18.24050002]

[29]   Vora LK, Gholap AD, Jetha K, Thakur RRS, Solanki HK, Chavda VP. Artificial intelligence in pharmaceutical technology and drug delivery design. Pharmaceutics 2023; 15(7): 1916.
[http://dx.doi.org/10.3390/pharmaceutics15071916] [PMID: 37514102]

[30]   Hu J, Wan J, Xi J, Shi W, Qian H. AI-driven design of customized 3D-printed multi-layer capsules with controlled drug release profiles for personalized medicine. Int J Pharm 2024; 656: 124114.
[http://dx.doi.org/10.1016/j.ijpharm.2024.124114] [PMID: 38615804]

[31]   Kumar K, Rani V, Mishra M, Chawla R. New paradigm in combination therapy of siRNA with chemotherapeutic drugs for effective cancer therapy. Curr Res Pharmacol Drug Discov 2022; 3: 100103.
[http://dx.doi.org/10.1016/j.crphar.2022.100103] [PMID: 35586474]

[32]   Noorain SV, Srivastava V, Parveen B, Parveen R. Artificial intelligence in drug formulation and development: applications and future prospects. Curr Drug Metab 2023; 24(9): 622-34.
[http://dx.doi.org/10.2174/0113892002265786230921062205] [PMID: 37779408]

[33]   Chaurawal N, Quadir SS, Joshi G, Barkat MA, Alanezi AA, Raza K. Development of fucoidan/polyethyleneimine based sorafenib-loaded self-assembled nanoparticles with machine learning and DoE-ANN implementation: Optimization, characterization, and *in-vitro* assessment for the anticancer drug delivery. Int J Biol Macromol 2024; 279(Pt 1): 135123.
[http://dx.doi.org/10.1016/j.ijbiomac.2024.135123] [PMID: 39208886]

[34]   Khadela A, Popat S, Ajabiya J, Valu D, Savale S, Chavda VP. AI, ML and other bioinformatics tools

for preclinical and clinical development of drug products. Bioinform Tools Pharm Drug Prod Dev 2023; 255-84.
[http://dx.doi.org/10.1002/9781119865728.ch12]

[35]   Philip AK, Shahiwala A, Rashid M, Faiyazuddin M. A Handbook of Artificial Intelligence in Drug Delivery. Academic Press 2023.

[36]   Sabbagh F, Muhamad II, Niazmand R, Dikshit PK, Kim BS. Recent progress in polymeric non-invasive insulin delivery. Int J Biol Macromol 2022; 203: 222-43.
[http://dx.doi.org/10.1016/j.ijbiomac.2022.01.134] [PMID: 35101478]

[37]   Adwani G, Bharti S, Kumar A. Engineered nanoparticles in non-invasive insulin delivery for precision therapeutics of diabetes. Int J Biol Macromol 2024; 275(Pt 1): 133437.
[http://dx.doi.org/10.1016/j.ijbiomac.2024.133437] [PMID: 38944087]

[38]   Singh AV, Rosenkranz D, Ansari MHD, *et al.* Artificial intelligence and machine learning empower advanced biomedical material design to toxicity prediction. Adv Intell Syst 2020; 2(12): 2000084.
[http://dx.doi.org/10.1002/aisy.202000084]

[39]   Serov N, Vinogradov V. Artificial intelligence to bring nanomedicine to life. Adv Drug Deliv Rev 2022; 184: 114194.
[http://dx.doi.org/10.1016/j.addr.2022.114194] [PMID: 35283223]

[40]   Asif HM, Zafar F, Ahmad K, *et al.* Synthesis, characterization and evaluation of anti-arthritic and anti-inflammatory potential of curcumin loaded chitosan nanoparticles. Sci Rep 2023; 13(1): 10274.
[http://dx.doi.org/10.1038/s41598-023-37152-7] [PMID: 37355723]

[41]   Fereig SA, El-Zaafarany GM, Arafa MG, Abdel-Mottaleb MMA. Boosting the anti-inflammatory effect of self-assembled hybrid lecithin–chitosan nanoparticles *via* hybridization with gold nanoparticles for the treatment of psoriasis: elemental mapping and *in vivo* modeling. Drug Deliv 2022; 29(1): 1726-42.
[http://dx.doi.org/10.1080/10717544.2022.2081383] [PMID: 35635314]

[42]   Dawoud MHS, Mannaa IS, Abdel-Daim A, Sweed NM. Integrating artificial intelligence with quality by design in the formulation of lecithin/chitosan nanoparticles of a poorly water-soluble drug. AAPS PharmSciTech 2023; 24(6): 169.
[http://dx.doi.org/10.1208/s12249-023-02609-5] [PMID: 37552427]

[43]   Yacoub AS, Ammar HO, Ibrahim M, Mansour SM, El Hoffy NM. Artificial intelligence-assisted development of *in situ* forming nanoparticles for arthritis therapy *via* intra-articular delivery. Drug Deliv 2022; 29(1): 1423-36.
[http://dx.doi.org/10.1080/10717544.2022.2069882] [PMID: 35532141]

[44]   Zhang Q, Zhang Y, Wan Y, Carvalho W, Hu L, Serpe MJ. Stimuli-responsive polymers for sensing and reacting to environmental conditions. Prog Polym Sci 2021; 116: 101386.
[http://dx.doi.org/10.1016/j.progpolymsci.2021.101386]

[45]   Pontrelli G, Toniolo G, McGinty S, Peri D, Succi S, Chatgilialoglu C. Mathematical modelling of drug delivery from pH-responsive nanocontainers. Comput Biol Med 2021; 131: 104238.
[http://dx.doi.org/10.1016/j.compbiomed.2021.104238] [PMID: 33618104]

[46]   Singh S, Sahani H. Current advancement and future prospects: biomedical nanoengineering. Curr Radiopharm 2024; 17(2): 120-37.
[http://dx.doi.org/10.2174/0118744471027437623112306313] [PMID: 38058099]

[47]   Bin Abu Sofian ADA, Sun X, Gupta VK, *et al.* Advances, synergy, and perspectives of machine learning and biobased polymers for energy, fuels, and biochemicals for a sustainable future. Energy Fuels 2024; 38(3): 1593-617.
[http://dx.doi.org/10.1021/acs.energyfuels.3c03842]

[48]   Li M, Wang R, Bao Q. Hyper-spectra imaging analysis of PLGA microspheres *via* machine learning enhanced Raman spectroscopy. J Control Release 2024; 367: 676-86.

[http://dx.doi.org/10.1016/j.jconrel.2024.01.071] [PMID: 38309305]

[49]   De A, Kang JH, Sauraj , Lee OH, Ko YT. Optimizing long-term stability of siRNA using thermoassemble ionizable reverse pluronic-Bcl2 micelleplexes. Int J Biol Macromol 2024; 264(Pt 2): 130783.
[http://dx.doi.org/10.1016/j.ijbiomac.2024.130783] [PMID: 38471603]

[50]   Zhu X. AI and Robotic Technology in Materials and Chemistry Research. John Wiley & Sons 2025.
[http://dx.doi.org/10.1002/9783527848836]

[51]   Patel RA, Webb MA. Data-driven design of polymer-based biomaterials: high-throughput simulation, experimentation, and machine learning. ACS Appl Bio Mater 2024; 7(2): 510-27.
[http://dx.doi.org/10.1021/acsabm.2c00962] [PMID: 36701125]

[52]   Jurczak KM, van der Boon TAB, Devia-Rodriguez R, *et al.* Recent regulatory developments in eu medical device regulation and their impact on biomaterials translation. Bioeng Transl Med 2025; 10(2): e10721.
[http://dx.doi.org/10.1002/btm2.10721] [PMID: 40060767]

[53]   Silva FM, Queirós C, Pereira M, *et al.* Precision fertilization: a critical review analysis on sensing technologies for nitrogen, phosphorous and potassium quantification. Comput Electron Agric 2024; 224: 109220.
[http://dx.doi.org/10.1016/j.compag.2024.109220]

[54]   Sharma A, Mukhopadhyay T, Rangappa SM, Siengchin S, Kushvaha V. Advances in computational intelligence of polymer composite materials: machine learning assisted modeling, analysis and design. Arch Comput Methods Eng 2022; 29(5): 3341-85.
[http://dx.doi.org/10.1007/s11831-021-09700-9] [PMID: 35035213]

[55]   Asl ZR, Rezaee K, Ansari M, Zare F, Roknabadi MHA. A review of biopolymer-based hydrogels and IoT integration for enhanced diabetes diagnosis, management, and treatment. Int J Biol Macromol 2024; 280(Pt 3): 135988.
[http://dx.doi.org/10.1016/j.ijbiomac.2024.135988] [PMID: 39322132]

[56]   Shafiq M, Thakre K, Pandurangan R, Lalitha RVS. Generative AI designs the next generation of smart materials from pixels to products. Int J Adv Manuf Technol 2025; 1-12.
[http://dx.doi.org/10.1007/s00170-025-14999-w]

[57]   Joshi A, Yadav P, Yadav C, Kanthaliya B, Verma KK, Arora J. Valorization strategies for agriculture residue: an overview. In: Arora J, Joshi A, Ray RC eds, Transforming Agriculture Residues for Sustainable Development From Waste to Wealth 2024; 21-43.
[http://dx.doi.org/10.1007/978-3-031-61133-9_2]

[58]   Jena GK, Patra CN, Jammula S, Rana R, Chand S. Artificial intelligence and machine learning implemented drug delivery systems: a paradigm shift in the pharmaceutical industry. J Bio-X Res 2024; 7: 0016.
[http://dx.doi.org/10.34133/jbioxresearch.0016]

[59]   Aguilar-Gallardo C, Bonora-Centelles A. Integrating artificial intelligence for academic advanced therapy medicinal products: challenges and opportunities. Appl Sci (Basel) 2024; 14(3): 1303.
[http://dx.doi.org/10.3390/app14031303]

[60]   Lohita B, Srijaya M. Novel technologies for shelf-life extension of food products as a competitive advantage: a review. In: Chakraborty R, Mathur P, Roy S eds, Food Production, Diversity, and Safety Under Climate Change 2024; 285-306.
[http://dx.doi.org/10.1007/978-3-031-51647-4_24]

[61]   Gholap AD, Uddin MJ, Faiyazuddin M, Omri A, Gowri S, Khalid M. Advances in artificial intelligence for drug delivery and development: A comprehensive review. Comput Biol Med 2024; 178: 108702.
[http://dx.doi.org/10.1016/j.compbiomed.2024.108702] [PMID: 38878397]

[62] Zaslavsky J, Bannigan P, Allen C. Re-envisioning the design of nanomedicines: harnessing automation and artificial intelligence. Expert Opin Drug Deliv 2023; 20(2): 241-57.
[http://dx.doi.org/10.1080/17425247.2023.2167978] [PMID: 36644850]

[63] Akinsemolu AA, Idowu AM, Onyeaka HN. Recycling technologies for biopolymers: current challenges and future directions. Polymers (Basel) 2024; 16(19): 2770.
[http://dx.doi.org/10.3390/polym16192770] [PMID: 39408479]

[64] Gadade DD, Kulkarni DA, Raj R, Patil SG, Modi A. Pushing boundaries: the landscape of ai-driven drug discovery and development with insights into regulatory aspects. Artificial Intelligence and Machine Learning in Drug Design and Development 2024; pp. 533-61.
[http://dx.doi.org/10.1002/9781394234196.ch17]

CHAPTER 8

Biopolymers in Stimuli-triggered and Enzyme-activated Drug Delivery Systems

Surbhi Gupta[1] and **Anubhav Anand**[2,*]

[1] *Ashoka Institute of Technology and Management, Varanasi, Uttar Pradesh, India*

[2] *Shri Ramswaroop College of Engineering and Management (Pharmacy), Lucknow, Uttar Pradesh, India*

Abstract: Biopolymers have emerged as a crucial component in advanced drug delivery systems due to their low toxicity, biodegradability, and inherent biocompatibility. Their ability to respond to specific stimuli or enzymes offers enhanced control over the release of therapeutic agents, making them highly valuable in precision medicine. Stimuli-triggered systems utilize external or internal signals such as pH, magnetic fields, temperature, redox potential, and light to precisely control drug release at targeted sites. For example, pH-sensitive biopolymers can release drugs in the acidic environment of tumors, while temperature-responsive systems adapt to local heat variations in tissues. This method ensures efficient drug delivery while minimizing side effects by targeting specific diseased areas. Enzyme-activated systems, on the other hand, rely on the presence of specific enzymes in the body to trigger the degradation of biopolymers and release the encapsulated drugs. These systems are beneficial in diseases such as cancer or infections, where overexpressed enzymes can be exploited for localized drug delivery. Biopolymers such as chitosan, dextran, alginate, and hyaluronic acid have been widely used in these enzyme-responsive systems, showing promising results in selective drug release. The precision and adaptability of biopolymer-based systems provide numerous benefits, including reduced systemic toxicity, controlled drug release, and enhanced therapeutic efficacy. However, challenges such as ensuring the stability of biopolymers in physiological conditions and scaling up production remain key obstacles. Ongoing research into novel biopolymers and more specific response mechanisms continues to push the boundaries of personalized medicine. Overall, biopolymer-based delivery systems signify a cutting-edge approach to achieving precise and controlled therapeutic interventions.

Keywords: Anti-bacterial, Anti-inflammatory, Biopolymer, Cancer, Enzyme-responsive systems, pH-responsive, Stimuli-responsive systems, Targeted drug delivery system.

---

[*] **Corresponding author Anubhav Anand:** Shri Ramswaroop College of Engineering and Management (Pharmacy), Lucknow, Uttar Pradesh, India; E-mail: anubhavanand2000@gmail.com

INTRODUCTION

Drug delivery is an interdisciplinary field combining expertise from chemistry, pharmaceutical sciences, medicine, and engineering. It relies on drug formulation, dosage, and delivery methods. Drug delivery systems (DDS) address the limitations of traditional approaches by improving solubility, extending drug activity, reducing side effects, and maintaining bioactivity. These systems also enhance bioavailability, increase drug absorption, maintain concentration through controlled release, and reduce side effects by targeting specific cells. Advances in material chemistry have driven the development of biodegradable carriers and biocompatible carriers with increased responsiveness. The creation of molecularly engineered biomaterials aims to address the challenges of delivering hydrophobic drugs and large biomolecules, such as proteins and nucleic acids. Smart materials that react to biological signals offer promising solutions for more efficient therapeutic delivery [1].

Biopolymers are a diverse and versatile class of compounds derived from natural sources or synthesized from biological materials. Like other polymers, they consist of repeating monomer units linked together. Due to their distinct characteristics, including biodegradability, availability, and the ability to modify their physicochemical properties, biopolymers are increasingly utilized in novel formulations. As the shift toward sustainable living gains momentum, biopolymers offer a promising platform for eco-friendly solutions. They have garnered significant attention for designing DDS, which delivers therapeutic agents precisely to diseased sites with minimal side effects. An ideal DDS could be one that controls drug release and protects drugs from degradation during transport. Both natural and synthetic polymers are considered suitable materials for developing DDS due to their biocompatibility and biodegradability [2].

Responsive polymers are materials that adapt their chemical or physical properties to respond to external stimuli.

Responsive polymers can alter characteristics like:

a. **Chain dimensions/size:** The polymer chains might expand or contract.
b. **Secondary Structure:** Modifications to the polymer chains' coiling or folding.
c. **Solubility:** The polymer's ability to dissolve in solvents may vary.
d. **Intermolecular Association:** Changes in the degree to which polymer molecules associate with each other.

These changes are typically driven by external factors such as:

- **Secondary Forces:** Like electrostatic interactions, hydrophobic effects, or hydrogen bonding.
- **Simple Chemical Reactions:** Acid-base reactions involving functional groups attached to the polymer.
- **Osmotic Pressure Differentials:** Changes in osmotic conditions caused by the external environment [3].

To reduce adverse effects, these biomaterials are designed to release therapeutic agents in response to biological, physical, or chemical triggers [4]. Enzymes, which are specific to certain organs and sites within the body, play a crucial role in this process. In enzyme-activated systems, drug release is triggered by enzymatic reactions. Prodrugs, liposomes, nanoparticles, and microparticles are some of the several types of enzymatic drug delivery systems. The key components in these systems include the drug, carrier, moiety, coating polymer, and ligand [5]. Stimuli-responsive polymers are particularly appealing due to their reactivity to both endogenous triggers, namely pH changes, redox conditions, and biomolecule recognition, as well as external stimuli like temperature and light [6]. These triggers can be categorized into intrinsic (internal) and extrinsic (external) stimuli. Intrinsic stimuli are linked to the pathological characteristics of diseased tissues, such as altered pH, redox environments, temperature, and the overexpression of specific biomolecules. External stimuli, conversely, are applied externally, namely heat, light, ultrasound, or magnetic fields [7].

Biopolymers are polymers derived from natural sources, either through chemical processes or synthesized entirely by living organisms. They consist of chain-like molecules made up of repeating chemical units obtained from renewable resources that are environmentally degradable. Biopolymers have been widely researched for pharmaceutical and biomedical applications owing to their diverse compositions, adaptable physical properties, and broad spectrum of potential products. They are generally categorized into two types based on their origin: natural biopolymers which include polysaccharides (such as alginate, chitin/chitosan, starch, and hyaluronic acid derivatives) and proteins (like silk, collagen, fibrin, gelatin, and soy) and synthetic polymers examples include poly(vinyl alcohol), poly(lactic acid), poly(glycolic acid), poly(caprolactone), poly(urethane), poly(propylene fumarate), and poly(hydroxybutyrate) [8].

Biopolymers offer several advantages due to their natural origin and inherent properties:

a. **Biocompatibility:** Biopolymers are typically more compatible with biological systems compared to synthetic polymers. This makes them well-suited for medical and pharmaceutical purposes, including drug delivery systems and tissue engineering, as they pose a lower risk of triggering adverse reactions in the body [9].

b. **Biodegradability:** One of the significant advantages of biopolymers is their ability to degrade naturally in the environment. This helps in reducing the accumulation of waste and pollution, making them an environmentally friendly alternative to conventional plastics. Biodegradable biopolymers break down into non-toxic byproducts, which can be absorbed by or contribute to the ecosystem [10].

These properties make biopolymers a promising choice in various fields, including medicine, environmental management, and sustainable manufacturing.

Endogenous stimuli like (acidic pH, enzymes, and GSH), exogenous stimuli (temperature, light, magnetic fields, and ultrasound), and both endogenous and exogenous stimuli, can be engineered to change size, charge, and/or stability to prolong blood circulation, accumulate at the diseased site, penetrate in tissues, internalize to target cells, and finally deliver and control drug release on demand [11].

STIMULI- RESPONSIVE SYSTEM

Stimuli-responsive drug delivery systems provide controlled and targeted drug release based on specific biological or external triggers. pH-responsive systems are highly effective for cancer and inflammatory diseases but require careful tuning to prevent premature drug release. Temperature-responsive systems work well in hyperthermia treatments but face challenges in maintaining precise *in vivo* control. Redox-responsive systems utilize elevated glutathione levels in tumors for the selective release of drugs, although individual variability affects their consistency. Enzyme-responsive systems offer exceptional specificity for cancer and infections but depend on enzyme overexpression. Photoresponsive systems enable precise, light-triggered drug release but have limited tissue penetration, whereas magnetic-responsive systems allow for noninvasive targeting, albeit requiring specialized equipment [12]. The ideal system depends on the medical application, and hybrid approaches combining multiple stimuli (*e.g.*, pH-temperature or enzyme-redox) are emerging to enhance efficacy. Future research should focus on improving biocompatibility, stability, and large-scale production for clinical translation.

pH-responsive Polymeric System

The pH of the human body varies depending on the location. For instance, the physiological pH of blood and other contact areas is around 7.4, whereas in the stomach, it ranges from 1 to 2.5. The pH of the colon ranges from 7.9 to 8.5, while that of the small intestine is 7.2 to 7.5. Polymeric materials with acidic or basic residues can undergo specific physicochemical changes when exposed to different pH. pH-sensitive polymers often incorporate cleavable bonds or functional groups that respond to changes in pH [13].

The altered pH in pathological conditions, such as cancer and inflammation, is often utilized to trigger the targeted release of drugs in specific biological organs (*e.g.*, the gastrointestinal tract) or compartments (*e.g.*, lysosomes or endosomes). In diseased areas, such as tumors and inflammatory tissues, a lower pH, ranging from 6.4 to 7.0, is observed in the environment surrounding cancer cells. Therefore, the use of substances sensitive to cancerous pH levels prevents unexpected drug release in normal cells and enables the selective delivery of drugs to cancer cells. Upon changes in pH, the cleavage of chemical bonding and/or a change in the charge balance between components, such as drugs, polymers, and DNA, by protonation can be used to control the release of drugs in various smart systems [14].

For polymeric nano-carriers to react to acidic tumor microenvironments, two important methodologies have been devised. The first involves polymers with functional groups that act as proton donors or acceptors. These polymers are deprotonated at physiological pH but protonate in acidic environments, altering their structure and hydrophobicity, which triggers drug release. pH-responsive nano-carriers are frequently made from polymers having ionizable groups, including amines or carboxylic acids [15].

In drug delivery systems, these polymers allow the creation of nanogels or hydrogel networks that respond to environmental pH changes through mechanisms such as bond cleavage, ionic interactions, or swelling. pH-responsive hydrogels adjust their volume based on the external pH, making them useful for controlled drug release. For instance, incorporating pH-cleavable bonds, such as acetal or hydrazone, into the hydrogel enables it to degrade in acidic environments, releasing the drug [16].

Temperature-responsive Polymeric System

Temperature-responsive materials are widely studied for smart drug delivery due to their ability to undergo phase transitions with temperature changes. Temperature can serve as an external stimulus (*e.g.*, heat applied externally) or be internal, such as in pathological conditions like tumors or inflammation. For instance, tumor tissues are slightly warmer than normal tissues (1–3°C warmer than normal tissue). These polymers exhibit a temperature-dependent phase transition, known as the critical solution temperature (CST), at which they become either soluble or insoluble. The shift from a more soluble to a less soluble state is recognized as the lower critical solution temperature (LCST), a feature utilized in most temperature-responsive polymers [17].

In DDS, these polymers allow the creation of a hydrogel that swells at higher temperatures and shrinks at lower ones. However, polymers with an LCST behave differently. Above the LCST, their solubility decreases, causing the hydrogel to contract, demonstrating negative temperature dependence.

The swelling behavior of these polymers is controlled by the hydrophilic/hydrophobic balance of the nano-carrier material. When the local temperature exceeds the LCST, the polymer chains become more hydrophobic and collapse, causing the release of encapsulated drugs, making them effective for targeted drug delivery [18].

Negative thermosensitive hydrogels consist of polymer chains rich in hydrophobic groups. Above the LCST, hydrophobic interactions dominate, leading the hydrogel to contract and reduce its surface area exposed to water, lowering the system's energy. Below the LCST, hydrogen bonding with water becomes more significant, causing the hydrogel to swell. Adjusting the ratio of hydrophilic to hydrophobic groups in the polymer changes the LCST. More hydrophobic groups result in a lower LCST, while more hydrophilic groups raise it [19].

Temperature-responsive injectable pre-hydrogels are designed to be liquid or semisolid at room temperature. After injection, they undergo gelation under physiological conditions, transitioning from sol to gel. This property enables precise control over drug release in biomedical applications [20].

Redox Potential-responsive System

Redox-responsive (ROS) polymeric materials are a promising tool for controlled drug delivery. These systems take advantage of the redox potential differences between the oxidizing extracellular space and the reducing intracellular space. In particular, the abundance of reduced glutathione (GSH) in cells serves as a key trigger for this process. While GSH has an extracellular concentration of approximately 2 μM, its intracellular concentration is around 10 mM, with tumor tissues often exhibiting higher levels than healthy tissues. This difference in redox potential can be used for targeted intracellular drug delivery [21].

The design of redox-responsive nano-carriers typically involves the use of disulfide bonds, which remain stable in the oxidizing extracellular environment. However, in the reduced intracellular space, these bonds are reduced to form thiol groups, leading to the disassembly of the nano-carriers and the release of the drug.

Drugs can be either encapsulated or conjugated to these disulfide-containing nano-carriers. Numerous redox-responsive nanomaterials have been created, including biodegradable polymers, dendritic polymers, and block copolymers. Notably, biodegradable nano-carriers improve *in vivo* metabolism and clearance, making them efficient platforms for drug delivery by incorporating disulfide bonds into the polymer structure [22].

GSH levels in some cancer tissues are reported to be four times higher than in normal tissues. GSH, along with thioredoxin, plays a key role in reducing disulfide bonds. This enhanced reduction at tumor sites can be leveraged to develop redox-responsive nanogels that react to the higher GSH concentrations in cancer tissues. One approach is to use crosslinkers containing disulfide bonds to form the hydrogel structure. Additionally, anti-cancer drugs can be attached to the hydrogel network through disulfide bonds [23].

In the high-GSH environment of cancer tissues, the disulfide bonds in the crosslinker break down, causing the hydrogel network to degrade. This process facilitates the targeted release of the encapsulated drug directly into the tumor site. Alternatively, when anti-cancer drugs are bound to the hydrogel's polymeric backbone *via* disulfide bonds, the reductive environment of cancer triggers drug release, minimizing damage to healthy cells [24].

In addition to disulfide bonds, diselenide bonds have been explored for redox-responsive nanogels. Diselenide bonds, with lower binding energies, are cleavable in both reductive (GSH) and oxidative (ROS) environments, providing an additional method for targeted drug delivery [25].

Photoresponsive Polymeric System

Light is an ideal remote stimulus for drug delivery systems because it requires no external additives or sensitive moieties, making it a reliable and efficient option. Photoresponsive self-assembled nanostructures have gained significant attention due to their ability to trigger a response rapidly and selectively when exposed to specific light wavelengths, including visible light, ultraviolet (UV), and near-infrared (NIR). These systems typically incorporate photo-sensitive moieties in polymers that undergo light-induced degradation or conformational changes. Light is used as a trigger for intelligent drug delivery, where the release of therapeutic agents can be controlled in an on/off manner by modulating the light's wavelength, intensity, and exposure time. The NIR window (650–900 nm) is particularly attractive for biomedical applications due to its minimal absorption by skin and tissue [26].

These polymers can be finely tuned by adjusting light wavelength, intensity, and exposure time, which affects both their physical properties (such as stiffness, shape, and degradation rate) and chemical properties (like surface hydrophilicity). Light-responsive polymers offer several advantages, including a broad range of applicable wavelengths, 4D control over material behavior, adjustable therapeutic light doses, and the ability to regulate *in vivo* responses within the optical tissue window [27].

Potential applications for these polymers include reversible photomechanical transduction, protein bioactivity alteration, tissue engineering, optical storage, viscosity control, and controlled release of active compounds. In drug delivery, light-responsive polymers play a significant role in controlled release and encapsulation [28].

These polymers can be categorized by the type of change they undergo:

- **Photo-isomerization (reversible):** Involves ring opening/closing or cis-trans transformations, affecting polarity or ionic structures (*e.g.*, azobenzene, spiropyran).
- **Photo-dimerization (reversible):** Involves photo-cross-linking through the [2+2] cycloaddition mechanism (*e.g.*, cinnamic ester, coumarin derivatives).
- **Photo-regulation (irreversible):** Involves irreversible Wolff rearrangement (*e.g.*, diazonaphthoquinone derivatives).
- **Photo-fragmentation (irreversible):** Causes hydrophobic-hydrophilic transformation through structural separation (*e.g.*, o-nitrobenzyl ester, coumarinyl ester).

Light-responsive nanomaterials are particularly attractive for biomedical applications due to their non-invasiveness, precise control for on-demand drug release, and high spatiotemporal resolution. In these systems, therapeutic agents are released only when exposed to external light sources (UV/visible or near-infrared) with the appropriate wavelength. Light-responsive linkers, which break upon irradiation, are often used in nano-carriers [29].

Photo-sensitive units in hydrogels can undergo cleavage, isomerization, or dimerization when exposed to light, leading to physical and chemical changes such as degradation, swelling, contraction, and modifications in the hydrogel's network. These changes enable photoresponsive hydrogels to alter their volume, either swelling or shrinking, allowing controlled drug release through water uptake or release [30].

These hydrogels adjust their properties in response to external light stimuli through three primary mechanisms. Firstly, hydrogels grafted with photo-sensitive groups can undergo phase transitions when exposed to photons of a certain energy, triggering the desired response. This is the most common mechanism for light-responsive behavior. Secondly, hydrogels containing photoactive molecules can generate ions upon light exposure, which interact with the hydrogel network or alter osmotic pressure, leading to swelling. Lastly, hydrogels with photo-sensitive compounds can change their properties by absorbing photon energy and responding to environmental changes [31].

Magnetic Field-responsive Polymeric System

Magnetic field-responsive systems hold significant promise for biomedical applications, including therapeutics, imaging, and diagnostics, due to their noninvasive nature, high penetration capability, absence of energy dissipation, and ease of control. These systems are typically synthesized by integrating ferromagnetic or paramagnetic materials into self-assembling structures and are extensively studied for applications like magnetically triggered drug delivery and magnetic resonance imaging. Magnetic field-responsive systems can be influenced by magnetic guidance, induced temperature increases, or a combination of both [32].

Magnetic nanoparticles are widely utilized in various biomedical fields, such as magnetic resonance imaging contrast agents, targeted drug delivery, hyperthermia, tissue repair, and magnetic field-assisted radionuclide therapy. Magnetic nanoparticles can be synthesized using several techniques, including microemulsion, solvothermal/hydrothermal methods, electrochemical approaches, and laser pyrolysis. Among the materials, iron oxide nanoparticles like maghemite (γ-Fe_2O_3) and magnetite (Fe_3O_4) are particularly favored due to their high biocompatibility. Optimal performance is achieved with nanoparticles smaller than 20 nm, at which point they exhibit superparamagnetism. These superparamagnetic nanoparticles, often composed of iron oxides and encased in biocompatible polymeric coatings, can be further functionalized with drugs, antibodies, proteins, or plasmids to enhance their therapeutic efficacy [33].

ENZYME-RESPONSIVE POLYMERIC SYSTEMS

Enzymes play key roles in biological and metabolic processes, and their upregulation is associated with several diseases, including cancer, thrombosis, inflammation, diabetes, and infections. This makes enzyme-responsive systems ideal for drug delivery, as they can encapsulate both hydrophilic and hydrophobic drugs. Their ability to respond to the elevated enzyme levels in disease environments offers a promising approach to targeted therapy, minimizing off-target effects and improving therapeutic outcomes [34].

A growing area of research in DDS involves the design of nanomaterials that respond to the selective catalytic action of enzymes. Enzymes, which are proteins, play a crucial role in biochemical reactions by catalyzing or initiating physiological processes. Enzyme concentrations and compositions vary significantly in different diseases, providing a promising approach for developing enzyme-sensitive DDS. Notably, inflammation and cancer cells overexpress enzymes like proteases, esterases, and glycosidases compared to normal cells. Enzymes are highly selective toward their substrates, enabling complex and precise chemical reactions under the moderate conditions found *in vivo*. Phospholipids, polymers, and inorganic nanomaterials are commonly used as carriers in enzyme-sensitive DDS. These nanomaterials contain enzyme-cleavable groups in their main chains or side groups [35].

Enzyme-responsive nano-carriers can undergo structural changes, such as swelling or degradation, when exposed to specific enzymes. Azo units, acting as cleavable linkers, provide various chemical pathways to enable enzyme-responsive behavior. Many cancer tissues overexpress certain enzymes, which can break specific bonds and serve as a trigger for drug release [36].

Enzyme-responsive hydrogels change their structure or properties when exposed to specific enzymes in their environment. These systems are valuable for the controlled release of bioactive components in targeted areas, such as the human gut, where enzymes like proteases and amylases are concentrated. They also have applications in the food industry for regulating fermentation processes, as microbes release different enzymes during various growth stages. Enzyme-responsive hydrogels are typically designed using polymers that can be hydrolyzed by specific enzymes, like proteases, which digest proteins, or amylases, which break down starches. Alternatively, the hydrogel may contain encapsulated substrates that react to enzymes in the environment. Common enzymes used in these hydrogels include phosphatases and matrix metalloproteinases. Compared to physically or chemically responsive hydrogels, enzyme-responsive hydrogels offer advantages such as high catalytic efficiency and excellent substrate specificity, making them effective for precise, controlled drug release or industrial applications [37].

DUAL-RESPONSIVE POLYMERIC SYSTEM

Stimuli-responsive polymeric nano-carriers have gained significant attention in biomedical fields for controlled drug delivery. They are designed to encapsulate therapeutic agents while ensuring specific delivery to targeted sites and times [38].

Single-stimuli-responsive hydrogels and advanced hydrogels with dual and multi-stimuli responsiveness have been developed. These hydrogels can be engineered to respond to two or more types of stimuli, enhancing the selectivity and efficacy of drug release. Recent advancements include dual and multi-responsive systems, including pH-temperature, pH-redox, temperature-redox, pH-light, and protease-redox-pH responsive systems [39]. These innovations aim to improve the response rate and efficacy of drug delivery by leveraging multiple stimuli for precise control over drug release.

APPLICATION OF BIOPOLYMERS

Many papers have been published on the use of biopolymers in different stimuli-responsive and enzyme-responsive systems for Cancer, anti-microbial, and anti-inflammatory.

For Cancer

This delivery system offers several advantages over conventional therapies, as nanomedicine can sometimes lead to off-target drug distribution, potentially harming non-target areas. Tumor microenvironment-responsive nanomedicine has shown tremendous promise for the efficient, safe, and precise delivery of therapeutics by leveraging stimuli specific to the tumor microenvironment.

This chapter aims to enhance understanding of the various strategies employed to integrate stimuli-responsive elements into polymeric nano-carriers for cancer therapy, rather than providing a comprehensive overview of the entire field of stimuli-responsive polymeric nano-carriers. Particularly, it focuses on biopolymers with stimuli-responsive mechanisms related to pH, redox potential, enzymes, light, and temperature, with discussions assisted by highlight recognition from the literature [40].

Different types of polymers, delivery systems, techniques, and mechanisms are employed in cancer treatment to enhance drug targeting, bioavailability, and therapeutic efficacy. These include polymer-based nanoparticles, micelles, hydrogels, and conjugates that facilitate controlled and targeted drug delivery. Table **1** summarizes the key polymers, delivery systems, and their mechanisms used in cancer therapy.

For Anti-microbial/Anti-bacterial

Smart polymer-based carriers, designed to respond to endogenous bacterial microenvironments such as acidic pH, bacterial toxins, specific enzymes, and elevated ROS levels, as well as external stimuli like light, have been extensively utilized in wound dressings. These carriers enable the controlled release of anti-bacterial agents or activation of anti-microbial properties, significantly enhancing treatment outcomes. This progress is attributed to the unique physicochemical characteristics of nanomaterials and the rapid advancements in nanotechnology [69].

Table 1. Different types of polymers, delivery systems, techniques, and mechanisms are utilized in the treatment of Cancer disease.

Sr. No.	API	Polymer	Delivery System	Techniques/Methods for Drug Encapsulation/Entrapment	Mechanism of Cancer-Targeted Drug Delivery	Description	Site of Cancer	References
1	5-Fluorouracil	Alginate	3,4-dihydroxybenzaldehyde metal organic framework 5-Fluorouracil @hydrogel as a dual drug delivery system (DHBD@MOF/5-FU@hydrogel as a DDDS)	-	pH-responsive	-UiO-66-NH$_2$ and DHBD underwent a Schiff base reaction to produce a MOF-based prodrug. -With just 1.31% DHBD release at pH 1.2 after 2.5 hours, the hydrogel-encapsulated DHBD@MOF and 5-FU ensured controlled release without burst under acidic stomach conditions. -At different pH values, including simulated stomach fluid at pH 1.2, 4.5, 7.4, and 6.5, the hydrogel successfully regulated the release of DHBD and 5-FU. After 8.5 and 24 hours, respectively, DHBD@MOF/5-FU@hydrogel released 10.60% of DHBD in the gut (pH 7.4) and 41.68% in the colorectum (pH 6.5). After 8.5 and 24 hours, the residual DHBD@MOF prodrug had an inactive rate of 89.40% and 58.32% in acidic tumours, respectively. - Significant concentration-dependent cytotoxicity was demonstrated by the DHBD@MOF@hydrogel against SW480 colon cancer cells.	Colon cancer	Binaeian *et al.* (2024) [41]
2	Quercetin	-	Hyaluronic acid-decorated pH and redox dual-stimuli responsive poly (methacrylic acid)/mesoporous organosilica nanoparticles with a core-shell structure for controlled drug release of quercetin. (QUE-loaded HA-MON/PMAA)	Polymerization method	pH/redox dual-stimuli responsive	*In vitro* assays showed that the nanomaterials are biocompatible, with drug-loaded targeted nanoparticles exhibiting higher cytotoxicity against MCF-7 breast cancer cells compared to free drugs and non-targeted carriers. This is attributed to enhanced cellular uptake *via* CD44 receptors.	Breast cancer	Ghalehkhondabi *et al.* (2024) [42]
3	Doxorubicin	-	Self-assembly of a cysteine-cored diphenylalanine-appended tetrapeptide (SN) into hollow spheres	-	Redox-responsive	In the presence of elevated glutathione (GSH) concentrations, which are common in cancer cells, the disulfide bond promotes selective breakdown. Doxorubicin (Dox), an anti-cancer medication, was successfully encapsulated (68.72%), and its redox-responsive, GSH-dependent release within malignant cells was demonstrated. When compared to non-malignant cells (50 mM), SN-Dox demonstrated a 20-fold lower effective dose (2.5 mM) for impairing breast cancer cell survival. SN-Dox was able to initiate DNA damage signalling and induce apoptosis, just like the unencapsulated drug.	Breast cancer cell	Nayak *et al.* (2024) [43]
4	5-Fluorouracil	Gelatin	Graphitic carbon nitride (g-C$_3$N$_4$) coated with Gelatin (G)/Polyethylene glycol (PEG) biopolymers was prepared by W/O/W double emulsion technique using olive oil and Span 80 surfactants. 5-FU was successfully loaded into the nanovehicle.	Double emulsion technique	pH-sensitive	-They created a new graphitic carbon nitride/gelatin/polyethylene glycol nano-carrier. -High encapsulation and drug loading efficiency levels were attained. -A pH-sensitive and regulated release of the medication 5-FU was noted. -Compared to the free medication, the loaded biocompatible nano-carrier displayed more apoptotic cells. -The treatment of breast cancer may benefit greatly from this sustainable nano-carrier. -While anomalous transport predominates at neutral pH, kinetic investigations indicated a diffusion-controlled mechanism of 5-FU release from the nano-carrier in an acidic environment. On MCF-7 breast cancer cells, the 3-(4,5-dimethylthiazol-2-yl)-2,5-diphenyltetrazolium bromide (MTT) assay demonstrated that G/PEG/g-C$_3$N$_4$ significantly cytotoxically affected the cells. According to flow cytometry, G/PEG/g-C$_3$N$_4$/5-FU generated a notably greater proportion of apoptotic cells than free 5-FU, while the free drug showed a higher proportion of necroptotic cells. As a pH-sensitive DDS against breast cancer cells, this innovative bio-nanocomposite exhibits a lot of promise.	Breast cancer	Rahmani *et al.* (2024) [44]

Sr. No.	API	Polymer	Delivery System	Techniques/Methods for Drug Encapsulation/Entrapment	Mechanism of Cancer-Targeted Drug Delivery	Description	Site of Cancer	References
5	5-Fluorouracil	Chitosan	Composed of graphitic carbon nitride (g-C₃N₄), hydroxyapatite (HAp), and chitosan (CS) and loaded with 5-FU. Nanosheets of g-C₃N₄ were added to CS/HAp hydrogel.	Emulsification method	pH sensitivity	The MTT technique demonstrated the biocompatibility of CS/HAp/g-C₃N₄ against MCF-7 cells, and flow cytometry verified this. In MCF-7 cells, CS/HAp/g-C₃N₄@5-FU produced the highest apoptosis rate, demonstrating the effectiveness of the nano-carrier in destroying cancer cells. These findings suggest that CS/HAp/g-C₃N₄@5-FU may be a promising medication for the treatment of cancer cells.	Breast cancer	Ahmari *et al.* (2024) [45]
6	5-Fluorouracil	Chitosan/ Agarose/ γ-Alumina	The addition of γ-alumina nanoparticles to the polymeric network enhanced the hydrogel's mechanical qualities and made drug encapsulation much easier. To further stabilise the nano-carrier's structure and produce a prolonged release profile for 5-FU, the nanocomposite was also trapped in a water-in-oil-in-water emulsion system.	Crosslinking agent (glyoxal 0.02% (v/v))	pH-sensitive feature	-Electrostatic interactions between 5-FU and γ-alumina, which enhance drug entrapment. -The nano-carrier releases more 5-FU in acidic conditions compared to neutral ones, showing potential for reducing chemotherapy side effects. - Cellular experiments, including MTT assays and flow cytometry with the MCF-7 cell line, demonstrated that 5-FU-loaded nanoemulsions are more effective in eliminating cancer cells and inducing apoptosis than free 5-FU.	Breast cancer	Bayat *et al.* (2023) [46]
7	Doxorubicin	Gelatin-alginate	Gelatin-alginate nanocomposites containing doxorubicin-loaded niosome (Nio-DOX@GT-AL)	-	pH-responsive	-The anti-cancer drug regulated gene expression level in MCF-7 cells, - Acidic pH causes a higher rate of DOX release than neutral pH. -The enhanced anti-cancer effect resulted from gene regulation and increased reactive oxygen species production in cancer cells.	Breast cancer	Zaer *et al.* (2023) [47]
8	Doxorubicin	Gelatin	The doxorubicin-loaded gelatin-based nanocluster (DOX-icluster) was electrostatically assembled from folic acid and dimethylmaleic anhydride modified gelatin (FA-GelDMA) and small-sized DOX-loaded NH₂ modified hollow mesoporous organosilicon nanoparticles (DOX-HMO--NH₂).	-	Stimuli-responsive drug delivery	At neutral pH, the DOX-icluster was initially around 199 nm in size. 48 nm of positively charged DOX-HMON-NH₂ was released when matrix metalloproteinase (MMP-2) dissolved the DMA bond of FA-GelDMA and broke down the gelatin after it had accumulated in tumour tissue. This allowed for easier penetration and cell internalisation.	Antitumor therapy	Xiao *et al.* (2023) [48]
9	5-Fluorouracil	-	Polyvinylpyrrolidone (PVP), carboxymethyl cellulose (CMC), and γ-alumina were combined to create an H-sensitive nanocomposite utilising the water in oil in water (W/O/W) double emulsion technique. 5-Fluorouracil has been used with the manufactured emulsion.	-	pH-sensitive	PVP /CMC/CMC/γ-alumina/5-FU exhibits late apoptosis and increased cytotoxicity. Based on the results of the MTT experiment on normal cell lines (L929), which showed cell viability above 90%, the biocompatibility and biosafety of the synthesized nano-carrier have been verified. Additionally, due to the porosity of PVP/CMC/γ-alumina, this nano-carrier has a large specific surface area and is more susceptible to environmental variables like pH. These results suggest that the new pH-sensitive PVP/CMC/γ-alumina nanocomposite may be a viable option for drug administration, particularly in the treatment of cancer.	Cancer therapy	Shamsabadipour *et al.* (2023) [49]
10	5-Fluorouracil	Chitosan/agarose/graphene oxide	Chitosan/agarose/graphene oxide 5-fluorouracil (CS/AG/GO/5-FU)	Emulsification technique	pH-sensitive	- A sustained drug release profile was observed in the acidic medium, with nearly 100% of the drug released within two days. -The cell viability against MCF-7 breast cancer cell (BCC) lines was approximately 23% as well. -The suggested drug delivery system's anti-cancer performance, pH sensitivity, and prolonged drug release profile were all found to be enhanced by the encapsulation of 5-FU medication within the nano-carrier. -For the treatment of breast cancer, the developed nano-carrier offers a great potential for pH-selective release of 5-FU (as a model medication) in a regulated manner.	Breast cancer therapy	Rajaei *et al.* (2023) [50]
11	5-Fluorouracil	-	Halloysite nanotube (HNT) coated with carboxymethyl cellulose (CMC)/polyethylene glycol (PEG) hydrogel for controlled delivery of 5-Fluorouracil (CMC/PEG/HNT nanocomposite loaded with 5-FU)	Emulsification technique	pH-responsive	When compared to the physiological medium, *in-vitro* drug release data demonstrated better and longer-lasting 5-FU distribution in an acidic environment, supporting the created nano-carrier's pH sensitivity. The 5-FU-loaded nanocomposite was shown to have significant cytotoxicity on MCF-7 breast cancer cells by flow cytometry and MTT assays; however, it was not toxic to L929 fibroblast cells. The pH-sensitive release of anti-cancer medications may be facilitated by the nanocomposite created here.	Breast cancer	Ghasemizadeh *et al.* (2023) [51]

(Table 1) cont.....

Sr. No.	API	Polymer	Delivery System	Techniques/Methods for Drug Encapsulation/Entrapment	Mechanism of Cancer-Targeted Drug Delivery	Description	Site of Cancer	References
12	5-Fluorouracil and Gemcitabine hydrochloride	Chitosan	5-Fluorouracil@ Materials Institute Lavoisier (MIL)-100 and 5-Fluorouracil-GEM@MIL-100 were then coated with chitosan, sequentially chelated with iron(II) and conjugated with quercetin, eventually obtaining a multifunctional MIL-100 nanocarrier.	Impregnation approach	pH-sensitive releases	5-FU and gemcitabine are more hydrophilic in acidic conditions, leading to significantly higher drug release efficiency in the acidic tumor microenvironment compared to neutral tissues at pH 5.0.	Breast cancer	Resen *et al.* (2022) [52]
13	Doxorubicin hydrochloride, mitoxantrone	-	di(ethylene glycol) methyl ether methacrylate Poly(ethylene glycol) methyl ether methacrylate hyaluronic acid methacrylate di(ethylene glycol) diacrylate Nanogels (MEO₂MA-OEGMA-MeHa-DEGDA NGs)	Simple precipitation polymerization, crosslinks	pH and Enzyme-responsive	Increasing the amount of the hyaluronidase enzyme (HAdase), which is typically overexpressed in cancerous environments, could improve the controlled enzymatic degradation of NGs. According to the results of the MTT experiment, the NGs showed cytotoxic action against human MCF-7 breast cancer cells, A278 ovarian cancer cells, and cytocompatibility towards MCF-10A and HOF healthy cells.	Human MCF-7 breast cancer cells, the A278 ovarian cancer cells	Liwinska *et al.* (2022) [53]
14	Violacein	Chitosan	Lipase Encapsulated in 3D-Bioprinted Chitosan-Hydroxypropyl Methylcellulose Matrix	-	Enzymatic Active Release, pH responsive	- The printing's final violacein encapsulation efficiency exceeded 90%. Violacein rapid release was around 20% higher in the mesh with NLC-lipase than in the mesh without the enzyme, according to the kinetic release of the biodye at pH = 7.4. Nonetheless, the profiles of the two Violacein kinetic releases were similar at pH = 5.0, where the lipase is inactive. -Cytotoxic investigations in A549 and HCT-116 cancer cell lines demonstrated the biological synergistic activities of lipase and violacein, as well as the strong anti-cancer activity of the matrix mediated by mucoadhesive chitosan.	Anti-cancer activity	Berti *et al.* (2022) [54]
15	5-Fluorouracil	-	5-Fluorouracil-loaded azobenzene-capped mesoporous silica (5-Fu loaded and azo-capped MS)	-	Enzyme-responsive drug delivery	-To function as an enzyme-responsive drug delivery system, the pores were gated by an azobenzene derivative. -The 5-Fu release from MS and the azo-capped molecule was evaluated, and a regulated release was noted when sodium dithionite was employed as a reducing agent and an azoreductase enzyme mimicker.	Anti-cancer drug	Farjadian *et al.* (2022) [55]
16	Doxorubicin	-	DOX-loaded and DOX + indocyanine green (ICG)-loaded hydrogels	-	Reactive oxygen species (ROS) can be generated *in situ* by exposing near-infrared (NIR) light to a photosensitizer	-By taking advantage of ROS reversible cleavage of diselenide bonds, diselenide-based drug delivery devices can be designed to react to NIR light. When a photosensitiser, such as ICG, is exposed to near-infrared light, ROS are produced. NIR light is useful because it can target without damaging healthy tissues due to its deep tissue penetration and noninvasive nature. -ICG-containing hydrogels generated ROS when exposed to NIR light, which resulted in a burst release of more than 50% doxorubicin during the first four hours and a steady release after that. -These hydrogels' biocompatibility was validated by *in vitro* experiments, which also revealed that they were not harmful to HT-29 colorectal cancer cells or HFF-1 fibroblast cells. Both DOX-loaded and DOX + ICG-loaded hydrogels showed comparable anti-cancer effects to free DOX and efficiently suppressed HT-29 cell metabolism, suggesting that they could be used in minimally invasive cancer treatment.	Colon cancer	Gulfam *et al.* (2022) [56]
7	Indocyanine green (ICG) and doxorubicin	Gelatin	Gelatin nanoparticle indocyanine green (ICG) and doxorubicin GNP-DOX/ICG	-	Photothermal effects	- A photothermal/MMP-2 dual-responsive nanosystem was created by combining the chemotherapeutic doxorubicin (DOX) with the photothermal agent indocyanine green (ICG) in gelatin nanoparticles (GNP-DOX/ICG). In the fight against breast cancer, this chemophotothermal therapy provides "on-demand" medication release and synergistic therapeutic efficacy.	Breast cancer	Chen *et al.* (2021) [57]

(Table 1) cont.....

Sr. No.	API	Polymer	Delivery System	Techniques/Methods for Drug Encapsulation/Entrapment	Mechanism of Cancer-Targeted Drug Delivery	Description	Site of Cancer	References
18	Doxorubicin hydrochloride	-	-	-	pH Responsive	-Tamarind gum-co-poly (acrylamidoglycolic acid) (TMGA)-based semi-interpenetrating polymer hydrogels were created *via* straightforward free radical polymerisation with potassium persulfate acting as an initiator and bis [2-(methacryloyloxy)ethyl] phosphate acting as a crosslinker. Furthermore, these hydrogels served as templates for the environmentally friendly synthesis of silver nanoparticles (13.4 ± 3.6 nm in diameter, TMGA-Ag) utilising Terminalia bellirica leaf extract as a reducing agent. -Cell viability and cell cycle analysis to evaluate the impact of therapy on HCT116 human colon cancer cells. The anti-bacterial efficacy of TMGA-Ag hydrogels against *Klebsiella pneumonia* and *Staphylococcus aureus* was investigated. Lastly, the results show that TMGA and TMGA-Ag are potential options for the inactivation of harmful bacteria and the delivery of anti-cancer drugs, respectively.	Colon cancer	Nagaraja *et al.* (2021) [58]
19	Curcumin	-	Polymer–curcumin conjugates (PCC$_8$) based on glycidyl azide polymer (GAP)	-	pH-responsive	By comparing the medication's release under acidic (pH 5.6) and normal physiological settings (pH 7.4 and pH 5.6), it was shown that the PCCs released the drug quickly under acidic conditions, reaching a 64% release rate in the first 30 hours. It was anticipated that the stimuli-responsive PCCs would exhibit a moderate degree of drug release and satisfactory stability at a pH of 7.4. These findings demonstrated that the PCCs can avoid drug leakage over undesirable tissues and organs and transfer and release conjugated pharmaceuticals in sufficient concentrations to the targeted site of action, such as tumour tissue. Therefore, PCCs may be used as an intelligent drug delivery system to deliver curcumin or other anti-cancer drugs into cells efficiently.	Cancer therapy	Rashidzadeh *et al.* (2021) [59]
20	Cabazitaxel	-	-	-	Enzyme responsive	MMP-2 is overexpressed in the tumour microenvironment of prostate cancer and can cleave enzyme-responsive peptides.	Prostate cancer	Barve *et al.* (2020) [60]
21	Cisplatin	-	Matrix metalloproteinase-2 (MMP-2)-responsive mesoporous silica nanoparticles (MSNs)	-	Enzyme-triggered drug release	This delivery strategy produced collagen-coated MSNs (Cis-col-MSN) by coating collagen on the surface of Cis-loaded MSNs (Cis-MSN) to create a capping layer. Collagen is freed from MSN holes due to the overexpression of the MMP-2 enzyme in the cancer microenvironment, which leads to both controlled and enhanced medication release. MMP-responsive experiments have shown increased enzyme-triggered medication release. Research on cellular uptake and cytocompatibility in A549 adenocarcinomic lung cancer cell lines indicates that this nano-carrier may be endocytosed successfully in a day and has good biocompatibility with the cells. The cytotoxicity results of cis-col-MSN demonstrated dose-dependent toxicity. When the MMP-2 enzyme was added to the cell growth media at gradually higher levels, the efficiency of Cis-col-MSN was considerably boosted. The formulation's effectiveness was credited with greatly increasing apoptosis, cell cycle arrest, and reactive oxygen species. Cis-col-MSN is anticipated to offer a practical method for building an "on-demand" smart drug delivery system that will only administer a therapeutic payload at the tumour location.	Cancer Therapy (lung cancer cells)	Vaghasiya *et al.* (2020) [61]
22	Doxorubicin	-	Casein and *N*-isopropylacrylamide (NIPA) can self-assemble into biodegradable micelles of approximately 80 nm at physiological conditions.	-	Enzymatic and pH-responsive	-Free radical polymerisation was used to create casein/NIPA graft copolymers. -Graft polymers can form micelles through thermal self-assembly at physiological pH. -Casein acted as an enzymatic and pH-responsive block in the micelles. -Doxorubicin (Dox) delivery from micelles was activated by trypsin and acidic pH. -Dox was successfully transported by microbes to the MD231 cells' nucleus, where it was destroyed. In the breast cancer cell line MDA231, cellular uptake and cytotoxicity tests demonstrated Dox's effective transport to the nucleus and antiproliferative properties.	Breast cancer	Cuggino *et al.* (2020) [62]

(Table 1) cont.....

Sr. No.	API	Polymer	Delivery System	Techniques/Methods for Drug Encapsulation/Entrapment	Mechanism of Cancer-Targeted Drug Delivery	Description	Site of Cancer	References
23	5-Fluorouracil	Dextran	Dextran and Dextran aldehyde (Dex and Dex-CHO) coated silica aerogels	-	Enzyme-triggered drug delivery systems	The quantity of 5-FU released from uncoated silica aerogels in these media was 86.4%, but the amount released from Dex and Dex-CHO coated silica aerogels in simulated gastric and intestinal fluids was 1.7% and 3.4%, respectively. However, in the colonic medium containing dextranase, 5-FU release within 12 hours, immediately following dextranase degradation, was 24% and 13.4% from Dex and Dex-CHO coated silica aerogels, respectively. MTT assay findings of untreated, amine-modified, Dex, and Dex-CHO coated silica aerogels did not exhibit any significant cytotoxic effect on Caco-2 cells. However, MTT assay findings of 5-FU loaded silica aerogels (unmodified, amine-modified, Dex, and Dex-CHO coated) demonstrated a decrease in the viability of Caco-2 cells. These findings show that Dex and Dex-CHO coated silica aerogels are biocompatible nanoparticles that are effective at delivering drugs to the colon area through an enzyme-triggered mechanism that is not impacted by the upper gastrointestinal tract.	Colorectal adenocarcinoma	Tiryaki *et al.* (2020) [63]
24	Erlotinib	Chitosan	Erlotinib loaded Methotrexate chitosan copolymer magnetic nanoparticles (ETB-loaded MTX CSC@MNPs)	-	pH- and thermo-responsive	The rate of drug release may decrease because the steric and hydrophobic interactions increase at 40'C as opposed to 37'C. In contrast, pH-responsive DDS minimises drug release and, consequently, cytotoxicity in healthy tissue with normal physiological pH while producing effective drug release in the acidic tumour microenvironment. The rate of drug release may decrease because the steric and hydrophobic interactions increase at 40'C as opposed to 37'C. In contrast, pH-responsive DDS minimises drug release and, consequently, cytotoxicity in healthy tissue with normal physiological pH while producing effective drug release in the acidic tumour microenvironment.	Ovarian cancer	Fathi *et al.* (2020) [64]
25	Doxorubicin	Chitosan	Biodegradable chitosan (CS) was bonded to hollow mesoporous silica spheres (HMSS) *via* cleavable azo linkages (HMSS-N=N-CS). Doxorubicin (DOX) (HMSS-N=N-CS/ DOX)	-	Enzyme-responsive	Following incubation of HMSS-N=N-CS/DOX with the colonic enzyme mixture, there was a noticeable increase in the cellular uptake of DOX. The results of the cellular uptake showed that the preincubation of HMSS- N=N-CS/DOX with a colonic enzyme combination clearly boosted the uptake of DOX. The cytotoxicity of the HMSS-N=N-CS/DOX group incubated with colon enzymes was higher, and its IC50 value was three times lower than that of the group without colon enzymes.	Colon-specific delivery	Cai *et al.* (2020) [65]
26	Curcumin	Starch	Hydroxyethyl starch curcumin nanoparticles (HES-CUR NPs)	-	Acid-responsive release	Curcumin and food-derived hydroxyethyl starch were combined to form nano-micelles, which greatly enhanced curcumin's solubility and stability. They also released curcumin in an acid-responsive manner, demonstrating greater antioxidant and anti-cancer activity than curcumin. They also had excellent colloidal and storage stability in addition to their acid-responsive release mechanism.	antioxidant and anti-cancer	Chen *et al.* (2020) [66]
27	Doxorubicin	-	Doxorubicin (DOX) was conjugated onto PEGylated keratin (PK) using both bioreducible disulphide linkage and acid-cleavable hydrazone bond in a series connection manner to create keratin-based drug-protein conjugate prodrug nanoparticles (PK-SS-Hy-D NPs).	-	pH and reduction dual-responsive triggered release	-Prodrug nanoparticles based on keratin from chicken feathers were created for DOX delivery. Disulphide and hydrazone linkages were used to conjugate DOX in a sequential connection mode. -The PK-SS-Hy-D NPs demonstrated outstanding dual-triggered DOX release related to pH and reduction. -The PK-SS-Hy-D NPs were more effective against tumours than the free DOX. -The controlled release profiles showed that the suggested drug-protein conjugate nanoparticles induced the release of DOX in a pH and reduction dual-responsive manner, with a modest premature drug leakage of 5.5% in the physiological medium simulation. According to the *in vitro* tests, the suggested prodrug nanoparticles may carry DOX into the cell nucleus with a higher level of antitumor activity. It is anticipated to be a viable option for tumour chemotherapy in the future with fewer toxic side effects.	Tumor	Zhang *et al.* (2019) [67]

(Table 1) cont.....

Sr. No.	API	Polymer	Delivery System	Techniques/Methods for Drug Encapsulation/Entrapment	Mechanism of Cancer-Targeted Drug Delivery	Description	Site of Cancer	References
28	Doxorubicin	-	Doxorubicin loaded poly-cystine- dopamine-Mesoporous silica nanoparticles (DOX-loaded poly-Cy-DA-MSNs)	-	Enzyme-/Redox-Responsive	Poly-Cy-DA-MSN was found to be highly biocompatible, non-toxic, and appropriate for use as drug carriers in controlled release drug delivery systems (CDDS) based on *in vitro* cellular cytotoxicity experiments conducted on HeLa cells and normal Marc-145 cells. The DOX-loaded poly-Cy-DA-MSNs displayed improved enzyme- and redox-responsive release behavior, attributed to the presence of peptide and disulfide linkages in Cy-DA. Confocal laser scanning microscopy revealed that the anti-cancer drug DOX was primarily released into the cytoplasm and nucleus of the cells, with DOX@poly-Cy-DA-MSNs being internalized through an endocytosis mechanism. These findings highlight poly-Cy-DA-MSN as a promising dual-responsive controlled drug delivery system (CDDS) for sustained cancer therapy.	Cancer therapy	Zhu *et al.* (2019) [68]

Antibiotics that are entrapped are shielded from degradation and clearance by delivery systems, which also distribute them to affected locations, tissues, or pathogens in a controlled way without endangering other tissues [70].

Various polymers, delivery systems, techniques, and mechanisms are employed to enhance anti-bacterial treatment outcomes (Table **2**). Polymers like natural, synthetic, and biodegradable types serve as carriers to improve drug stability and release. Advanced delivery systems, including nanoparticles, liposomes, and hydrogels, are utilized to ensure targeted and controlled drug delivery. Techniques such as encapsulation and conjugation improve drug penetration, while mechanisms like sustained release and bioadhesion enhance drug retention and therapeutic efficacy.

For Anti-inflammatory

Due to its significance, this chapter section discusses the new therapeutic possibilities of stimuli-responsive "Smart" biomaterials that target the inflammatory molecular signatures (Table **3**) linked to intricately inflammatory arthritis to produce negligible or non-existent adverse effects [79].

CHALLENGES IN THE STABILITY OF BIOPOLYMERS IN PHYSIOLOGICAL CONDITIONS

Biopolymers face significant stability challenges in physiological conditions due to enzymatic degradation, hydrolysis, oxidation, temperature fluctuations, and ionic interactions, which can impact their effectiveness in drug delivery. To enhance their stability, various strategies have been developed, including chemical modifications such as crosslinking, PEGylation, and grafting hydrophobic groups; physical encapsulation using nanoparticles, hydrogels, and layer-by-layer coatings; and the use of stabilizing agents like enzyme inhibitors, antioxidants, and pH buffers. These approaches help protect biopolymers from premature degradation, ensuring controlled and targeted drug release. Ongoing

Table 2. Different types of polymers, delivery systems, techniques, and mechanisms are utilized in the improvement of anti-bacterial treatment outcomes.

Sr. No.	API	Polymer	Delivery System	Techniques/Methods for Drug Encapsulation/Entrapment	Mechanism of Targeted Drug Delivery	Description	References
1	Amoxicillin trihydrate and ornidazole	Dextrin	Polysaccharide-dextrin-based hybrid hydrogelator (Dxt-PMAA)	Cross-linking	Significant variations in the swelling research revealed a stimulus-sensitive unit.	-The MCF7 cell line was used to perform the Dxt-PMAA hydrogel, showing over 98% cell viability. -Both cytocompatible and pH-sensitive. When taken orally, Dxt-PMAA hydrogel may act as a sustained-release matrix for amoxicillin and ornidazole.	Das *et al.* (2023) [71]
2	Capsaicin	Chitosan	Poly(lactic-co-glycolic acid) @ chitosan @ capsaicin (CAP@CS@PLGA)	-	Enzyme and pH dual-responsive	-To enable phase-responsive release of the loaded antifoulant, a composite structure consisting of "large spheres wrapped in small spheres" was built. To achieve the regulated release of antifouling chemicals, PLGA@CS@CAP microparticles with dual pH and enzyme responsiveness were created. -The uniqueness of single-responsive materials is compensated for by the creation of multi-responsive materials. The antibacterial rate is nearly five times greater at pH 4 than it is at pH 8. When the pH drops as a result of bacterial reproduction, the protonation product of CS is triggered to release internal CAP. This keeps the bacterial inhibition rate above 90% and causes controlled release sterilisation of the pH response in the second stage. - In order to achieve the smart controlled release of enzymatic response, the PLGA shell rapidly breaks down and ruptures in the first stage due to its exceptional biodegradability, releasing the internal CAP and CAP@CS.	Guo *et al.* (2023) [72]

Drug Delivery Systems

(Table 2) cont.....

Sr. No.	API	Polymer	Delivery System	Techniques/Methods for Drug Encapsulation/Entrapment	Mechanism of Targeted Drug Delivery	Description	References
3	Fusidic acid	-	-	Nanoprecipitation technique	Enzymatic reactions	The lipase enzyme showed site-specific fusidic acid delivery at multiple pH levels, including 5.5 (pH at normal skin), 6.5 (pH at wound site), and 7.4 (physiological pH).	Ullah *et al.* (2023) [73]
4	Gentamicin sulphate	Sodium hyaluronate and gelatin	Fabricated sodium HA (hyaluronic acid sodium salt), gelatin, PVA, and sorbitol-based microarray patches (including gentamicin sulphate)	-	Enzyme responsive	The formation of stable enzyme-substrate complexes would include non-covalent interactions. These interactions would cause the microarray patch to release gentamicin sulphate, which would allow the microbial enzymes (gelatinase and hyaluronate lyase) to break down the matching substrates (sodium HA and gelatin). The bacterial enzymes hyaluronate lyase and gelatinase hydrolysed sodium HA and gelatin, which resulted in liquification. These findings suggested that after enzyme action, the encapsulated anti-microbial ingredient would be released promptly. In the presence of *S. aureus* and *P. aeruginosa*, sodium HA and gelatin were hydrolyzed, indicating that the processed patches would release the encapsulated antibiotic in response to a pathologic stimulus (enzymes hyaluronate lyase and gelatinase). During the *in vitro* study, the produced patches quickly (within 150 minutes) released the loaded antibiotic.	Arshad *et al.* (2023) [74]
5	Doxycycline hydrochloride	Chitosan	Hyaluronic acid chitosan Doxycycline gelatin nanoparticles (HACS-Doxy-GNPs)	-	Both pH and bacterial enzyme-responsive drug release	-In the acidic biofilm microenvironment, chitosan protonation and swelling control the degradation, which is regulated by hyaluronidase and gelatinase.	Wang *et al.* (2022) [75]

(Table 2) cont.....

Sr. No.	API	Polymer	Delivery System	Techniques/Methods for Drug Encapsulation/Entrapment	Mechanism of Targeted Drug Delivery	Description	References
6	Ciprofloxacin	Alginate	The model drug, ciprofloxacin, was conjugated with alginate and poly-L-lysine using a copper-free 1,3-dipolar cycloaddition (click reaction).	Ionotropic gelation/ conjugation	Enzyme-responsive	Ionotropic gelation of the alginate was performed using longer lysine sequences with an azide function as the end group that were enzymatically cleavable. When an enzyme is present, the nanogels and the layers break down, releasing the ciprofloxacin. The linker residues adversely impair the released drug's ability to combat *Staphylococcus aureus*.	Bourga *et al.* (2021) [76]
7	Nisin	Chondroitin sulfate	Nanogels of chondroitin sulfate-nisin (CS-N NGs)	-	Enzyme and pH-responsive, thermal-responsive nanogel	CS-N NGs were pH-responsive and enzyme-responsive due to the vulnerable bonds in chondroitin sulphate, which resulted in a controlled and effective release of nisin in the infectious media simulation.	Tayeferad *et al.* (2021) [77]

Drug Delivery Systems

(Table 2) cont.....

Sr. No.	API	Polymer	Delivery System	Techniques/Methods for Drug Encapsulation/Entrapment	Mechanism of Targeted Drug Delivery	Description	References
8	Enrofloxacin	Chitosan/ cyclodextrin	Hyaluronic acid/chitosan (HA/CS) and enrofloxacin-cyclodextrin (β-CD) inclusion complexes (IC) that self-assemble.	–	pH/hyaluronidase dual-responsive	- The complexes were entrapped inside F68-covered nanogels to create the nanosystems. - The electrostatic interaction between HA and CS produced the nanogels. -Excellent sustained release is seen by the complexes that contain nanosystems coated with F68. -The nanosystems exhibited a dual response to hyaluronidase and pH. -The nanosystems increased bacterial activity and sensitivity to their surroundings. - The ideal nanosystems at 118.8 ± 30.7 nm demonstrated remarkable stability and responsive release when *S. aureus* was present in the LB broth medium, hyaluronidase-containing medium, and acid media. Strong surface adsorption and increased activity against *S. aureus* were demonstrated by the nanosystems. -Compared to the single HA/CS nanogels and the polymeric nanoparticles created by incorporating IC into F68, it exhibited a more sustained release. -A promising multifunctionalized nanosystem is presented in this study to address the therapeutic problem of *S. aureus* and other bacterial infections.	*Liu et al.* (2021) [78]

research into biomimetic engineering and smart polymers continues to advance the field, improving the durability and therapeutic efficacy of biopolymer-based drug delivery systems [81].

DRUG-BIOPOLYMER INTERACTION TOXICITY

The interaction between drugs and biopolymers is a critical factor influencing the stability, release kinetics, and biological activity of drug delivery systems. Some drugs may form strong electrostatic, hydrophobic, or covalent bonds with biopolymers, altering their degradation profile and potentially leading to uncontrolled or delayed drug release. Additionally, certain biopolymer-drug interactions may affect the bioavailability of the therapeutic agent, either enhancing or reducing its intended effect. In some cases, degradation byproducts of biopolymers or drugs can induce cytotoxicity, inflammation, or immune responses, raising safety concerns. To mitigate these risks, rigorous biocompatibility assessments, *in vitro* and *in vivo* toxicity studies, and computational modeling of drug-polymer interactions are essential. Strategies such as optimizing polymer composition, using inert or biocompatible polymer coatings, and designing stimuli-responsive systems that minimize unintended interactions can help enhance the safety and efficacy of biopolymer-based drug delivery platforms [82].

Table 3. Polymer, delivery system, techniques, and mechanisms utilized in the treatment of inflammation.

Sr. No.	API	Polymer	Delivery system	Techniques/ Methods for drug encapsulation/ entrapment	Mechanism of targeted drug delivery	Description	Reference
1	Ibuprofen	Beta-Cyclodextrin	Magnetic mesoporous silica nanoparticle-beta-cyclodextrin (MMSN-β-CD) nano-carrier	-	Redox and light-sensitive	Following stimulation with a redox system for disulphide bonds or UV light, or by combining these two treatments to achieve greater therapeutic efficiency, MMSN-β-CD demonstrated outstanding, dependent-regulated drug delivery capabilities.	Hegazy *et al.* (2019) [80]

CONCLUSIONS

This chapter discusses stimuli-responsive polymeric systems, highlighting their macroscopic reactions to external stimuli, driven by intricate modifications in their dynamically reversible non-covalent interactions at a fundamental level. A detailed review of the literature reveals that research has made significant progress beyond the initial focus on single-stimulus-responsive polymer systems. It now includes the development of dual- and multi-stimuli-sensitive systems, featuring well-defined structures and precisely controllable properties within a single polymer, tailored for various applications as previously outlined.

Various approaches have been explored, with particular emphasis on polymeric materials exhibiting stimuli-responsive mechanisms triggered by factors such as pH, redox potential, enzymes, temperature, light, and dual stimuli. Key aspects of the mechanisms underlying drug release, along with examples of their primary biomedical applications, are highlighted. A comprehensive review of the literature reveals extensive research focused on cancer. Significant advancements have been achieved in synergistic cancer treatments using biopolymers in stimuli-triggered drug delivery systems that specifically respond to unique changes in cancer cells, such as altered pH gradients and elevated secretion of specific enzymes, rather than targeting normal cells. Additionally, dual- and enzyme-responsive systems have been employed to deliver anti-bacterial agents to targeted sites, while redox- and light-sensitive systems have been utilized to deliver anti-inflammatory drugs.

LIST OF ABBREVIATIONS

CST   Critical Solution Temperature

DDS   Drug Delivery Systems

Dox   Doxorubicin

GSH   Glutathione

LCST Lower Critical Solution Temperature

NIR   Near-infrared

ROS   Redox-responsive

UV   Ultraviolet

5-FU   5-Fluorouracil

REFERENCES

[1] Gopi S, Amalraj A. Effective drug delivery system of biopolymers based on nanomaterials and hydrogels - a review. Drug Des Open Access 2016; 5(2)
[http://dx.doi.org/10.4172/2169-0138.1000129]

[2] Fazal T, Murtaza BN, Shah M, *et al.* Recent developments in natural biopolymer based drug delivery systems. RSC Advances 2023; 13(33): 23087-121.
[http://dx.doi.org/10.1039/D3RA03369D] [PMID: 37529365]

[3]   Roy D, Cambre JN, Sumerlin BS. Future perspectives and recent advances in stimuli-responsive materials. Oxford: Progress in Polymer Science 2010; 35: 278-301.
[http://dx.doi.org/10.1016/j.progpolymsci.2009.10.008]

[4]   Wells CM, Harris M, Choi L, Murali VP, Guerra FD, Jennings JA. Stimuli-responsive drug release from smart polymers. Vol. 10. J Funct Biomater 2019; 10(3): 34.
[http://dx.doi.org/10.3390/jfb10030034] [PMID: 31370252]

[5]   Ahad HA, Haranath C, Vikas SS, Varam NJ, Ksheerasagare T, Gorantla SPR. A Review on Enzyme Activated Drug Delivery System. Research Journal of Pharmacy and Technology. 2021; 14: 516-22.
[http://dx.doi.org/10.5958/0974-360X.2021.00094.9]

[6]   Preman NK, Barki RR, Vijayan A, Sanjeeva SG, Johnson RP. Recent developments in stimuli-responsive polymer nanogels for drug delivery and diagnostics: A review. Eur J Pharm Biopharm 2020; 157: 121-53.
[http://dx.doi.org/10.1016/j.ejpb.2020.10.009] [PMID: 33091554]

[7]   Wang Y, Shim MS, Levinson NS, Sung HW, Xia Y. Stimuli-responsive materials for controlled release of theranostic agents. Adv Funct Mater 2014; 24(27): 4206-20.
[http://dx.doi.org/10.1002/adfm.201400279] [PMID: 25477774]

[8]   Christy PN, Basha SK, Kumari VS, *et al.* Biopolymeric nanocomposite scaffolds for bone tissue engineering applications – A review. J Drug Deliv Sci Technol 2020; 55: 101452.
[http://dx.doi.org/10.1016/j.jddst.2019.101452]

[9]   Jabeen N, Atif M. Polysaccharides based biopolymers for biomedical applications: A review. Polym Adv Technol 2024; 35(1): e6203.
[http://dx.doi.org/10.1002/pat.6203]

[10]   Samir A, Ashour FH, Hakim AAA, Bassyouni M. Recent advances in biodegradable polymers for sustainable applications. npj Mater Degrad. 2022; 6.
[http://dx.doi.org/10.1038/s41529-022-00277-7]

[11]   Xiao R, Zhou G, Wen Y, Ye J, Li X, Wang X. Recent advances on stimuli-responsive biopolymer-based nanocomposites for drug delivery. Compos, Part B Eng 2023; 266: 111018.
[http://dx.doi.org/10.1016/j.compositesb.2023.111018]

[12]   Karimi M, Sahandi Zangabad P, Ghasemi A, *et al.* Temperature-responsive smart nanocarriers for delivery of therapeutic agents: applications and recent advances. ACS Appl Mater Interfaces 2016; 8(33): 21107-33.
[http://dx.doi.org/10.1021/acsami.6b00371] [PMID: 27349465]

[13]   Rizzo F, Kehr NS. Recent advances in injectable hydrogels for controlled and local drug delivery. Adv Healthc Mater 2021; 10(1): 2001341.
[http://dx.doi.org/10.1002/adhm.202001341] [PMID: 33073515]

[14]   Koçak G, Tuncer C, Bütün V. Stimuli-responsive polymers providing new opportunities for various applications. Hacet J Biol Chem 2020; 48(5): 527-74.
[http://dx.doi.org/10.15671/hjbc.811267]

[15]   Mozafari M. Nanoengineered Biomaterials for Advanced Drug Delivery. 2020.
https://www.elsevier.com/books-and-journals

[16]   Nanoengineered biomaterials for advanced drug delivery. Nanoeng Biomater Adv Drug Deliv 2020.
[http://dx.doi.org/10.1016/C2018-0-02205-7]

[17]   Kenchegowda M, Rahamathulla M, Hani U, *et al.* Smart nano-carriers as an emerging platform for cancer therapy: a review. Vol. 27. Molecules 2022.
[PMID: 35011376]

[18]   Klouda L, Mikos AG. Thermoresponsive hydrogels in biomedical applications. Eur J Pharm Biopharm 2008; 68(1): 34-45.

[http://dx.doi.org/10.1016/j.ejpb.2007.02.025] [PMID: 17881200]

[19]   Buwalda SJ, Vermonden T, Hennink WE. Hydrogels for therapeutic delivery: current developments and future directions. Biomacromolecules 2017; 18(2): 316-30.
[http://dx.doi.org/10.1021/acs.biomac.6b01604] [PMID: 28027640]

[20]   Arif ZU, Khalid MY, Tariq A, Hossain M, Umer R. 3D printing of stimuli-responsive hydrogel materials: Literature review and emerging applications. Giant (Oxf) 2024; 17: 100209.
[http://dx.doi.org/10.1016/j.giant.2023.100209]

[21]   Sies H, Jones DP. Reactive oxygen species (ROS) as pleiotropic physiological signalling agents. Nat Rev Mol Cell Biol 2020; 21(7): 363-83.
[http://dx.doi.org/10.1038/s41580-020-0230-3] [PMID: 32231263]

[22]   Lee CG, Kwon TH. Controlling morphologies of redox-responsive polymeric nanocarriers for a smart drug delivery system. Chemistry 2023; 29(34): e202300594.
[http://dx.doi.org/10.1002/chem.202300594]

[23]   Kennedy L, Sandhu JK, Harper ME, Cuperlovic-Culf M. Role of glutathione in cancer: From mechanisms to therapies. Vol. 10. Biomolecules 2020; 10(10): 1429.
[http://dx.doi.org/10.3390/biom10101429] [PMID: 33050144]

[24]   Abed HF, Abuwatfa WH, Husseini GA. Redox-responsive drug delivery systems: a chemical perspective. Nanomaterials (Basel) 2022; 12(18): 3183.
[http://dx.doi.org/10.3390/nano12183183] [PMID: 36144971]

[25]   Shi Z, Liu J, Tian L, *et al.* Insights into stimuli-responsive diselenide bonds utilized in drug delivery systems for cancer therapy. Biomed Pharmacother 2022; 155: 113707.
[http://dx.doi.org/10.1016/j.biopha.2022.113707]

[26]   Xing Y, Zeng B, Yang W. Light responsive hydrogels for controlled drug delivery. Front Bioeng Biotechnol 2022; 10: 1075670.
[http://dx.doi.org/10.3389/fbioe.2022.1075670]

[27]   Teasdale I. Stimuli-responsive phosphorus-based polymers. Eur J Inorg Chem 2019; 2019(11-12): 1445-56.
[http://dx.doi.org/10.1002/ejic.201801077] [PMID: 30983876]

[28]   Adekoya OC, Adekoya GJ, Sadiku ER, Hamam Y, Ray SS. Application of dft calculations in designing polymer-based drug delivery systems: an overview. Pharmaceutics 2022; 14(9): 1972.
[http://dx.doi.org/10.3390/pharmaceutics14091972] [PMID: 36145719]

[29]   Sia CS, Tey BT, Low LE. Light-responsive nanoassemblies: advancing biomedical innovation. Adv Funct Mater 2024; 34(23): 2314278.
[http://dx.doi.org/10.1002/adfm.202314278]

[30]   Li L, Scheiger JM, Levkin PA. Design and applications of photoresponsive hydrogels. Adv Mater 2019; 31(26): 1807333.
[http://dx.doi.org/10.1002/adma.201807333] [PMID: 30848524]

[31]   Ghalehkhondabi V, Soleymani M, Fazlali A. Synthesis of quercetin-loaded hyaluronic acid-conjugated pH/redox dual-stimuli responsive poly(methacrylic acid)/mesoporous organosilica nanoparticles for breast cancer targeted therapy. Int J Biol Macromol 2024; 263(Pt 1): 130168.
[http://dx.doi.org/10.1016/j.ijbiomac.2024.130168] [PMID: 38365162]

[32]   Bossmann SH, Payne MM, Kalita M, Bristow RMD, Afshar A, Perera AS. Iron-based magnetic nanosystems for diagnostic imaging and drug delivery: towards transformative biomedical applications. Pharmaceutics 2022; 14(10): 2093.
[http://dx.doi.org/10.3390/pharmaceutics14102093] [PMID: 36297529]

[33]   Sun C, Lee JSH, Zhang M. Magnetic nanoparticles in MR imaging and drug delivery. Adv Drug Deliv Rev 2008; 60(11): 1252-65.
[http://dx.doi.org/10.1016/j.addr.2008.03.018] [PMID: 18558452]

[34] Godfrey WH, Kornberg MD. The role of metabolic enzymes in the regulation of inflammation. Vol. 10. Metabolites 2020; 10(11): 426.
[http://dx.doi.org/10.3390/metabo10110426] [PMID: 33114536]

[35] Kuperkar K, Atanase L, Bahadur A, Crivei I, Bahadur P. Degradable polymeric bio(nano)materials and their biomedical applications: a comprehensive overview and recent updates. Polymers (Basel) 2024; 16(2): 206.
[http://dx.doi.org/10.3390/polym16020206] [PMID: 38257005]

[36] Kapalatiya H, Madav Y, Tambe VS, Wairkar S. Enzyme-responsive smart nanocarriers for targeted chemotherapy: an overview. Drug Deliv Transl Res 2022; 12(6): 1293-305.
[http://dx.doi.org/10.1007/s13346-021-01020-6] [PMID: 34251612]

[37] Sobczak M. Enzyme-responsive hydrogels as potential drug delivery systems—state of knowledge and future prospects. Int J Mol Sci 2022; 23(8): 4421.
[http://dx.doi.org/10.3390/ijms23084421]

[38] Das SS, Bharadwaj P, Bilal M, et al. Stimuli-responsive polymeric nano-carriers for drug delivery, imaging, and theragnosis. Vol. 12. Polymers (Basel) 2020; 12(6): 1397.
[http://dx.doi.org/10.3390/polym12061397]

[39] El-Husseiny HM, Mady EA, Hamabe L, et al. Smart/stimuli-responsive hydrogels: Cutting-edge platforms for tissue engineering and other biomedical applications. Mater Today Bio 2022; 13: 100186.
[http://dx.doi.org/10.1016/j.mtbio.2021.100186] [PMID: 34917924]

[40] Yao Y, Zhou Y, Liu L, et al. Nanoparticle-based drug delivery in cancer therapy and its role in overcoming drug resistance. Front Mol Biosci 2020; 7: 193.
[http://dx.doi.org/10.3389/fmolb.2020.00193]

[41] Binaeian E, Rohani S. pH-responsive prodrug containing Zr-based MOF and aldehyde-based drug; anionic hydrogel coating as a smart delivery system. Adv Powder Technol 2024; 35(1): 104316.
[http://dx.doi.org/10.1016/j.apt.2023.104316]

[42] Sharma R, Basist P, Alhalmi A, Khan R, Noman OM, Alahdab A. Synthesis of quercetin-loaded silver nanoparticles and assessing their anti-bacterial potential. Micromachines (Basel) 2023; 14(12): 2154.
[http://dx.doi.org/10.3390/mi14122154] [PMID: 38138323]

[43] Nayak S, Das K, Sivagnanam S, et al. Cystine-cored diphenylalanine appended peptide-based self-assembled fluorescent nanostructures direct redox-responsive drug delivery. iScience 2024; 27(4): 109523.
[http://dx.doi.org/10.1016/j.isci.2024.109523] [PMID: 38577103]

[44] Rahmani M, Pourmadadi M, Abdouss M, Rahdar A, Díez-Pascual AM. Gelatin/polyethylene glycol/g-C_3N_4 hydrogel with olive oil as a sustainable and biocompatible nanovehicle for targeted delivery of 5-fluorouracil. Ind Crops Prod 2024; 208: 117912.
[http://dx.doi.org/10.1016/j.indcrop.2023.117912]

[45] Ahmari A, Pourmadadi M, Yazdian F, Rashedi H, Khanbeigi KA. A green approach for preparation of chitosan/hydroxyapatite/graphitic carbon nitride hydrogel nanocomposite for improved 5-FU delivery. Int J Biol Macromol 2024; 258(Pt 2): 128736.
[http://dx.doi.org/10.1016/j.ijbiomac.2023.128736] [PMID: 38101677]

[46] Bayat F, Pourmadadi M, Eshaghi MM, Yazdian F, Rashedi H. Improving release profile and anticancer activity of 5-fluorouracil for breast cancer therapy using a double drug delivery system: chitosan/agarose/γ-alumina nanocomposite@double emulsion. J Cluster Sci 2023; 34(5): 2565-77.
[http://dx.doi.org/10.1007/s10876-023-02405-y]

[47] Zaer M, Moeinzadeh A, Abolhassani H, et al. Doxorubicin-loaded Niosomes functionalized with gelatine and alginate as pH-responsive drug delivery system: A 3D printing approach. Int J Biol Macromol 2023; 253(Pt 2): 126808.

[http://dx.doi.org/10.1016/j.ijbiomac.2023.126808] [PMID: 37689301]

[48]   Xiao R, Ye J, Li X, Wang X. Dual size/charge-switchable and multi-responsive gelatin-based nanocluster for targeted anti-tumor therapy. Int J Biol Macromol 2023; 238: 124032.
[http://dx.doi.org/10.1016/j.ijbiomac.2023.124032] [PMID: 36921812]

[49]   Shamsabadipour A, Pourmadadi M, Rashedi H, Yazdian F, Navaei-Nigjeh M. Nanoemulsion carriers of porous γ-alumina modified by polyvinylpyrrolidone and carboxymethyl cellulose for pH-sensitive delivery of 5-fluorouracil. Int J Biol Macromol 2023; 233: 123621.
[http://dx.doi.org/10.1016/j.ijbiomac.2023.123621] [PMID: 36773864]

[50]   Rajaei M, Rashedi H, Yazdian F, Navaei-Nigjeh M, Rahdar A, Díez-Pascual AM. Chitosan/agarose/graphene oxide nanohydrogel as drug delivery system of 5-fluorouracil in breast cancer therapy. J Drug Deliv Sci Technol 2023; 82: 104307.
[http://dx.doi.org/10.1016/j.jddst.2023.104307]

[51]   Ghasemizadeh H, Pourmadadi M, Yazdian F, *et al.* Novel carboxymethyl cellulose-halloysite-polyethylene glycol nanocomposite for improved 5-FU delivery. Int J Biol Macromol 2023; 232: 123437.
[http://dx.doi.org/10.1016/j.ijbiomac.2023.123437] [PMID: 36708898]

[52]   Resen AK, Atiroğlu A, Atiroğlu V, *et al.* Effectiveness of 5-Fluorouracil and gemcitabine hydrochloride loaded iron-based chitosan-coated MIL-100 composite as an advanced, biocompatible, pH-sensitive and smart drug delivery system on breast cancer therapy. Int J Biol Macromol 2022; 198: 175-86.
[http://dx.doi.org/10.1016/j.ijbiomac.2021.12.130] [PMID: 34973989]

[53]   Liwinska W, Waleka-Bagiel E, Stojek Z, Karbarz M, Zabost E. Enzyme-triggered- and tumor-targeted delivery with tunable, methacrylated poly(ethylene glycols) and hyaluronic acid hybrid nanogels. Drug Deliv 2022; 29(1): 2561-78.
[http://dx.doi.org/10.1080/10717544.2022.2105443] [PMID: 35938558]

[54]   Rivero Berti I, Rodenak-Kladniew BE, Katz SF, *et al.* Enzymatic active release of violacein present in nanostructured lipid carrier by lipase encapsulated in 3d-bioprinted chitosan-hydroxypropyl methylcellulose matrix with anticancer activity. Front Chem 2022; 10: 914126.
[http://dx.doi.org/10.3389/fchem.2022.914126] [PMID: 35873038]

[55]   Farjadian F, Moghadam M, Monfared M, Mohammadi-Samani S. Mesoporous silica nanostructure modified with azo gatekeepers for colon targeted delivery of 5-fluorouracil. AIChE J 2022; 68(12): e17900.
[http://dx.doi.org/10.1002/aic.17900]

[56]   Gulfam M, Jo SH, Jo SW, Vu TT, Park SH, Lim KT. Highly porous and injectable hydrogels derived from cartilage acellularized matrix exhibit reduction and NIR light dual-responsive drug release properties for application in antitumor therapy. NPG Asia Mater 2022; 14(1): 8.
[http://dx.doi.org/10.1038/s41427-021-00354-4]

[57]   Chen X, Zou J, Zhang K, *et al.* Photothermal/matrix metalloproteinase-2 dual-responsive gelatin nanoparticles for breast cancer treatment. Acta Pharm Sin B 2021; 11(1): 271-82.
[http://dx.doi.org/10.1016/j.apsb.2020.08.009] [PMID: 33532192]

[58]   Nagaraja K, Krishna Rao KSV, Zo S, Soo Han S, Rao KM. Synthesis of novel tamarind gum-c--poly(Acrylamidoglycolic acid)-based pH responsive semi-IPN hydrogels and their Ag nanocomposites for controlled release of chemotherapeutics and inactivation of multi-drug-resistant bacteria. Gels 2021; 7(4): 237.
[http://dx.doi.org/10.3390/gels7040237] [PMID: 34940297]

[59]   Rashidzadeh H, Rezaei SJT, Zamani S, Sarijloo E, Ramazani A. pH-sensitive curcumin conjugated micelles for tumor triggered drug delivery. J Biomater Sci Polym Ed 2021; 32(3): 320-36.
[http://dx.doi.org/10.1080/09205063.2020.1833815] [PMID: 33026298]

[60]   Barve A, Jain A, Liu H, Zhao Z, Cheng K. Enzyme-responsive polymeric micelles of cabazitaxel for

prostate cancer targeted therapy. Acta Biomater 2020; 113: 501-11.
[http://dx.doi.org/10.1016/j.actbio.2020.06.019] [PMID: 32562805]

[61]   Vaghasiya K, Ray E, Sharma A, Katare OP, Verma RK. Matrix metalloproteinase-responsive mesoporous silica nanoparticles cloaked with cleavable protein for "self-actuating" on-demand controlled drug delivery for cancer therapy. ACS Appl Bio Mater 2020; 3(8): 4987-99.
[http://dx.doi.org/10.1021/acsabm.0c00497] [PMID: 35021676]

[62]   Cuggino JC, Ambrosioni FE, Picchio ML, *et al.* Thermally self-assembled biodegradable poly(casein-g-N-isopropylacrylamide) unimers and their application in drug delivery for cancer therapy. Int J Biol Macromol 2020; 154: 446-55.
[http://dx.doi.org/10.1016/j.ijbiomac.2020.03.138] [PMID: 32194104]

[63]   Tiryaki E, Başaran Elalmış Y, Karakuzu İkizler B, Yücel S. Novel organic/inorganic hybrid nanoparticles as enzyme-triggered drug delivery systems: Dextran and Dextran aldehyde coated silica aerogels. J Drug Deliv Sci Technol 2020; 56: 101517.
[http://dx.doi.org/10.1016/j.jddst.2020.101517]

[64]   Fathi M, Barar J, Erfan-Niya H, Omidi Y. Methotrexate-conjugated chitosan-grafted pH- and thermo-responsive magnetic nanoparticles for targeted therapy of ovarian cancer. Int J Biol Macromol 2020; 154: 1175-84.
[http://dx.doi.org/10.1016/j.ijbiomac.2019.10.272] [PMID: 31730949]

[65]   Cai D, Han C, Liu C, *et al.* Chitosan-capped enzyme-responsive hollow mesoporous silica nanoplatforms for colon-specific drug delivery. Nanoscale Res Lett 2020; 15(1): 123.
[http://dx.doi.org/10.1186/s11671-020-03351-8] [PMID: 32488526]

[66]   Chen S, Wu J, Tang Q, Xu C, Huang Y, Huang D, *et al.* Nano-micelles based on hydroxyethyl starch-curcumin conjugates for improved stability, antioxidant and anti-cancer activity of curcumin. Carbohydr Polym 2020; 228.
[http://dx.doi.org/10.1016/j.carbpol.2019.115398]

[67]   Zhang H, Pei M, Liu P. Keratin-based drug-protein conjugate with acid-labile and reduction-cleavable linkages in series for tumor intracellular DOX delivery. J Ind Eng Chem 2019; 80: 739-48.
[http://dx.doi.org/10.1016/j.jiec.2019.05.041]

[68]   Zhu D, Hu C, Liu Y, Chen F, Zheng Z, Wang X. Enzyme-/redox-responsive mesoporous silica nanoparticles based on functionalized dopamine as nanocarriers for cancer therapy. ACS Omega 2019; 4(4): 6097-105.
[http://dx.doi.org/10.1021/acsomega.8b02537]

[69]   Pang Q, Jiang Z, Wu K, Hou R, Zhu Y. Nanomaterials-based wound dressing for advanced management of infected wound. Antibiotics (Basel) 2023; 12(2): 351.
[http://dx.doi.org/10.3390/antibiotics12020351]

[70]   Skwarczynski M, Bashiri S, Yuan Y, *et al.* Antimicrobial activity enhancers: towards smart delivery of antimicrobial agents. Antibiotics (Basel) 2022; 11(3): 412.
[http://dx.doi.org/10.3390/antibiotics11030412] [PMID: 35326875]

[71]   Das D, Roy A, Pal S. A polysaccharide-based ph-sensitive hybrid hydrogel as a sustained release matrix for antimicrobial drugs. ACS Appl Polym Mater 2023; 5(5): 3348-58.
[http://dx.doi.org/10.1021/acsapm.2c02256]

[72]   Guo Y, Feng H, Li W, Wang W, Yu M, Chen S. Enzyme and pH dual-responsive CAP@CS@PLGA microcapsules for controlled release anti-bacterial application. Biochem Eng J 2023; 196: 108956.
[http://dx.doi.org/10.1016/j.bej.2023.108956]

[73]   Ullah N, Khan D, Ahmed N, Zafar A, Shah KU. ur Rehman A. Lipase-sensitive fusidic acid polymeric nanoparticles based hydrogel for on-demand delivery against MRSA-infected burn wounds. J Drug Deliv Sci Technol 2023; 80: 104110.
[http://dx.doi.org/10.1016/j.jddst.2022.104110]

[74]   Arshad MS, Zafar S, Rana SJ, Nazari K, Chang MW, Ahmad Z. Fabrication of gentamicin sulphate laden stimulus responsive polymeric microarray patches for the treatment of bacterial biofilms. J Drug Deliv Sci Technol 2023; 84: 104504.
[http://dx.doi.org/10.1016/j.jddst.2023.104504]

[75]   Wang Y, Shukla A. Bacteria-responsive biopolymer-coated nanoparticles for biofilm penetration and eradication. Biomater Sci 2022; 10(11): 2831-43.
[http://dx.doi.org/10.1039/D2BM00361A] [PMID: 35441624]

[76]   Bourgat Y, Mikolai C, Stiesch M, Klahn P, Menzel H. Enzyme-responsive nanoparticles and coatings made from alginate/peptide ciprofloxacin conjugates as drug release system. Antibiotics (Basel) 2021; 10(6): 653.
[http://dx.doi.org/10.3390/antibiotics10060653] [PMID: 34072352]

[77]   Tayeferad M, Boddohi S, Bakhshi B. Dual-responsive nisin loaded chondroitin sulfate nanogel for treatment of bacterial infection in soft tissues. Int J Biol Macromol 2021; 193(Pt A): 166-72.
[http://dx.doi.org/10.1016/j.ijbiomac.2021.10.116] [PMID: 34688678]

[78]   Liu Y, Chen D, Zhang A, *et al.* Composite inclusion complexes containing hyaluronic acid/chitosan nanosystems for dual responsive enrofloxacin release. Carbohydr Polym 2021; 252: 117162.
[http://dx.doi.org/10.1016/j.carbpol.2020.117162] [PMID: 33183613]

[79]   Hegazy M, Zhou P, Rahoui N, Wu G, Taloub N, Lin Y, *et al.* A facile design of smart silica nano-carriers *via* surface-initiated RAFT polymerization as a dual-stimuli drug release platform. Colloids Surf A Physicochem Eng Asp 2019; 581: 123797.
[http://dx.doi.org/10.1016/j.colsurfa.2019.123797]

[80]   Pourjavadi A, Tehrani ZM. Poly(N-isopropylacrylamide)-coated β-cyclodextrin–capped magnetic mesoporous silica nanoparticles exhibiting thermal and pH dual response for triggered anticancer drug delivery. Int J Polym Mater 2017; 66(7): 336-48.
[http://dx.doi.org/10.1080/00914037.2016.1217531]

[81]   Abotbina W, Sapuan SM, Ilyas RA, *et al.* Recent developments in cassava (*manihot esculenta*) based biocomposites and their potential industrial applications: a comprehensive review. Materials (Basel) 2022; 15(19): 6992.
[http://dx.doi.org/10.3390/ma15196992] [PMID: 36234333]

[82]   Gautam S, Lakhanpal I, Sonowal L, Goyal N. Recent advances in targeted drug delivery using metal-organic frameworks: toxicity and release kinetics. Next Nanotechnology 2023; 3-4: 100027.
[http://dx.doi.org/10.1016/j.nxnano.2023.100027]

CHAPTER 9

Biopolymer-based Nanofibers in Tissue Engineering

Aarti Tiwari[1], **Ajay Kumar Shukla**[1,*], **Vimal Kumar Yadav**[1], **Kunal Agam Kanujia**[1], **Vishnu Prasad Yadav**[1], **Rama Sankar Dubey**[2] and **Manoj Kumar Mishra**[3]

[1] *Institute of Pharmacy, Dr Rammanohar Lohia Avadh University, Ayodhya, Uttar Pradesh, India*

[2] *Department of Pharmacy, MMM University Gorakhpur Uttar Pradesh, India*

[3] *Shambhunath Institute of Engineering and Technology, Prayagraj, Uttar Pradesh, India*

Abstract: Natural materials such as wood, shells, fungi, bacteria, and plants can be used to make biopolymer nanofibers (BPNFs), which are natural polymeric materials. Nanofibers (NFs) are the class of nanostructured materials that are widely used in tissue engineering (TE) and regenerative medicine (RM). These biomaterials aim to promote bone tissue regeneration at the defect location, whereupon they will eventually degrade naturally and be replaced by freshly produced bone tissue. Nanocomposite biomaterials are a relatively new class of materials that combine readily resorbable, bioactive fillers that are nanoscale in size with biopolymeric and biodegradable matrix architectures. The biocompatibility, tissue regeneration, and incorporation of nanomaterials have been assessed with alginate, fucoidan, chitosan (CS), collagen (Col), cellulose, and silk fibroin (SF). Examples of synthetic polymer-based nanocomposites in this chapter include polyethylene glycol (PEG), polycaprolactone (PCL), poly (lactic-co-glycolic) acid (PLGA), poly (lactic acid) (PLA), and polyurethane (PU) based nanocomposites. In bone tissue regeneration research, a wide range of nanofillers are used, such as graphene oxide (GO), nano titanium dioxide ($nTiO_2$), nano silica (nSi), nano zirconia (nZr), nano-hydroxyapatite (nHA), and nano silver nanoparticles (AgNPs). Biopolymer-based nanofibers have unique properties that replicate the extracellular matrix (ECM) of natural tissues, making them a promising tool for tissue engineering. These nanofibers, which can be made from synthetic or natural biopolymers, have customizable mechanical properties, biocompatibility, and biodegradability, making them ideal scaffolding materials for tissue regeneration. Cell attachment, proliferation, and differentiation are essential for the successful use of nanofibers in tissue engineering applications because of their high surface area-to-volume ratio. Further improving the functionality of biopolymer-based nanofibers and encouraging targeted tissue regeneration and healing is the addition of growth factors, medications, and bioactive compounds. The latest developments in biopolymer-based nanofibers for tissue engineering emphasize their properties, techniques of production,

---

* **Corresponding author Ajay Kumar Shukla:** Institute of Pharmacy, Dr Rammanohar Lohia Avadh University, Ayodhya, Uttar Pradesh, India; E-mail: ashukla1007@gmail.com

and uses in the regeneration of various tissues, such as skin, bone, cartilage, and neural tissues. The necessity for multidisciplinary research to enhance nanofiber-based scaffolds for therapeutic applications is highlighted by the exploration of the possible obstacles and future prospects in this quickly developing sector. This chapter discusses a few biomaterials that have the potential to regenerate bone tissue in the form of polymeric nanocomposites.

Keywords: Biodegradability, Biopolymer-based nanofibers, Natural or synthetic biopolymers, Offer biocompatibility, Tissue engineering.

INTRODUCTION

The naturally occurring substances found in natural sources are known as biopolymers. The Greek terms bio and polymer, which stand for nature and living things, are the roots of the term biopolymer. Biopolymers are large macromolecules composed of many repeating units. A macromolecule, according to the IUPAC definition, is a single molecule. Because the biopolymers are biocompatible and biodegradable, they can be used in a variety of applications, including the food industry for edible films and emulsions, as well as in the pharmaceutical industry for wound healing, tissue scaffolds, dressing materials, drug transport materials, and medical implants such as organs [1].

The most common macromolecules are biopolymers (Fig. **1**), which include proteins, carbohydrates, lipids, nucleic acids, and huge non-polymeric molecules like macrocycles and lipids. Synthetic macromolecules include plastics, synthetic fibers, and experimental materials like carbon nanotubes [2]. Their molecular backbones may include repeating units of amino acids, saccharides, or nucleic acids as well as a range of chemical side chains that support the molecules' functions. Using conventional chemical techniques, biopolymers such as polylactic acid (PLA) and polyhydroxyalkanoates (PHAs) are identified in microbes or genetically engineered organisms. These consist of proteins from milk or collagen and carbohydrates from cellulose. The genetic manipulation of microorganisms enables the biotechnological manufacture of biopolymers with specific properties appropriate for high-value medicinal applications, such as tissue engineering and drug delivery. The classification of biopolymers based on their origin is displayed in Table **1**

Some synthetic polymers based on chemicals are harmful to microbiological organisms, plants, animals, and people. To make them suitable for use in biomedical applications, biopolymers have been proposed as a substitute. Although biopolymers have useful biomedical uses, they can be harmful in some situations and are unstable in biological fluids. Recent developments in nanotechnology have increased its practical relevance across a range of fields, particularly in the development of biopolymers to create nanoparticles for a wide range of biomedical uses. To improve their biological qualities and be used for specific biomedical applications, biopolymers are specifically created as nanofibers [3].

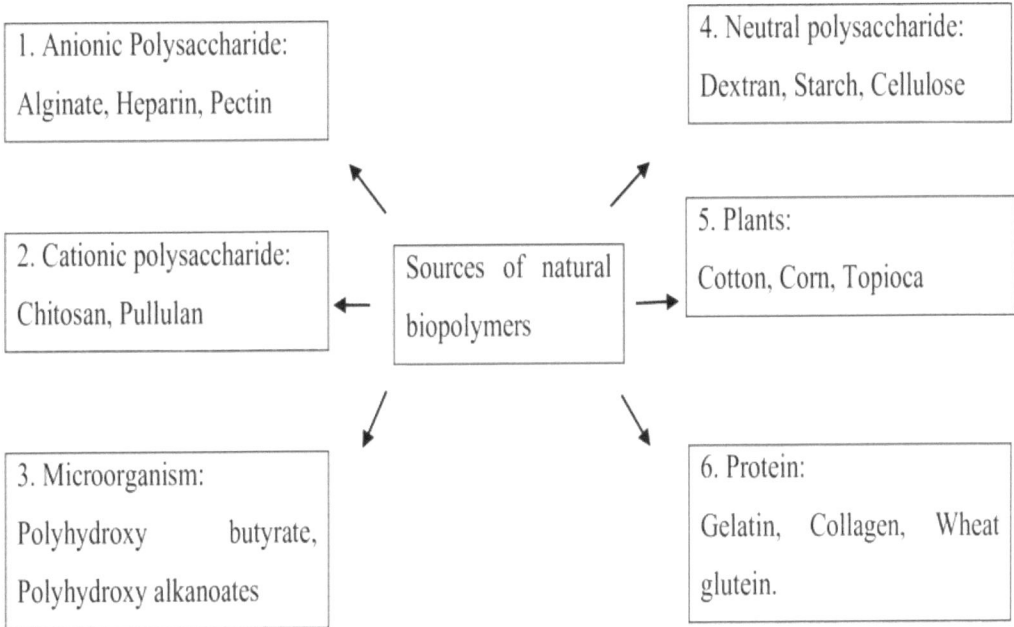

| 1. Anionic Polysaccharide: Alginate, Heparin, Pectin |
| 4. Neutral polysaccharide: Dextran, Starch, Cellulose |
| 2. Cationic polysaccharide: Chitosan, Pullulan |
| Sources of natural biopolymers |
| 5. Plants: Cotton, Corn, Topioca |
| 3. Microorganism: Polyhydroxy butyrate, Polyhydroxy alkanoates |
| 6. Protein: Gelatin, Collagen, Wheat glutein. |

Fig. (1). Sources of natural biopolymers.

Table 1. A list of types of biopolymers, advantages and disadvantages.

Type	Advantages	Disadvantages	References
Natural biopolymers	Biodegradable, biocompatible, non-toxic, bioadhesive, biofunctional, and biologically renewable	Structurally more complicated, low melting point, high surface tension, and less stable.	[2]
Synthetic biopolymers	Improved mechanical and chemical stability, increased repeatability, and biocompatibility	Costly, toxic, and non-biodegradable synthesis process.	[2]

A nanofibrous structure is designed for a variety of medicinal applications using synthetic and/or natural polymers (biopolymers). The biopolymers are more biocompatible and have fewer toxicities and immunogenic effects on the body than synthetic polymers. Biopolymer-based electrospun nanofibrous materials have found extensive applications in drug delivery, tissue engineering, regenerative medicine, and wound dressing. To create composite nanofibers with ideal properties that are essential to their operation, biopolymers are occasionally electrospun with synthetic polymers.

The possibilities for creating scaffolds that might be able to handle this difficulty have significantly increased with the introduction of nanofibers. Nanofibers can currently be created using three different methods: phase separation, self-assembly, and electrospinning. Out of all of these methods, electrospinning has been investigated the most and has shown the most promise for tissue engineering applications. The opportunities for creating nanofibrous scaffolds have expanded due to the abundance of natural and synthetic biomaterials, particularly when employing the electrospinning method. Nanofiber-based three-dimensional synthetic biodegradable scaffolds are a great way to promote cell adhesion, proliferation, and differentiation. Thus, regardless of how they are made, nanofibers have been utilized as scaffolds for musculoskeletal tissue engineering (which includes bone, cartilage, ligaments, and skeletal muscle), skin, vascular, and neural tissue engineering, as well as carriers for the regulated delivery of medications, proteins, and DNA [4].

Tissue engineering (TE) develops efficient tissue substitutes using techniques from engineering, materials science, and biology. The features of the natural EM are frequently not well replicated by common scaffolding materials, which calls for the investigation of alternative materials. Natural sources of biopolymers offer a great substitute because of their advantageous biological properties [5]. Nanofibers are highly designed fibers made of various polymers with varying physical characteristics and potential applications. They have a diameter of less than 500 nm. The typical diameter of a nanofiber is roughly 200 times smaller than that of a human hair, which is about 80 micrometers in diameter. They possess unique characteristics that traditional physics finds challenging to account for. The three techniques available for producing nanofibers are phase separation, self-assembly, and electrospinning. However, electrospinning is the most studied technique and seems to provide the most promising results for exceptional applications in a range of fields of interest [6].

The basic idea behind tissue engineering (TE) is to combine living cells with a natural or synthetic support structure, known as a scaffold, to create a three-dimensional, living construct. This construct is designed to be similar to or even better than the tissue it is meant to replace in terms of function, structure, and mechanics. TE involves placing cells onto the scaffold, allowing them to grow in a lab setting before they are implanted into the injured area of the body. The scaffold initially provides support for cell attachment, growth, and specialization, helping to form a new extracellular matrix (ECM) (Fig. **2**) [7, 8].

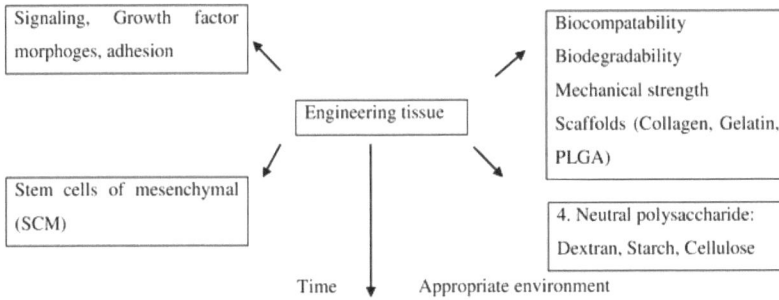

The general principles of Tissue Engineering

Fig. (2). Basic principle of tissue engineering.

Stem cells are unique, unspecialized cells found in many organisms. They have two key characteristics: self-renewal and potency. Self-renewal allows them to divide multiple times, creating identical daughter cells, while potency refers to their ability to develop into specific types of mature cells. The two main types of stem cells in mammals are embryonic stem (ES) cells, which come from blastocysts, and adult stem cells, which are found in mature tissues. ES cells can differentiate into almost any cell type except placental cells. This is done outside of the body in the formation of cell clusters referred to as embryoid bodies (EBs) that initiate pathways of development and yield cell types from all three germ layers [9].

ES cells are of great value in cell-replacement therapies because they have a wide differentiation potential, although they have ethical issues and the risk of immune rejection. Therapeutic cloning would assist in overcoming this problem of rejection [10].

Stem cells Mesenchymal (SCM), a particular cell type of stem cell, may differentiate into many cell types, such as bone cells, fat cells, and cartilage cells, as well as into cells from other layers of the tissue. SCM are non-hematopoietic, multipotent stem cells that can differentiate into endodermal and ectodermal lineages as well as mesodermal lineages such as osteocytes, adipocytes, and chondrocytes. Stem cell growth may be supported using different natural as well as synthetic biomaterials. The biomaterials stimulate cell attachment, proliferation, and migration and develop conditions that promote EM formation as well as repair of tissue. Numerous synthetic and natural biomaterials can be used to enhance stem cell proliferation. Biomaterials are suitable for cell adhesion, migration, proliferation, and growth. Biomaterials can also create an environment that is favorable for cells to form and replace the extracellular matrix (ECM) [11].

TISSUE SCAFFOLD

A biomaterial-based support system called a tissue scaffold promotes cell attachment, migration, and the growth of functional tissues. This scaffold often has a porous structure that permits nutrients to be transported and breaks down gradually over time. Tissue engineering places significant emphasis on creating efficient scaffolds because they give cells a short-term framework to develop and create new tissue before it is transferred into the body. A scaffold's performance is significantly influenced by its design, with specifications changing according to its intended usage. In tissue engineering, stem cells must be organized in a matrix that permits the passage of nutrients and oxygen while eliminating waste. Tissues that can be placed in an injured location to promote natural tissue regeneration can be created in a lab setting using scaffolds. Different materials are used to make scaffolds, but their suitability for tissue engineering depends on a few key parameters. These include the capacity to transport nutrients, mechanical strength, biocompatibility, and biodegradability. Advanced biomaterials and technology are therefore necessary to establish a natural environment in which tissue can grow [12, 13].

BIOPOLYMERS FOR THE PRODUCTION OF NANOFIBERS

Biopolymer Types

Proteins, nucleic acids, and polysaccharides are the three types of biopolymers; each has special qualities that make it suitable for tissue engineering applications.

Natural Polysaccharides

Chitosan, alginate, hyaluronic acid, and cellulose are examples of common polysaccharides. Excellent water retention, biocompatibility, and the capacity to encourage cell adhesion and proliferation are characteristics of these materials.

Polysaccharides offer significant potential in cartilage tissue engineering due to their biocompatibility, structural resemblance to the extracellular matrix, and adaptability to chemical modifications. While challenges remain, particularly in electrospinning techniques due to solubility, viscosity, and surface tension issues, various solutions such as the incorporation of carrier polymers and the optimization of solvent systems are promising. Polysaccharides like alginate, chitosan, hyaluronic acid, chondroitin sulfate, and cellulose stand out for their potential in cartilage regeneration. Continued research and technological advancements will be essential to fully realize the benefits of polysaccharide-based nanofibers in tissue engineering applications [14].

Ocular disorders pose significant treatment challenges due to the complex anatomy and unique physiological barriers of the eye. However, polysaccharides offer promising alternatives for enhancing drug delivery systems, thanks to their biocompatibility, biodegradability, and adhesive properties. Novel developments in polysaccharide-based technologies have the potential to revolutionize the management of various ocular conditions. By examining the anatomy of the eye and the specific obstacles faced in drug transport, we provide insights into different administration methods that can address these challenges effectively. The emphasis on biologically adhesive polymers such as chitosan, hyaluronic acid, cellulose, cyclodextrin, and poloxamer illustrates their role in improving drug retention and bioavailability. Furthermore, the evaluation of various ophthalmic formulation designs, including gels, lenses, eye drops, nanofibers, microneedles, microspheres, and nanoparticles, reveals their respective benefits and limitations in therapeutic applications. The exploration of novel polysaccharides opens new avenues for innovative ocular drug delivery solutions, instilling hope for enhanced treatment options in the future [15].

This chapter presents a novel approach to enhancing the properties of conductive hydrogels, particularly those based on polysaccharides like chitosan. By integrating dissolved chitosan and solid chitosan nanofibers, it exhibits heightened sensitivity and low detection limits when utilized as a strain sensor. Its remarkable durability, stability, and temperature tolerance further enhance its potential for practical applications in wearable sensor technology. Overall, our findings contribute to the advancement of conductive hydrogels, paving the way for their expanded use in various fields related to human activity monitoring and health diagnostics [16].

Proteins

Well-known protein-based biopolymers include collagen, gelatin, and silk fibroin. They have bioactive qualities that can improve tissue regeneration and cellular responses. The development of a silk-elastin-like protein (SELP) nanofiber membrane, combined with bacterial cellulose (BC), presents a significant advancement in creating effective skin substitutes. This innovative design successfully mimics the gradient structure of the skin, allowing for optimal drug penetration while offering robust protection against bacterial invasion. The impressive permeation efficiency and the conducive environment for cell growth further highlight the potential of SELP-based biomaterials in enhancing wound healing processes. Overall, this study underscores the significance of using silk-elastin-like protein constructs in clinical applications for treating full-thickness skin injuries and ulcers, paving the way for future research and development in regenerative medicine [17].

This study highlights the significant role of surface modifications on electrospun poly(ε-caprolactone) nanofibers in promoting the chondrogenic differentiation of bone marrow-derived mesenchymal stromal cells (BMSCs). By employing two distinct surface modification techniques, plasma polymerization and chondroitin sulfate immobilization, we demonstrated that the introduction of polar functional groups can enhance the cells' proliferation, matrix production, and expression of key extracellular matrix proteins. Notably, the carboxylic acid-rich nanofibers emerged as the most effective in facilitating chondrogenesis, evidenced by elevated levels of aggrecan, Sox9, and collagen II, alongside a reduction in hypertrophic markers. These findings suggest that tailored surface properties can be leveraged to engineer nanofibers with optimal chondro-inductive characteristics, paving the way for innovative strategies in cartilage tissue engineering and regeneration [18].

Nucleic Acid

Although less prevalent, nucleic acids are being studied for their potential in tissue engineering and gene therapy applications.

Biopolymer Nanofiber Properties

Nanofibers made of biopolymers have several important characteristics, such as:

- High surface-area-to-volume ratio
- Interconnected pore structure and porosity
- Flexibility and mechanical strength
- Bioactivity and biodegradability
- They are perfect candidates for tissue scaffolds because of their qualities.

METHODS OF FABRICATION

A variety of techniques are used to create biopolymer nanofibers:

The Process of Electrospinning

The method most frequently employed to create nanofibers is electrospinning. To create continuous fibers, a high voltage is applied to a polymer melt or solution. Viscosity, electrospinning parameters, and polymer content are some of the factors that affect fiber morphology.

One method that is interesting for turning polymeric biomaterials into nanofibers is electrospinning. By using a very straightforward experimental setup, this technique also provides the ability to adjust the porosity of the nanofiber meshes as well as the thickness and composition of the nanofibers [19].

The concept of electrospinning or electro-spraying has been around for over a century; however, polymeric nanofibers by electrospinning have garnered significant attention during the past ten years. Potential candidates for tissue engineering applications, electrospun nanofibers' high surface area and high porosity enable advantageous cell interactions [20].

Electrospinning technology has revolutionized the development of nanofibers, particularly with the incorporation of polysaccharides like starch, chitosan, and cellulose. These polysaccharide-based nanofiber membranes exhibit great potential across various sectors, including filtration, wound care, food preservation, and electronic monitoring, due to their unique properties, like high surface area and porosity. While the current advancements are promising, challenges persist, requiring further research to optimize their applications and unlock their full potential for improving modern living standards [21].

Herbal extracts, such as *Nigella Sativa*, curcumin, chamomile, neem, and nettle, have demonstrated potential in promoting wound healing due to their antibacterial, antioxidant, and anti-inflammatory effects. However, the inherent instability of these extracts highlights the need for advanced wound dressings to optimize their efficacy. The development of nanofibrous materials through electrospinning offers an effective platform for creating dressings with enhanced biocompatibility, moisture retention, and mechanical strength. By integrating herbal extracts into these innovative materials, modern wound care can potentially achieve better healing outcomes while reducing complications like adhesion and inflammation [22].

The keratin-based nanofiber scaffolds developed through electrospinning exhibit significant promise for skin tissue engineering applications. The combination of keratin and PVA resulted in uniformly interconnected nanofibers with diameters ranging from 100 to 250 nm, as confirmed by SEM analysis. FTIR (Fourier Transform Infrared Spectroscopy) and XRD (X-Ray Diffraction) analyses indicated that hydrogen bonding was crucial in the interaction between the two materials. The scaffolds supported cell adhesion, infiltration, and growth without cytotoxic effects, as shown in *in vitro* cell culture studies. Furthermore, the co-culture study demonstrated the scaffold's ability to replicate the natural structure of the epidermal and dermal layers, making it a viable option for skin regeneration applications [23].

This study demonstrates the effective production of nanofiber membranes using polycaprolactone (PCL) and PCL-gelatin blends *via* electrospinning. By fine-tuning polymer concentrations and selecting appropriate solvent combinations, the solution viscosity was controlled, leading to nanofibers with favorable physical, mechanical, and thermal characteristics. The incorporation of gelatin into PCL significantly enhanced the bioactivity of the scaffolds, fostering cell growth, which is essential for tissue engineering and other biomedical applications. These findings contribute to improving the functionality of PCL-based biomaterials, paving the way for innovative advancements in the biomedical field [24].

Polysaccharides offer significant potential in cartilage tissue engineering due to their biocompatibility, structural resemblance to the extracellular matrix, and adaptability to chemical modifications. While challenges remain, particularly in electrospinning techniques due to solubility, viscosity, and surface tension issues, various solutions, such as the incorporation of carrier polymers and the optimization of solvent systems, have shown promise. Polysaccharides like alginate, chitosan, hyaluronic acid, chondroitin sulfate, and cellulose stand out for their potential in cartilage regeneration. Continued research and technological advancements are essential to fully realize the benefits of polysaccharide-based nanofibers in tissue engineering applications [25].

Electrospinning has proven to be a highly effective method for creating nanofibers from various polymeric materials, including chitin. The unique properties of chitin, such as its biodegradability, non-toxicity, and biocompatibility, have made it an attractive option for a wide range of industrial and biomedical applications. Despite its potential, chitin's high crystallinity and limited solubility present challenges that require appropriate solvent systems for efficient electrospinning. This review has highlighted the importance of choosing the right solvents and optimizing processing parameters to achieve high-quality chitin nanofibers. Future research and development will likely focus on expanding the applications of chitin nanofibers in fields such as wound care, drug delivery, and sustainable packaging [26].

Collagen and gelatin are pivotal natural biopolymers extensively applied in tissue engineering and biomaterials due to their remarkable physicochemical and biocompatibility properties. Electrospinning technology allows the transformation of these materials into nanofibrous structures with high surface area, mimicking the extracellular matrix and making them ideal for biomedical applications. However, challenges such as rapid degradation in aqueous environments limit their use, making cross-linking essential for enhancing stability and functionality. This review has highlighted key research on electrospun collagen and gelatin, exploring their structural, mechanical, and biological characteristics. The advancements in electrospinning processes and cross-linking techniques have broadened their potential, despite ongoing challenges, positioning these biofibers as valuable candidates for a range of biomedical applications [27].

The development of CE-Lalb nanofibers represents a significant advancement in addressing the challenge of multidrug-resistant bacterial infections that impede wound healing. By integrating α-lactalbumin with cephalexin and epigallocatechin through electrospinning, we have created a scaffold that not only exhibits potent antimicrobial properties but also enhances key processes in wound healing, such as fibroblast migration, proliferation, and collagen synthesis. The promising *in vitro* and *in vivo* results highlight the efficacy of CE-Lalb NFs in reducing bacterial infections, promoting tissue regeneration, and minimizing scar formation. These findings underscore the potential of such tailored scaffolds to serve as innovative and personalized wound dressings, offering noninvasive and effective therapeutic solutions for patients with infected wounds. The success of this approach paves the way for further research and development in the field of wound care, addressing an urgent need in contemporary healthcare [28].

This study successfully demonstrated the fabrication of superparamagnetic-fluorescent bioactive glasses in various forms, including particles, nanofibers, and 3D scaffolds, by integrating maghemite nanoparticles and photoluminescent rare earth ions through sol-gel, electrospinning, and robocasting techniques. The investigation into the *in vitro* cytotoxicity and hemolytic activity revealed a concentration-dependent cytotoxic response for particle and nanofiber forms, while 3D scaffolds exhibited biocompatibility with no cytotoxic effects on the tested cell lines. Furthermore, the drug loading and release profiles indicated that 3D scaffolds had lower loading rates but demonstrated significant release capabilities, varying based on the morphology of the bioactive glasses and the pH of the medium. These findings highlight the potential of these bioactive glasses for innovative applications in tissue engineering and cancer therapy, paving the way for future research and development in this promising area [29].

The rising incidence of diabetes underscores the need for effective wound-healing strategies, particularly for diabetic ulcers. This study successfully presents novel hybrid nanofibrous scaffolds created from PVA/CS and Gel/PCL polymers using a double-nozzle electrospinning technique. By examining various Gel/PCL blend ratios, we identified the optimal combination of PVA/CS (80:20)-Gel/PCL (80:20), which demonstrated superior mechanical properties, a favorable contact angle, and excellent cytocompatibility with L-929 fibroblast cells. These results indicate the scaffold's potential to promote cell proliferation and integration, making it a promising candidate for improving diabetic wound healing outcomes. The advancements achieved in this research mark a significant contribution toward developing effective solutions for the complexities of diabetic wound care, offering hope for enhanced treatment options in clinical practice [30].

The mechanical characteristics of various tissues are intricately linked to the composition and structural arrangement of the nanofibrous extracellular matrix. The crimped microstructure of collagen nanofibers, prevalent in tissues like blood vessels, tendons, and heart valves, contributes to the distinctive non-linear 'J-shaped' stress-strain behavior observed in these tissues. By employing a nanofabrication technique based on electrospinning, this study successfully developed two-component hybrid electrospun fibrous materials that replicate the microstructure and mechanical properties of vascular tissue. The fabrication parameters can be adjusted to achieve precise optimization of these properties. Furthermore, the creation of tubular grafts with biomimetic structures underscores the feasibility of this fabrication method for applications in vascular graft replacement, showcasing the potential to closely match the geometry and compliance of natural blood vessels through careful optimization of graft microstructure [31].

This study demonstrates the successful transformation of jute microfiber into Holo and Alpha-cellulose nanofibers (NFs), achieving a significant reduction in size through electrospinning techniques. The characterization of these materials confirmed their unique morphological, physicochemical, and thermal properties. Notably, while both Holo and Alpha cellulose nanofibers displayed excellent mechanical performance and liquid absorption capabilities, Alpha CNF exhibited superior morphological stability and slower biodegradation rates. Importantly, the biocompatibility assays indicated that neither type of cellulose nanofiber was cytotoxic to COS-7 cells, promoting cell viability and proliferation. These findings underscore the potential of jute-derived cellulose as a promising candidate for developing advanced cellulose-based nano biomaterials, opening new avenues for applications in various fields, including biomedicine and material science [32].

Phase Partition

Porous nanofibers can be produced using phase separation processes, including solvent casting and non-solvent-induced phase separation. This technique is useful for producing fibers with regulated pore diameters.

Freeze Drying

Freeze drying, also known as lyophilization, is a technique used to turn solutions containing sensitive materials into solids, making them stable and usable across fields like food technology, pharmaceuticals, and enzyme preservation. In this method, the solution is first frozen at very low temperatures, typically between -70°C and -80°C. The frozen sample is then placed in a low-pressure chamber under partial vacuum, where ice within the material is removed by sublimation (changing directly from solid to gas). Next, any remaining unfrozen water is removed through a process called desorption. A significant limitation of scaffolds created by freeze drying is that they form solid, non-interconnected pore walls, which can restrict cell growth and the flow of nutrients [33].

Foaming

Foaming uses safe, soluble gases like CO_2 or N_2 to create pores in polymers. This technique can be applied to composite materials made of polymer and bio-ceramics, as well as single polymers, especially for solid tissue engineering structures. One major advantage of foaming is that it does not require solvents, which avoids the risk of leftover solvent residue. It also uses low temperatures, so the polymer is less likely to degrade during processing. However, this method produces scaffolds with closed surfaces and isolated pores, which limits the transport of nutrients through the scaffold. In certain cases, open pores can be achieved, but they tend to be too small for tissue engineering applications [34, 35].

Self-Assembly

Self-assembly refers to the self-organization of components into specific patterns or structures to create different types of nanofibers. This process can happen through either covalent or non-covalent interactions in biological molecules. Many tiny protein filaments, or peptides, can join together to form nanofibers that mimic the physical environment of cells in the body. These nanofibers cover cell surfaces, acting like cables that connect and support neighboring cells by forming 3D networks. In bone tissue engineering, scientists explore certain peptide amphiphiles to create nanofibers using a pH-driven self-assembly process [36, 37].

Scaling Up Biopolymer Nanofiber Production for Quality and Reproducibility

1. Challenges Associated with Materials

Variability of the Source: Biopolymers that are naturally occurring, such as polysaccharides and proteins, frequently display batch-to-batch variability in molecular weight, content, and purity, which can have an impact on the development of fibers [38].

Processing and Solvent Stability

The selection of a suitable solvent system is of the utmost importance. Many biopolymers require solvents that are either aqueous or environmentally friendly, which might have an impact on the fiber morphology and yield. Biopolymers face the challenges related to biodegradability and storage stability. They have the potential to decay or undergo unwelcome structural changes while they are being stored, which can result in inconsistencies in the characteristics of the fibers [39].

2. Challenges in Process Reproducibility

Electrospinning Parameters: It is difficult to achieve repeatability on a wide scale since even minute changes in parameters like voltage, flow rate, and humidity can have a major impact on the diameter and porosity of the fiber.

Shear and Thermal Sensitivity: During the process of extrusion-based approaches, the creation of fibers might be affected by the fact that many biopolymers are sensitive to heat and shear forces.

Scalability of Equipment: The electrospinning setups used in laboratories do not necessarily translate well to the production processes used in industrial settings. This is because there are changes in the nozzle arrangements, spinneret designs, and collecting procedures [40].

3. Quality Control Challenges

Morphological Uniformity: Production on a large scale has several challenges, including the control of fiber diameter, porosity, and alignment. The mechanical strength and bioactivity of a substance can be affected by variation. It is challenging to ensure that fibers are cross-linked or functionalized in a uniform manner with medicines, proteins, or nanoparticles when you are manufacturing in bulk.

Sterility and Contamination Management: Biopolymer nanofibers are also commonly used in biomedical fields, which require very high standards of sterility. This enhances the complexity of production.

4. Mechanical and Structural Stability

Differences in the molecular weight of the polymer, the process conditions for processing the nanofibers, and environmental conditions can all be responsible for the inconsistency of their mechanical properties. Numerous biopolymers are hygroscopic, indicating that they would be able to distort or break down when put into contact with excessive humidity, and this will affect the functionality of the product.

5. Regulatory and Economic Challenges

Due to variances in the fiber qualities, compliance with regulatory agencies like the FDA, EMA, or other regulatory agencies for biomedical uses may be a difficult task. Due to the possibility of huge costs involved in biopolymer extraction, purification, and processing, mass production can be very expensive.

Potential Solutions for Scale-Up

Process Optimization: Creating a reliable and automated electrospinning process or developing other techniques of producing fibers (such as centrifugal spinning or solution blow spinning, for example). Achieving and maintaining fiber uniformity through the utilization of real-time monitoring tools such as artificial intelligence-based image analysis, spectroscopy, and rheology measurements.

Standardization of Raw Materials: Utilising standardized biopolymer purification and modification procedures to limit the amount of variation that occurs from batch to batch [41, 42].

MOLECULAR MECHANISMS OF BIOPOLYMER-BASED NANOFIBERS

Integrin-mediated adhesion, controlled release of bioactive chemicals, intracellular signaling cascades, and immunomodulatory effects are some of the ways that biopolymer-based nanofibers interact with cells and play a critical role in tissue engineering. Because they cooperate to promote tissue regeneration and healing in a range of biomedical applications, these processes are promising strategies in regenerative medicine. They mimic the natural extracellular matrix (ECM) and support cell adhesion, proliferation, differentiation, and tissue regeneration.

Cell Adhesion: By interacting with integrins and other cell surface receptors, the biopolymers give functional groups (such as hydroxyl, carboxyl, and amine) that make it easier for cells to adhere to one another. It is important to investigate the function that biomimetic alterations, such as RGD peptides, play in the process of increasing adhesion.

Cell Proliferation: To control the proliferation of cells, biopolymers use bioactive signals, mechanical qualities, and breakdown byproducts as their mechanisms of action. Consideration should also be given to the impact that the porosity and stiffness of the polymer have on the development of cells.

Cell Differentiation: The biopolymer nanofibers serve as scaffolds to influence the fate of stem cells through the transmission of biochemical cues (such as the supply of growth factor) and biophysical features (such as fiber alignment and mechanical stiffness) [43 - 45].

Tissue Engineering's Common Nanomaterials

Nanomaterials are frequently employed in tissue engineering to foster better cellular responses, maintain structural integrity, and speed up the healing process.

The following are some of the nanomaterials that are most commonly used:

1. Using Nanofibers

For skin, cartilage, and bone tissue engineering, nanofibers are perfect because of their large surface area and ability to replicate the extracellular matrix, which promotes cell adhesion, proliferation, and differentiation. Nanofibers include things like silk fibroin, collagen, chitosan, polycaprolactone (PCL), polylactic acid (PLA), and polyglycolic acid (PGA) [46, 47]

2. Using Small Particles.

Bioactive chemicals can be delivered by nanoparticles, which can also improve imaging and encourage osteogenesis. For example, silver has antibacterial qualities that aid in wound healing, while HA is frequently utilized for bone regeneration. Carbon nanotubes (CNTs), hydroxyapatite (HA), silica, gold, silver, and iron oxide are a few examples of nanoparticles [48, 49].

3. Use of Nanocomposites

They are appropriate for applications like cartilage and bone scaffolding that demand both flexibility and strength because they improve mechanical qualities, bioactivity, and cell contact. Collagen and hydroxyapatite, chitosan and graphene oxide, or PLA and nano-silica are a few examples of nanocomposites [50].

4. Nanogels

Since nanogels create a moist environment that promotes cell growth, they are hydrophilic networks perfect for controlled release and medication administration in wound healing and cartilage regeneration. Hyaluronic acid, polyacrylamide, chitosan, and polyethylene glycol (PEG) are a few examples of nanogels.

5. Graphene and Graphene Oxide

Materials based on graphene are useful for tissue engineering of the heart and nerves because they are conductive. They are even more useful because of their strong mechanical strength and biocompatibility.

6. Carbon Nanotubes (CNTs)

Apart from their mechanical strength, CNTs can enhance electrical conductivity, which facilitates muscle and nerve regeneration. Nevertheless, potential cytotoxicity concerns limit their utilization [51, 52].

7. Peptides That Self-assemble

Many studies are being conducted on these peptides for nerve, cartilage, and skin tissue engineering because they create nanofiber structures that resemble the extracellular matrix and offer a favorable environment for cell proliferation. The functionality and integration of engineered tissues are improved by nanomaterials' adaptability, which allows their properties to be modified to match the particular needs of different tissue types [53].

New Materials in Biopolymer Nanofibers

a. Natural Polymers

Silk Fibroin: Offers excellent mechanical properties and biodegradability.

Chitosan: Antimicrobial, bioactive, and supports cell adhesion.

Collagen and Gelatin: Mimic native ECM, enhancing cell growth.

Alginate: Provides hydration and is used in wound healing scaffolds.

Polysaccharides (Tamarind, Fenugreek, Cellulose derivatives): Biocompatible and sustain drug delivery.

b. Synthetic Polymers (Blended with Biopolymers)

Polycaprolactone (PCL): Slow degradation, excellent mechanical strength.

Poly(lactic-co-glycolic acid) (PLGA): FDA-approved, tunable degradation.

Polyurethane (PU): Enhances elasticity and mechanical stability.

c. Hybrid Nanocomposites

Carbon-based Materials (Graphene Oxide, Carbon Nanotubes): Improve conductivity, cell differentiation.

Hydroxyapatite (HA)-infused Fibers: Used for bone tissue engineering.

Metallic Nanoparticles (Ag, ZnO, CuO): Impart antimicrobial properties [54, 55].

NOVEL FABRICATION TECHNIQUES

a.Electrospinning Innovations

Coaxial Electrospinning: Encapsulates drugs and enhances controlled release.

Emulsion Electrospinning: Integrates hydrophilic and hydrophobic phases.

Magnetically Assisted Electrospinning: Aligns fibers and enhances cell guidance [56].

b.3D Bioprinting

Electrohydrodynamic Jet Printing: Produces hierarchical nanofiber structures.

Hybrid 3D Printing: Combines micro/nanofibers with scaffolds.

c.Self-assembly and Phase Separation

Peptide-based Nanofibers: Improve cell signaling and tissue regeneration.

Thermally Induced Phase Separation (TIPS): Creates porous structures [57].

APPLICATIONS IN TISSUE ENGINEERING

Wound Healing: Antimicrobial chitosan/Ag nanofibers [58].

Bone Regeneration: HA/PCL nanofibers mimic mineralized ECM [59].

Cartilage Tissue Engineering: Collagen/PLGA nanofibers provide mechanical support.

Neural Tissue Regeneration: Conductive polymeric nanofibers with graphene.

Cardiac Tissue Engineering: Electroactive scaffolds for myocardium repair [60].

Nanofibers based on biopolymers have demonstrated enormous promise in a range of tissue engineering applications:

Regeneration of the Skin

Collagen or chitosan-based nanofibrous scaffolds have shown promise in accelerating skin regeneration. Their shape facilitates cell migration and proliferation by imitating the natural extracellular matrix.

The development of sulfated hyaluronic acid (SHA)/collagen-based nanofibrous biomimetic skins represents a significant advancement in addressing diabetic foot ulcers (DFUs) and enhancing wound healing in diabetic patients. The successful synthesis of SHA and its integration into a collagen matrix, along with the incorporation of polyurethane, resulted in hybrid nanofiber scaffolds that exhibited improved mechanical properties and favorable morphological characteristics. The findings indicate that these hybrid scaffolds not only promote cell proliferation and maintain normal cellular phenotypes but also significantly accelerate wound healing and skin remodeling in diabetic models. Hence, SHA/COL-based hybrid scaffolds show immense promise as efficient biomimetic treatments for diabetic wounds and provide new hope for patients of this debilitating complication of diabetes [61].

The hybrid dressing made from silk fibroin/polyvinyl alcohol nanofibers and sodium alginate/gum tragacanth hydrogel, which is modified with cardamom extract, shows immense promise for wound healing applications. The successful production of uniform nanofibers, along with desirable physical and mechanical properties, reflects the compatibility and structural integrity of the hybrid system. The favorable swelling behavior, controlled drug release, and excellent antibacterial activity against major pathogens further support its potential for skin tissue engineering. This novel approach reflects the potential of natural-synthetic polymer composites in the development of wound care technologies [62].

This research efficiently proves the application of electrospun nanofibers of polyurethane and hydroxypropyl methylcellulose, especially the extract-loaded ones, as novel wound dressings. The results demonstrate that the ideal mat composition (PU90/HPMC10) possesses outstanding characteristics such as higher biodegradability, water vapor permeability, and porosity with non-toxicity and desirable hemocompatibility. Furthermore, the notable increase in cell proliferation and viability and strong antibacterial activity against causative bacteria reinforces the effectiveness of these nanofiber mats. These findings confirm that the novel wound dressings developed herein are likely to find significant applications in healing injured skin tissues, where they may prove to be particularly useful in alleviating serious impediments in the management and cure of wounds [63].

The creation of an efficient antibacterial wound dressing is still a major challenge in clinical environments, especially with the rise of antibiotic resistance. The newly created two-layer wound dressing in this work, which couples decellularized bovine skin tissue with an antibacterial nanofiber layer, shows substantial potential for overcoming these challenges. Our results show that the DBS-PVA/CS/Abs scaffold not only replicates the natural extracellular matrix but also has better mechanical properties, high porosity, and good biocompatibility. Notably, the scaffold's high antibacterial activity against both normal and drug-resistant bacterial strains indicates its potential to significantly diminish the risk of post-wound infections. This novel dressing establishes a good platform for future preclinical and clinical studies to further optimize the treatment of infectious skin wounds and improve patient outcomes [64].

Engineering Bone Tissue

To enhance bone regeneration, bioactive materials such as hydroxyapatite have been blended with biopolymers such as gelatin and hyaluronic acid. These composites facilitate osteoconduction and osteoinduction. The synthesis of thermosensitive bioinks based on chitosan enriched with self-assembled aggregates of nanofibers and nanohydroxyapatite offers a bright future in the field of bone tissue engineering. Traditional chitosan bioinks have drawbacks of poor printability and low mechanical strength, whereas the enhancements developed in this study dramatically improved both properties. The best formulation, having a printability rate of 10% nanohydroxyapatite and showing more than 88% cell viability, emphasizes the potential of the bioink for efficient use in tissue regeneration. Additionally, the low water uptake and controlled degradation in lysozyme indicate its applicability in maintaining cellular environments, with an elastic modulus of 15.5 kPa and significant alkaline phosphatase (ALP) activity with osteogenic properties, this bio-ink composite meets not only the mechanical and biocompatibility properties but also elevates cellular performance, making it a strong contender for future applications in bone tissue engineering [50].

Repairing Cartilage

Chitosan and gelatin nanofibers have demonstrated promising outcomes in cartilage tissue engineering, as they provide a suitable setting for chondrocyte development and differentiation. This research emphasizes the effectiveness of surface modifications on electrospun poly(ε-caprolactone) nanofibers in facilitating the chondrogenic differentiation of bone marrow-derived mesenchymal stromal cells (BMSCs). Utilizing two different surface modification methods—plasma polymerization and chondroitin sulfate immobilization—showed that incorporation of polar functional groups can accelerate the cells' proliferation, matrix deposition, and production of important extracellular matrix proteins. Interestingly, the carboxylic acid-enriched nanofibers proved to be the most potent in supporting chondrogenesis, as indicated by high aggrecan, Sox9, and collagen II levels, accompanied by a decrease in hypertrophic markers. The results indicate that surface properties can be tailored to engineer nanofibers with the best chondro-inductive properties, opening up avenues for novel approaches in cartilage tissue engineering and regeneration [18].

Regeneration of Nerves

The application of silk fibroin (SF) and polycaprolactone (PCL) electrospun nanofibers for peripheral nerve regeneration has been investigated. Their composition supports the growth and alignment of neurons. The creation of the Janus nanofibrous scaffold, composed of SF and PCL, represents a significant advancement in bone tissue engineering. With the incorporation of selenium nanoparticles (SeNPs) and nano-hydroxyapatite (nHA), the scaffold successfully overcomes the limitations of conventional two-dimensional scaffolds, ensuring improved cell penetration and differentiation. The scaffold's unique framework, where the outer layer ensures cell guidance and the inner layer facilitates cell adhesion, demonstrates its multifunctionality. Moreover, the antibacterial properties of SeNPs and the osteogenic potential of the scaffold make it a promising candidate for clinical applications in bone reconstruction. Overall, the SF/PCL-based Janus nanofibrous scaffold serves as an innovative biomaterial for bone regeneration, making it highly valuable in regenerative medicine [48].

This research further highlights the versatility of electrospun nanofibers composed of polycaprolactone, collagen, and bioactive components such as hydroxyapatite (HA) in the development of advanced drug delivery systems. By leveraging the biocompatibility and mechanical properties of these materials, particularly concerning Cetirizine, key challenges such as limited solubility and rapid drug re-crystallization can be addressed. The incorporation of hydroxyapatite into the nanofibers not only enhances bioactivity but also modulates drug release kinetics, enabling a controlled and prolonged release of Cetirizine. Characterization techniques such as FESEM, HRTEM, and FTIR provided critical insights into the structural and functional properties of the nanofibers, confirming their suitability for tissue engineering applications. Ultimately, this study underscores the potential of these sophisticated nano scaffolds in developing effective drug delivery systems, paving the way for further research and clinical applications in topical disease treatments [65].

Additionally, the development of cost-effective triple-layered nanofibrous bandages presents a viable solution for addressing challenges associated with chronic wound healing. These bandages are fabricated using a layer-by-layer strategy, consisting of:

1. A hydrophilic polyvinyl alcohol (PVA) outermost layer,
2. A middle antibacterial cellulose acetate layer embedded with silver nanoparticles, and

3. A hydrophobic polycaprolactone (PCL) bottom layer.

This specific nanofibrous morphology exhibits excellent mechanical properties, enhanced wettability, and sustained drug release, demonstrating potent antibacterial activity against both Gram-negative and Gram-positive bacteria. Furthermore, *in vitro* and *in vivo* studies confirm the hemocompatibility, biocompatibility, and effectiveness of these bandages in achieving full-thickness wound healing. Future research is necessary to assess their clinical efficacy and refine their therapeutic potential for wound treatment [66].

OBSTACLES AND PROSPECTS

Biopolymer-based nanofibers hold much promise for therapeutic applications, but several hindrances limit their widespread use:

Lower Mechanical Strength-Biopolymer-based nanofibers are generally less mechanically strong compared to synthetic polymers.

Variability-Natural biopolymers exhibit inherent variability, leading to inconsistencies in performance.

Scalability and Fabrication Consistency-The large-scale production and reproducibility of biopolymer-based nanofibers remain challenging.

To enhance mechanical properties without compromising biocompatibility, future research should focus on developing hybrid scaffolds that combine synthetic and biopolymer-based materials. Additionally, incorporating growth factors and bioactive molecules could further improve cellular response [67].

LONG-TERM EFFECTS OF BIOPOLYMER NANOFIBERS ON TISSUE REGENERATION

Enhanced Cell Adhesion and Proliferation

Nanofibers serve as a framework that facilitates the attachment of cells, as well as proliferation and differentiation [68]. The enhancement of cellular responses can be achieved through the functionalization of bioactive compounds, such as growth factors and peptides.

Controlled Degradation and Remodeling

With time, natural biopolymers such as chitosan, collagen, gelatin, and silk fibroin break down, enabling progressive tissue remodeling. To prevent early scaffold collapse or an excessive accumulation of byproducts, the rate of degradation should be proportional to the rate at which tissue is formed.

Improved Vascularization

Certain biopolymer nanofibers stimulate angiogenesis, which is necessary for the delivery of oxygen and nutrients to tissues that are in the process of regeneration. Enhanced tissue function and prevention of fibrosis are both benefits of Long-term vascularization.

Immune Response and Biocompatibility

Nanofibers made of biopolymers that have been developed correctly reduce the risk of chronic inflammation and immunological rejection. Certain naturally occurring biopolymers, such as alginate and hyaluronic acid, for example, have immunomodulatory properties that facilitate the healing of tissues.

Functional Tissue Regeneration

In applications involving the musculoskeletal system (bone, cartilage, and muscle), nanofibers enhance mechanical characteristics and tissue integration, respectively. Electrospun nanofibers can direct the formation of axons and strengthen synaptic connections during the process of brain regeneration.

Long-term Stability and Performance

Over extended periods, certain biopolymer nanofibers manifest mechanical instability, necessitating the implementation of reinforcement strategies such as crosslinking and composite formulations. Nanofiber scaffolds that have been optimized have been shown to sustain steady tissue function without causing any harmful effects, according to Long-term research [69].

CHALLENGES AND FUTURE PERSPECTIVES

1. Standardization and Scalability: To bring clinical translation into practice, reproducible fabrication processes are required.

2. Degradation Products: An in-depth analysis of the effects that degradation byproducts have on the tissues in the surrounding area is required.

3. Personalized Approaches: Improvements in treatment outcomes are possible thanks to developments in bioprinting and patient-specific scaffold design.

CONCLUSION

Nanofibers derived from biopolymers present significant advantages in tissue engineering due to their excellent biocompatibility and bioactivity. These fibers mimic the natural extracellular matrix, promoting cell adhesion, proliferation, and differentiation, which are crucial for tissue regeneration. Despite these benefits, challenges such as scalability, mechanical stability, and controlled degradation remain obstacles to their widespread clinical application. Overcoming these limitations requires extensive research into advanced fabrication techniques, optimized material compositions, and innovative crosslinking strategies to enhance their structural and functional properties. Recent advancements in electrospinning, 3D bioprinting, and self-assembly techniques have shown promise in improving nanofiber production, allowing for better control over fiber morphology and alignment.

Additionally, incorporating bioactive agents, growth factors, or nanoparticles into biopolymer-based nanofibers can further enhance their therapeutic potential. By addressing current limitations and refining fabrication methods, these nanofibers can pave the way for revolutionary applications in regenerative medicine, including wound healing, bone regeneration, and nerve repair. Continued interdisciplinary collaboration among material scientists, biomedical engineers, and clinicians is essential to accelerate their clinical translation, ultimately transforming the landscape of tissue engineering and regenerative therapies.

REFERENCES

[1] Sánchez-Téllez DA, Baltierra-Uribe SL, Vidales-Hurtado MA, Valdivia-Flores A, García-Pérez BE, Téllez-Jurado L. Novel PVA–hyaluronan–siloxane hybrid nanofiber mats for bone tissue engineering. Polymers (Basel) 2024; 16(4): 497.
[http://dx.doi.org/10.3390/polym16040497] [PMID: 38399875]

[2] Baranwal J, Barse B, Fais A, Delogu GL, Kumar A. Biopolymer: A sustainable material for food and medical applications. Polymers (Basel) 2022; 14(5): 983.
[http://dx.doi.org/10.3390/polym14050983] [PMID: 35267803]

[3] Jeevanandam J, Pan S, Rodrigues J, Elkodous MA, Danquah MK. Medical applications of biopolymer nanofibers. Biomater Sci 2022; 10(15): 4107-18.
[http://dx.doi.org/10.1039/D2BM00701K] [PMID: 35788587]

[4] Chen Y, Bera H, Guo X, Cun D, Yang M. Engineering of biopolymer-based nanofibers for medical uses. Tailor-Made and Functionalized Biopolymer Systems for Drug Delivery and Biomedical Applications. Woodhead Publishing Series in Biomaterials 2021; pp. 383-424.
[http://dx.doi.org/10.1016/B978-0-12-821437-4.00012-8]

[5] Vasita R, Katti DS. Nanofibers and their applications in tissue engineering. Int J Nanomedicine 2006; 1(1): 15-30.
[http://dx.doi.org/10.2147/nano.2006.1.1.15] [PMID: 17722259]

[6] Yamashita Y, Tanaka A, Miyake H, Higashiyama A, Kato H. Establishment of nano fiber preparation technique for nanocomposite. 16th International Conference on Composite Materials 2007.

[7] Stock UA, Vacanti JP. Tissue engineering: current state and prospects. Annu Rev Med 2001; 52(1): 443-51.
[http://dx.doi.org/10.1146/annurev.med.52.1.443] [PMID: 11160788]

[8] Sachlos E, Czernuszka JT. Making tissue engineering scaffold work: review on the application of SFF technology to the production of tissue engineering scaffolds. Eur Cell Mater 2003; 5: 29-40.
[http://dx.doi.org/10.22203/eCM.v005a03] [PMID: 14562270]

[9] Thomson JA, Itskovitz-Eldor J, Shapiro SS, *et al.* Embryonic stem cell lines derived from human blastocysts. Science 1998; 282(5391): 1145-7.
[http://dx.doi.org/10.1126/science.282.5391.1145] [PMID: 9804556]

[10] Winston R. Embryonic Stem cell research - The case for.... Nat Med 2001; 7(4): 396-7.
[http://dx.doi.org/10.1038/86442] [PMID: 11283652]

[11] Zvaifler NJ, Marinova-Mutafchieva L, Adams G, *et al.* Mesenchymal precursor cells in the blood of normal individuals. Arthritis Res Ther 2000; 2(6): 477-88.
[http://dx.doi.org/10.1186/ar130] [PMID: 11056678]

[12] Shimojo AAM, Perez AGM, Galdames SEM, Brissac ICS, Santana MHA. Performance of PRP associated with porous chitosan as a composite scaffold for regenerative medicine. Sci World J 2015; 2015(1): 396131.
[http://dx.doi.org/10.1155/2015/396131] [PMID: 25821851]

[13] Tabata Y. Biomaterial technology for tissue engineering applications. J R Soc Interface 2009; 6(Suppl 3)
[http://dx.doi.org/10.1098/rsif.2008.0448.focus]

[14] Almajidi YQ, Ponnusankar S, Chaitanya MVNL, *et al.* Chitosan-based nanofibrous scaffolds for biomedical and pharmaceutical applications: A comprehensive review. Int J Biol Macromol 2024; 264(Pt 2): 130683.
[http://dx.doi.org/10.1016/j.ijbiomac.2024.130683] [PMID: 38458289]

[15] Wang TJ, Rethi L, Ku MY, Nguyen HT, Chuang AEY. A review on revolutionizing ophthalmic therapy: Unveiling the potential of chitosan, hyaluronic acid, cellulose, cyclodextrin, and poloxamer in eye disease treatments. Int J Biol Macromol 2024; 273(Pt 2): 132700.
[http://dx.doi.org/10.1016/j.ijbiomac.2024.132700] [PMID: 38879998]

[16] Wang X, Wang B, Liu W, *et al.* Using chitosan nanofibers to simultaneously improve the toughness and sensing performance of chitosan-based ionic conductive hydrogels. Int J Biol Macromol 2024; 260(Pt 1): 129272.
[http://dx.doi.org/10.1016/j.ijbiomac.2024.129272] [PMID: 38211925]

[17] Feng Z, Wang S, Huang W, Bai W. A potential bilayer skin substitute based on electrospun silk-elastin-like protein nanofiber membrane covered with bacterial cellulose. Colloids Surf B Biointerfaces 2024; 234: 113677.
[http://dx.doi.org/10.1016/j.colsurfb.2023.113677] [PMID: 38043505]

[18] Asadian M, Tomasina C, Onyshchenko Y, *et al.* The role of plasma-induced surface chemistry on polycaprolactone nanofibers to direct chondrogenic differentiation of human mesenchymal stem cells. J Biomed Mater Res A 2024; 112(2): 210-30.
[http://dx.doi.org/10.1002/jbm.a.37607] [PMID: 37706337]

[19] Jayaraman K, Kotaki M, Zhang Y, Mo X, Ramakrishna S. Recent advances in polymer nanofibers. J Nanosci Nanotechnol 2004; 4(1-2): 52-65.
[PMID: 15112541]

[20] Doshi J, Reneker DH. Electrospinning process and applications of electrospun fibers. J Electrost 1995; 35(2-3): 151-60.
[http://dx.doi.org/10.1016/0304-3886(95)00041-8]

[21] Su W, Chang Z, e Y, *et al.* Electrospinning and electrospun polysaccharide-based nanofiber

membranes: A review. Int J Biol Macromol 2024; 263(Pt 2): 130335.
[http://dx.doi.org/10.1016/j.ijbiomac.2024.130335] [PMID: 38403215]

[22] Sharifi M, Bahrami SH. Review on application of herbal extracts in biomacromolecules-based nanofibers as wound dressings and skin tissue engineering. Int J Biol Macromol 2024; 277(Pt 2): 133666.
[http://dx.doi.org/10.1016/j.ijbiomac.2024.133666] [PMID: 38971295]

[23] Aadil KR, Nathani A, Rajendran A, Sharma CS, Lenka N, Gupta P. Investigation of human hair keratin-based nanofibrous scaffold for skin tissue engineering application. Drug Deliv Transl Res 2024; 14(1): 236-46.
[http://dx.doi.org/10.1007/s13346-023-01396-7] [PMID: 37589816]

[24] Rodríguez-Martín M, Aguilar JM, Castro-Criado D, Romero A. Characterization of gelatin-polycaprolactone membranes by electrospinning. Biomimetics (Basel) 2024; 9(2): 70.
[http://dx.doi.org/10.3390/biomimetics9020070] [PMID: 38392116]

[25] Arash A, Dehgan F, Zamanlui Benisi S, Jafari-Nodoushan M, Pezeshki-Modaress M. Polysaccharide base electrospun nanofibrous scaffolds for cartilage tissue engineering: Challenges and opportunities. Int J Biol Macromol 2024; 277(Pt 1): 134054.
[http://dx.doi.org/10.1016/j.ijbiomac.2024.134054] [PMID: 39038580]

[26] Dzolkifle NAN, Wan Nawawi WMF. A review on chitin dissolution as preparation for electrospinning application. Int J Biol Macromol 2024; 265(Pt 1): 130858.
[http://dx.doi.org/10.1016/j.ijbiomac.2024.130858] [PMID: 38490398]

[27] Larue L, Michely L, Grande D, Belbekhouche S. Design of collagen and gelatin-based electrospun fibers for biomedical purposes: an overview. ACS Biomater Sci Eng 2024; 10(9): 5537-49.
[http://dx.doi.org/10.1021/acsbiomaterials.4c00948] [PMID: 39092811]

[28] Khan NU, Chengfeng X, Jiang MQ, *et al.* α-Lactalbumin based scaffolds for infected wound healing and tissue regeneration. Int J Pharm 2024; 663: 124578.
[http://dx.doi.org/10.1016/j.ijpharm.2024.124578] [PMID: 39153643]

[29] Deliormanlı AM, Rahman B, Atmaca H. *In vitro* cytotoxicity of magnetic-fluorescent bioactive glasses on SaOS-2, MC3T3-E1, BJ fibroblast cells, their hemolytic activity, and sorafenib release behavior. Biomaterials Advances 2024; 158: 213782.
[http://dx.doi.org/10.1016/j.bioadv.2024.213782] [PMID: 38377664]

[30] Ranjbar-Mohammadi M, Tajdar F, Esmizadeh E, Arab Z. Co electrospinning -poly (vinyl alcohol)-chitosan/gelatin-poly (ϵ -caprolacton) nanofibers for diabetic wound-healing application. Biomed Mater 2024; 19(4): 045017.
[http://dx.doi.org/10.1088/1748-605X/ad4df6] [PMID: 38768605]

[31] Beachley V, Kuo J, Kasyanov V, Mironov V, Wen X. Biomimetic crimped/aligned microstructure to optimize the mechanics of fibrous hybrid materials for compliant vascular grafts. J Mech Behav Biomed Mater 2024; 150: 106301.
[http://dx.doi.org/10.1016/j.jmbbm.2023.106301] [PMID: 38141364]

[32] Haider MK, Davood K, Kim IS. "Micro-to-nano": Reengineering of jute for constructing cellulose nanofibers as a next-generation biomaterial. Int J Biol Macromol 2024; 261(Pt 2): 129872.
[http://dx.doi.org/10.1016/j.ijbiomac.2024.129872] [PMID: 38302019]

[33] Whang K, Thomas CH, Healy KE, Nuber G. A novel method to fabricate bioabsorbable scaffolds. Polymer (Guildf) 1995; 36(4): 837-42.
[http://dx.doi.org/10.1016/0032-3861(95)93115-3]

[34] Singh L, Kumar V, Ratner BD. Generation of porous microcellular 85/15 poly (dl-lactide-co-glycolide) foams for biomedical applications. Biomaterials 2004; 25(13): 2611-7.
[http://dx.doi.org/10.1016/j.biomaterials.2003.09.040] [PMID: 14751747]

[35] Wang X, Li W, Kumar V. A method for solvent-free fabrication of porous polymer using solid-state

foaming and ultrasound for tissue engineering applications. Biomaterials 2006; 27(9): 1924-9.
[http://dx.doi.org/10.1016/j.biomaterials.2005.09.029] [PMID: 16219346]

[36]   Whitesides GM, Grzybowski B. Self-assembly at all scales. Science 2002; 295(5564): 2418-21.
[http://dx.doi.org/10.1126/science.1070821] [PMID: 11923529]

[37]   Zhang S, Gelain F, Zhao X. Designer self-assembling peptide nanofiber scaffolds for 3D tissue cell
cultures. Semin Cancer Biol 2005; 15(5): 413-20.
[http://dx.doi.org/10.1016/j.semcancer.2005.05.007] [PMID: 16061392]

[38]   de Lima Nascimento TR, de Amoêdo Campos Velo MM, Silva CF, *et al.* Current applications of
biopolymer-based scaffolds and nanofibers as drug delivery systems. Curr Pharm Des 2019; 25(37):
3997-4012.
[http://dx.doi.org/10.2174/1381612825666191108162948] [PMID: 31701845]

[39]   Benalaya I, Alves G, Lopes J, Silva LR. A review of natural polysaccharides: sources, characteristics,
properties, food, and pharmaceutical applications. Int J Mol Sci 2024; 25(2): 1322.
[http://dx.doi.org/10.3390/ijms25021322] [PMID: 38279323]

[40]   Abdulhussain R, Adebisi A, Conway BR, Asare-Addo K. Electrospun nanofibers: Exploring process
parameters, polymer selection, and recent applications in pharmaceuticals and drug delivery. J Drug
Deliv Sci Technol 2023; 90: 105156.
[http://dx.doi.org/10.1016/j.jddst.2023.105156]

[41]   Marjuban SMH, Rahman M, Duza SS, *et al.* Recent advances in centrifugal spinning and their
applications in tissue engineering. Polymers (Basel) 2023; 15(5): 1253.
[http://dx.doi.org/10.3390/polym15051253] [PMID: 36904493]

[42]   Paliwal R, Babu RJ, Palakurthi S. Nanomedicine scale-up technologies: feasibilities and challenges.
AAPS PharmSciTech 2014; 15(6): 1527-34.
[http://dx.doi.org/10.1208/s12249-014-0177-9] [PMID: 25047256]

[43]   Ruoslahti E, Pierschbacher MD. New perspectives in cell adhesion: RGD and integrins. Science 1987;
238(4826): 491-7.
[http://dx.doi.org/10.1126/science.2821619] [PMID: 2821619]

[44]   Han SB, Kim JK, Lee G, Kim DH. Mechanical properties of materials for stem cell differentiation.
Adv Biosyst 2020; 4(11): 2000247.
[http://dx.doi.org/10.1002/adbi.202000247] [PMID: 33035411]

[45]   Lim SH, Liu XY, Song H, Yarema KJ, Mao HQ. The effect of nanofiber-guided cell alignment on the
preferential differentiation of neural stem cells. Biomaterials 2010; 31(34): 9031-9.
[http://dx.doi.org/10.1016/j.biomaterials.2010.08.021] [PMID: 20797783]

[46]   Zhang R, Zhang Y, Zhang Q, Xie H, Qian W, Wei F. Growth of half-meter long carbon nanotubes
based on Schulz-Flory distribution. ACS Nano 2013; 7(7): 6156-61.
[http://dx.doi.org/10.1021/nn401995z] [PMID: 23806050]

[47]   Endo M, Strano MS, Ajayan PM. Potential applications of carbon nanotubes. In: Jorio A, Dresselhaus
G, Dresselhaus MS, Eds. Top Appl Phys. 2008; 13: p. 62.
[http://dx.doi.org/10.1007/978-3-540-72865-8_2]

[48]   Tang Z, Li J, Fu L, *et al.* Janus silk fibroin/polycaprolactone-based scaffold with directionally aligned
fibers and porous structure for bone regeneration. Int J Biol Macromol 2024; 262(Pt 1): 129927.
[http://dx.doi.org/10.1016/j.ijbiomac.2024.129927] [PMID: 38311130]

[49]   Zadegan S, Vahidi B, Nourmohammadi J, Shojaee A, Haghighipour N. Evaluation of rabbit adipose
derived stem cells fate in perfused multilayered silk fibroin composite scaffold for Osteochondral
repair. J Biomed Mater Res B Appl Biomater 2024; 112(3): e35396.
[http://dx.doi.org/10.1002/jbm.b.35396] [PMID: 38433653]

[50]   Bharadwaj T, Chrungoo S, Verma D. Self-assembled chitosan/gelatin nanofibrous aggregates
incorporated thermosensitive nanocomposite bioink for bone tissue engineering. Carbohydr Polym

2024; 324: 121544.
[http://dx.doi.org/10.1016/j.carbpol.2023.121544] [PMID: 37985063]

[51] Baughman RH, Zakhidov AA, de Heer WA. Carbon nanotubes-the route toward applications. Science 2002; 297(5582): 787-92.
[http://dx.doi.org/10.1126/science.1060928] [PMID: 12161643]

[52] Saito N, Usui Y, Aoki K, *et al.* Carbon nanotubes: biomaterial applications. Chem Soc Rev 2009; 38(7): 1897-903.
[http://dx.doi.org/10.1039/b804822n] [PMID: 19551170]

[53] Yao L, Ling B, Huang W, Shi S, Xiao J. Versatile triblock peptide self-assembly system to mimic collagen structure and function. Biomacromolecules 2024; 25(4): 2520-30.
[http://dx.doi.org/10.1021/acs.biomac.4c00033] [PMID: 38525550]

[54] Moohan J, Stewart SA, Espinosa E, *et al.* Cellulose nanofibers and other biopolymers for biomedical applications: A review. Appl Sci (Basel) 2019; 10(1): 65.
[http://dx.doi.org/10.3390/app10010065]

[55] Nandhini J, Karthikeyan E, Rajeshkumar S. Eco-friendly bio-nanocomposites: pioneering sustainable biomedical advancements in engineering. Discover Nano 2024; 19(1): 86.
[http://dx.doi.org/10.1186/s11671-024-04007-7] [PMID: 38724698]

[56] Yarin AL. Coaxial electrospinning and emulsion electrospinning of core–shell fibers. Polym Adv Technol 2011; 22(3): 310-7.
[http://dx.doi.org/10.1002/pat.1781]

[57] Abadi B, Goshtasbi N, Bolourian S, Tahsili J, Adeli-Sardou M, Forootanfar H. Electrospun hybrid nanofibers: Fabrication, characterization, and biomedical applications. Front Bioeng Biotechnol 2022; 10: 986975.
[http://dx.doi.org/10.3389/fbioe.2022.986975] [PMID: 36561047]

[58] Maliszewska I, Czapka T. Electrospun polymer nanofibers with antimicrobial activity. Polymers (Basel) 2022; 14(9): 1661.
[http://dx.doi.org/10.3390/polym14091661] [PMID: 35566830]

[59] Aidun A, Safaei Firoozabady A, Moharrami M, *et al.* Graphene oxide incorporated polycaprolactone/chitosan/collagen electrospun scaffold: Enhanced osteogenic properties for bone tissue engineering. Artif Organs 2019; 43(10): E264-81.
[http://dx.doi.org/10.1111/aor.13474] [PMID: 31013365]

[60] Ma X, Ge J, Li Y, Guo B, Ma PX. Nanofibrous electroactive scaffolds from a chitosan-grafted-aniline tetramer by electrospinning for tissue engineering. RSC Adv 2014; 4(26): 13652-61.
[http://dx.doi.org/10.1039/c4ra00083h]

[61] Zhou S, Wang Q, Yang W, *et al.* Development of a bioactive silk fibroin bilayer scaffold for wound healing and scar inhibition. Int J Biol Macromol 2024; 255: 128350.
[http://dx.doi.org/10.1016/j.ijbiomac.2023.128350]

[62] Irantash S, Gholipour-Kanani A, Najmoddin N, Varsei M. A hybrid structure based on silk fibroin/PVA nanofibers and alginate/gum tragacanth hydrogel embedded with cardamom extract. Sci Rep 2024; 14(1): 14010.
[http://dx.doi.org/10.1038/s41598-024-63061-4] [PMID: 38890349]

[63] Al-Naymi HAS, Mahmoudi E, Kamil MM, *et al.* A novel designed nanofibrous mat based on hydroxypropyl methyl cellulose incorporating mango peel extract for potential use in wound care system. Int J Biol Macromol 2024; 259(Pt 1): 129159.
[http://dx.doi.org/10.1016/j.ijbiomac.2023.129159] [PMID: 38181905]

[64] Alizadeh S, Majidi J, Jahani M, *et al.* Engineering of a decellularized bovine skin coated with antibiotics-loaded electrospun fibers with synergistic antibacterial activity for the treatment of infectious wounds. Biotechnol Bioeng 2024; 121(4): 1452-63.

[http://dx.doi.org/10.1002/bit.28659] [PMID: 38234099]

[65] Princy KD, Kaur D, Kaur A. Engineering of electrospun polycaprolactone/polyvinyl alcohol-collagen based 3D nano scaffolds and their drug release kinetics using cetirizine as a model drug. Int J Biol Macromol 2024; 268(Pt 2): 131847.
[http://dx.doi.org/10.1016/j.ijbiomac.2024.131847] [PMID: 38677678]

[66] Dugam S, Jain R, Dandekar P. Silver nanoparticles loaded triple-layered cellulose-acetate based multifunctional dressing for wound healing. Int J Biol Macromol 2024; 276(Pt 1): 133837.
[http://dx.doi.org/10.1016/j.ijbiomac.2024.133837] [PMID: 39009263]

[67] Rahmani K, Zahedi P, Shahrousvand M. Potential use of a bone tissue engineering scaffold based on electrospun poly (ε-caprolactone) - Poly (vinyl alcohol) hybrid nanofibers containing modified cockle shell nanopowder. Heliyon 2024; 10(10): e31360.
[http://dx.doi.org/10.1016/j.heliyon.2024.e31360] [PMID: 38813180]

[68] Patel DK, Won SY, Jung E, Han SS. Recent progress in biopolymer-based electrospun nanofibers and their potential biomedical applications: A review. Int J Biol Macromol 2025; 293: 139426.
[http://dx.doi.org/10.1016/j.ijbiomac.2024.139426] [PMID: 39753169]

[69] Dahlin RL, Kasper FK, Mikos AG. Polymeric nanofibers in tissue engineering. Tissue Eng Part B Rev 2011; 17(5): 349-64.
[http://dx.doi.org/10.1089/ten.teb.2011.0238] [PMID: 21699434]

CHAPTER 10

Role of Biopolymer in Bone Regeneration & Replacement

Shashi Kant Singh[1,*] and **Shreya Maddesiya**[1]

[1] *Faculty of Pharmaceutical Sciences, Mahayogi Gorakhnath University, Gorakhpur , Uttar Pradesh 273007, India*

Abstract: Bone regeneration and replacement are always challenging conditions, and with treatments, the chances of falling are often the least than expectations. This chapter emphasizes the function of biopolymers in bone replacement and regeneration, highlighting biopolymers that can be converted into scaffolds, hydrogels, and composites that mimic the extracellular matrix, facilitating the growth and repair of new bone. Their ability to promote cell adhesion, proliferation (rapid increase in the number or amount of cells), and differentiation makes them an alternative to traditional synthetic materials. Also, the biopolymer-based materials can be arranged in such a manner as to release bioactive molecules, enhancing osteogenesis,a genetic or heritable disease in which bones fracture or break easily, often without any major reason or minor injury. Angiogenesis is the process by which the body creates new capillaries from existing blood vessels. This chapter describes biopolymers in-depth and their role in the current biopolymer-based strategies for bone regeneration and replacement, including their mechanical properties, degradation rates, and osteoconductive potentials. It also focuses on the potential of biopolymers of composites and hybrid materials to enhance the regeneration the bones, combining the benefits of natural and synthetic materials. This chapter examines the potential of biopolymers for developing innovative solutions for bone replacement and regeneration, such as 3D-printed biopolymer-based implants and bioactive coatings. By harnessing the potential of biopolymers, the researchers can develop revolutionary treatments for bone regeneration and replacements, which lead to improvement in patient outcomes and enhances quality of life.

Keywords: Biocomposites, Biopolymer scaffolds, Bone cells, Bone healing, Bone replacement, *In vitro* testing, Novel biopolymer, Properties of biopolymer, Recent advancement, Stem cell therapy.

---

[*] **Corresponding author Shashikant Singh:** Faculty of Pharmaceutical Sciences, Mahayogi Gorakhnath University, Gorakhpur , Uttar Pradesh 273007, India; E-mail: Shashikantsingh59@gmail.com

INTRODUCTION TO BONE BIOLOGY AND STRUCTURE

Composition of Bone (Organic and Inorganic Component)

Bone is a complex organ with both organic and inorganic components that support its structure and functionality. Collagen, more especially type I collagen, which gives the organic matrix its flexibility and tensile strength, makes up the majority of the material. Because it enables energy dissipation and fracture resistance, this collagen framework is essential to the properties of bone mechanics [1]. The structure of hierarchy in the collagen fibers allows mineral crystals to be integrated into the collagen matrix, increasing bone toughness [2, 3]. Collagen cross-links stabilize the collagen network, which affects the overall strength and toughness of the skeletal tissue, which further adds to the mechanical integrity of bone [4, 5]. Hydroxyapatite, a primary inorganic constituent of bone, is primarily composed of crystalline calcium phosphate with the chemical formula $Ca_5[PO4]_3[OH]$. The compressive strength and stiffness of bone are attributed to this mineral phase [6]. At the nanoscale, the mineralization process entails the highly ordered deposition of HA crystals inside the collagenous framework. The degree of mineralization can influence the size and crystallinity of mineral crystals, and this can vary based on factors such as age and mechanical loading [7, 8]. Moreover, the mineral phase stores significant ions like calcium and phosphate, which are required for several physiological processes [9]. The interaction between the organic and inorganic components primarily examines the mechanistic features of bone. The collagen matrix shields the mineral crystals from mechanical stressors and serves as a framework for mineral deposition. Because of its composite nature, bone can tolerate a variety of stresses without losing its structural integrity or usefulness.

Role of Bone Cells

The following three primary cell types regulate the ongoing remodeling of bone:

Osteoblasts

The process of making new bone is carried out by specialized cells known as osteoblasts. They are derived from mesenchymal stem cells and play a crucial role in generating and secreting collagen type I and other vital components of the bone matrix, including osteocalcin and osteopontin. The mineralization of the bone matrix is dependent on osteoblasts, which facilitate the crystallization of hydroxyapatite, ultimately imparting strength and stiffness to the bone. Additionally, by secreting signaling molecules like Osteoprotegerin and RANKL, which influence the differentiation and function of osteoclasts, osteoblasts help control osteoclast activity. Osteoblasts and hematopoietic stem cells have a spatial relationship in the bone marrow that suggests a functional interdependence, with osteoblasts providing a niche that promotes the maintenance of hematopoietic stem cells [10, 11].

Osteoclasts

Macrophage lineage multinucleated cells and osteoclasts are principally responsible for bone resorption. Osteoclasts cling to the bone's surface and release an acidic microenvironment that encourages mineralized materials to break down bone and cause the deterioration of the organic matrix [12, 13]. The necessary components of the macrophage colony-stimulating factor (M-CSF) and RANKL are highly regulated processes of osteoclast precursor differentiation through mature osteoclasts [14, 15]. To remove deteriorated or outdated bone tissue, this resorption process is necessary for appropriate bone remodeling. Additionally, osteoclasts show dynamic behavior that can change their functional capability through fission and fusion processes [16, 17]. Osteoporosis and other metabolic bone diseases can result from excessive osteoclastogenesis, hence, controlling osteoclast activity is essential [18].

Osteocytes

Osteoblasts that become embedded in the mineralized matrix differentiate into osteocytes, the most prevalent cell type in bone tissue. These are essential for mechanotransduction because they detect mechanical loads and communicate with other bone cells to start remodeling processes. Through a network of dendritic processes that extend through canaliculi in the bone matrix, osteoblasts and osteoclasts can communicate with one another. Coordination of bone production and resorption depends on this communication, which guarantees that bone remodeling adapts to mechanical and metabolic demands. Additionally, osteoblasts and osteoclasts are influenced by the signaling molecules that these cells create. Fig. (1) shows the bone cells responsible for bone regeneration. As a result, osteoclasts and osteoblasts work together to preserve bone homeostasis.

BONE REMODELING AND REGENERATION PROCESSES

Bone remodeling and regeneration are complex biological processes that involve the coordinated function of several cell types, primarily osteoblasts, osteoclasts, and osteocytes [19].

Bone Remodeling Process

When stimulated, osteoclasts, which directly resorb bone, break down the mineralized matrix of bone and resorb aging or damaged bone tissue in this phase, which usually lasts for three weeks. Osteoclasts do so by secreting acids and proteolytic enzymes, which dissolve the mineral parts and degrade the collagen. The process is reversed from resorption to the formation phase, and in this, osteoblasts take a central stage. Osteoblasts deposit new bone matrix (osteoid) in these three to four months, which then gets mineralized to construct new bone [20 - 22]. Moreover, the osteoblasts secrete factors of signaling that control the activity of osteoclasts, maintaining a close coupling between the resorption and formation of bone [23, 24]. The resorption site is primed for new bone formation through the reversal phase, which lasts about five weeks, with the delivery of osteoblast precursors to the resorption site.

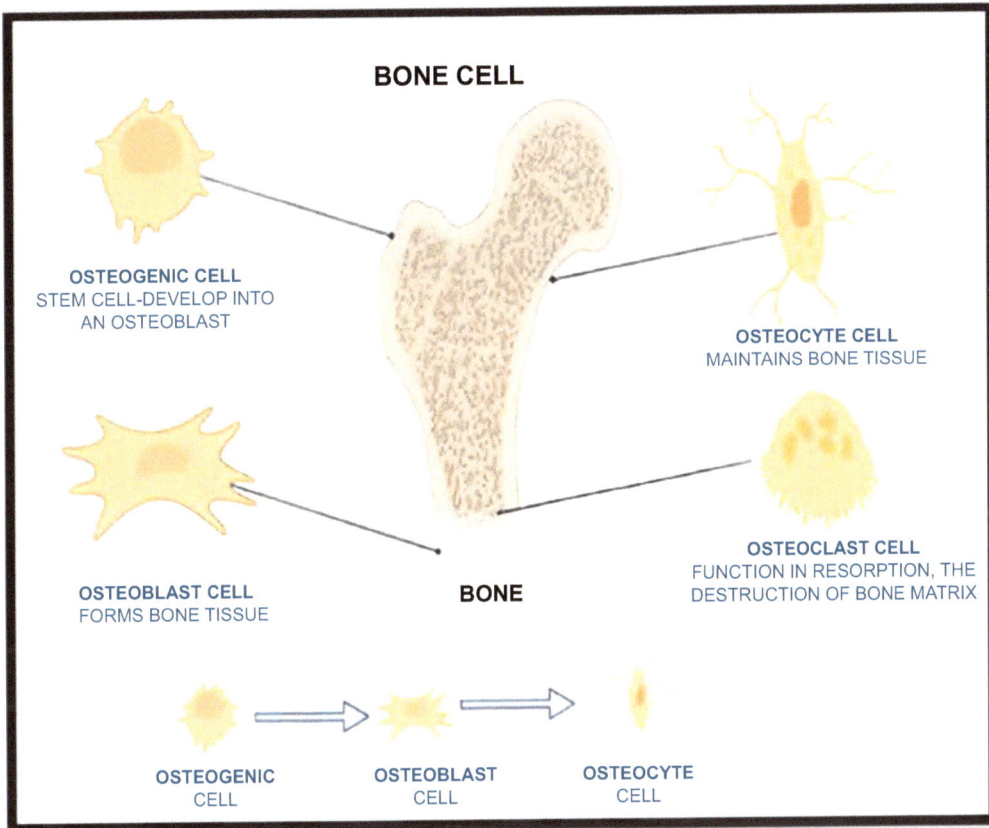

Fig. (1). Types of bone cells.

Bone Regeneration

A critical part of the remodeling cycle is bone regeneration, especially in response to traumas or fractures. Several structures, such as blood vessels and nerve fibers, which have a crucial function in the provision of nutrients, signaling molecules evoking healing, are involved in the process of regeneration [25]. There exists intricate interaction among osteoblasts and osteoclasts during bone tissue regeneration, with additional involvement of various cytokines and growth factors towards augmenting differentiation and activity of cells [26]. Osteoclast-derived extracellular vesicles were recently found to be pivotal in modulating osteoblast function and enhancing bone healing [27]. MicroRNAs and other signaling molecules contained in the vesicles may control osteoblast activity by enhancing osteoblast development and activity and suppressing aberrant osteoclastogenesis. Moreover, it has been demonstrated that molecules like bone morphogenetic proteins (BMPs) enhance bone regeneration by inhibiting the activity of osteoclasts and promoting osteoblast differentiation simultaneously [28].

Intercellular Communication

Osteoblasts and osteoclasts communicate with one another for effective bone remodeling and regeneration. Aside from producing osteoprotegerin (OPG) to inhibit osteoclastogenesis, osteoblasts also produce molecules such as RANKL (Nuclear Factor Kappa-B Ligand Receptor Activator) that activate osteoclast development and activity [29]. Such equilibrium is needed to achieve bone homeostasis, and its disturbance may cause osteoporosis, a condition whereby new bone decreases in production decreases while bone resorption increases [30]. In addition to this, mature osteocytes that reside within the mineralized matrix make use of signal transduction pathways to modulate osteoblasts and osteoclasts to mechanical stress [31].

BIOPOLYMERS AND THEIR PROPERTIES

The most prevalent type of biopolymers include: hyaluronic acid, chitosan, collagen, *etc.* Their central positions in biological mechanisms have rendered them especially appealing to scientists in areas like materials science, medicine, and tissue engineering. The escalating interest in naturally occurring compounds is due to their prospective use and their intrinsic significance in living organisms [32].

Collagen

The ubiquitous ECM protein is predominantly responsible for providing tissues with structural support. Collagen exists in numerous forms, the most prevalent of which is type 1 that is found in connective tissues, skin and bone. From its molecular structure to its fibrillar structure, the hierarchical structure of collagen has a profound influence on its mechanical properties. For example, research has shown that the level of hydration influences the mechanical properties of collagen microfibrils because collagen in a hydrated form possesses greater tensile strength and elasticity compared to dehydrated collagen [33]. Furthermore, the mechanical behavior of collagen gels is enhanced by glycosaminoglycans supplementation, *e.g.*, hyaluronic acid (HA). This indicates that interactions between collagen and GAGs are necessary for the structural stability of the extracellular matrix (ECM) [34, 35]. Another factor to consider in the use of collagen in tissue engineering is its bioactivity. Collagen scaffolds have been designed to facilitate cell adhesion and growth, an essential component of tissue regeneration. It has been demonstrated experimentally that crosslinking collagen using chemicals like Ethyl dimethylaminopropyl carbodiimide (EDC) increases the mechanical strength and stability of collagen scaffolds to make them more applicable for weight-bearing conditions. Collagen has also been investigated concerning the ability to establish biocompatible conditions for the proliferation of cells in hydrogels, particularly in corneal regeneration treatments [36]. The versatility of collagen is further illustrated in that it can also be utilized for 3D-printed models, which are adaptable for special tissue engineering techniques [37].

Chitosan

Due to its unique characteristics, *i.e.*, water-bonding property and antibacterial action, chitosan can potentially be used in a wide range of biomedical applications such as drug delivery systems and wound healing. Its cationic character, which enables it to bind with negatively charged biological molecules, increases its binding with these molecules. It is also a potential drug carrier due to its cationic character. Chitosan materials can induce cell proliferation and adhesion and thus help in their use for tissue engineering [38]. Due to the variety of chemical modifications, it is possible to adjust the mechanical properties of chitosan to be used optimally for specific applications. Hyaluronic acid is one of the naturally derived GAGs that is widely known for its highly hydrophilic as well as viscoelastic nature. It is essential for tissue hydration and the promotion of cellular migration when repair of the tissue takes place. Because it affects the mechanical properties of tissues, including tensile strength and elasticity, the interaction between hyaluronic acid and collagen is vital [39].

Hyaluronic Acid

It is applied in various areas such as in biomedicine, in dermal fillers, in osteoarthritis treatments, and as scaffolds in tissue regeneration. Its ability to form hydrogel enables it to emerge as a viable candidate for the construction of biocompatible matrices that support tissue regeneration and cell growth. Their use in composite materials has also been in the spotlight, as there is interest in leveraging the synergistic properties by incorporating them in biomedical applications to enhance performance. Hybrid hydrogels made up of both collagen and hyaluronic acid, for instance, have been found to possess improved mechanical performance and cellular behavior compared to their components.

Chemical Structure and Physical Properties of Biopolymers

The chemical and physical properties of biopolymers are represented in Table **1**.

Table 1. Biopolymer characteristics: physical and chemical properties.

Biopolymers	Physical Properties	Chemical Structure
Collagen	The protein known as natural collagen is very hydrophilic and insoluble in organic solvents [40].	Arabic numerals are used to identify the three polypeptide chains that make up collagen, referred to as collagen Alpha Chains. To produce homotrimers, the trio of chains may be identical or form, but the chains can differ. Collagen fibrils are formed from three polyproline II helices, each with a left-handed twist, which are then coiled together to form a right-handed triple helix. Additionally, amino acid residues are intertwined between adjacent chains, stabilizing the structure [41].
Chitosan	The polymer normally gels when the pH is higher than 7, yet it dissolves fully with a pH of less than 5. The composition of commercially available chitosan preparations can vary greatly, with molecular weights ranging from 300 to 1000 Kd and deacetylation levels ranging from 50 to 90% [42].	[43]
Hyaluronic Acid	The HA polymer's strong water-binding capacity and solubility combine to create a very viscous solution that gives cartilage and synovial fluid their crucial lubricating and viscoelastic qualities [44].	D-glucuronic acid N-acetyl-glucosamine [45]

Biocompatibility and Biodegradability of Biopolymers

The natural biocompatibility and biodegradability of biopolymers have created keen interest in the biomedical field. Through reducing unpleasant reactions and the facilitation of natural biological mechanisms, these substances provide the best solution for drug delivery systems, tissue engineering, and medical devices, where safety comes first and second. When one material is capable of performing what it was supposed to do without adversely affecting the adjoining tissues, then that substance is considered to be biocompatible. As biopolymers such as collagen, chitosan, and hyaluronic acid occur naturally and structurally resemble components of the extracellular matrix (ECM), they are also well recognized for their superior biocompatibility [46]. Gelatine, for instance, which is a derivative of collagen, is widely used as a material for tissue scaffolds and wound dressings owing to its desirable cell interaction and low antigenicity [47, 48]. The ability of biopolymers to enhance adhesion and cell proliferation, needed for effective incorporation into biological systems, also contributes to their increased biocompatibility [49]. Another key attribute of biopolymers is their ability to degrade naturally over time through the action of biological processes to non-toxic byproducts such as carbon dioxide and water, along with being biocompatible. This property is particularly useful in uses where the material is designed to be replaced by tissues in the body gradually [50]. An example of tissue engineering and drug delivery is two biomedical uses that can take advantage of the great biodegradability and biocompatibility of polyhydroxyalkanoates (PHAs) and microbial biopolymers [51, 52]. Additionally, chemical reactions can be applied to alter biodegradation in biopolymers to produce controlled rates of degradation that correlate with tissue healing processes [53]. For instance, it has been shown that the incorporation of nano clay fillers into films composed of maize starch biopolymer enhances barrier properties without degrading biodegradability. When developing biomaterials, it is imperative to consider the interaction between biodegradability and biocompatibility. Biopolymers with high rates of degradation may not provide sufficient support for tissue regeneration, while those with slow degradation rates could lead to foreign body responses or long-term inflammation [54].To enhance mechanical properties without compromising biocompatibility and biodegradability, scientists are focusing more on developing composite materials that combine biopolymers with synthetic polymers or nanoparticles [55, 56].

Nanotechnology-enhanced Biopolymers for Bone Regeneration

The characteristics of biopolymers used in tissue engineering have been greatly improved by the use of nanotechnology in bone regeneration.

Design and Synthesis of Nanomaterial

Nanostructured Biomaterials: These have been formulated to imitate the composition and nanoarchitecture of organic bone. Nanostructured biomaterials enhance host-cell interactions and bone regeneration. The materials integrate the mechanical properties of polymers with the osteoconductivity of artificial calcium phosphate ceramics [57, 58].

Drug Delivery Nanoparticles: Drugs can be delivered by nanoparticles using controlled release systems that augment healing [59]. For instance, electrospun nanofibers loaded with nanoparticles may be used to deliver BMP-2 sequentially and temporally in a manner to promote osteogenic differentiation.

Hybrid Scaffolds: Hybrid scaffolds combine natural polymers like chitosan with synthetic polymers like PCL to achieve the optimal benefit with the least disadvantage. The scaffolds are backed by the bioactivity of chitosan and are mechanically stronger as a result of the physicochemical properties of PCL [60].

Impact of Biopolymers on Scaffolds

Enhanced Mechanical Properties: Surface roughness and porosity can be modified by incorporating nanoparticles into biopolymer matrices to enhance their mechanical strength and load-bearing application suitability [61].

Osteoinductivity and Biocompatibility: Due to their rougher surface area and greater surface area, which enhance protein adsorption essential for cell attachment, nanocomposites are often found to be more biocompatible [62]. Bioactive glasses, like borate bioactive glasses, have been found to stimulate osteoblast growth through the release of ions that facilitate the expression of genes associated with bone.

Targeted Drug Delivery Systems:-They preserve bioactivity and encourage localized therapeutic responses, which are made possible by nanomaterials implanted in scaffolds. By offering sustained release profiles customized to meet particular therapeutic demands, this strategy is essential for maximizing healing processes.

Regulation of Cellular Behavior: The nanoscale components in scaffold design influence adhesion, migration, proliferation, and differentiation, among other aspects of cellular behavior. As per research, these properties can utilize enhanced integrin signaling pathways to direct stem cell fate toward osteogenesis.

Challenges and Opportunities

Despite challenges, nanotechnology has advanced significantly in terms of the improvement of the properties of biopolymers for bone regeneration,

- **Toxicological Concerns**: Degradation products or residual heavy metals in carbon nanotubes question the toxicity of nanoparticles.
- **Regulatory Hurdles**: To overcome the regulatory hurdles in applying these technologies to clinical practice, further research is needed.
- **Scalability Challenges:** Before widespread acceptance, it was still challenging

to scale up production while maintaining quality control.

Comparison of Biopolymer Properties

- **Poly(lactic acid) (PLA) and Poly(lactic-co-glycolic acid) (PLGA):** PLA and PLGA are widely used based on their capacity for encouraging bone development and regeneration through osteoinduction and osteogenesis When incorporated in hydroxyapatite (HA), they amplify bone formation due to encouraging adhesion and differentiation of cells [63, 64]. Both are biodegradable, and their degradation is rapid in PLGA compared to PLA. By altering the lactic to glycolic acid ratio in PLGA, their rates of breakdown can be maximized. Exceptionally biocompatible and have been used extensively in biological applications without causing any noticeable side effects [65]. Despite their relative stiffness, these polymers are fragile. Blending them with other chemicals, such as HA, can improve their mechanical properties.

- **Poly(caprolactone) (PCL):** Because of its slower rate of degradation, PCL can be used to create a scaffold for bone development that lasts longer. However, until it is combined with bioactive fillers like HA or bioglass, its osteogenic potential is frequently lower than that of PLA or PLGA. PCL can be used in conditions requiring a scaffold to be present for a longer duration because PCL degrades more slowly than PLA and PLGA. PCL is biocompatible and has found various uses in tissue engineering . PCL is softer than PLA and PLGA because its modulus of elasticity is lower. PCL is biocompatible and has various tissue engineering applications. PCL is more pliable than PLA and PLGA owing to the reason that its modulus of elasticity is lower. However, adding some fillers may compromise its mechanical properties.

- **Poly(3-hydroxybutyrate-co-3-hydroxyvalerate) (PHBV):** PHBV is osteogenically active, especially when mixed with bioactive inorganic particles like bioglass, which allows for cell proliferation and differentiation. PHBV is biodegradable, but degrades slowly compared to PLA. PHBV is well biocompatible, especially when mixed with bioactive fillers. PHBV is stiffer than PCL but not as stiff as PLA. Its composition can be adjusted to tailor its mechanical properties. It can be made to degrade through the adjustment of the copolymer's composition.

- **Silk Fibroin:** Silk fibroin possesses moderate mechanical properties and is useful for bone tissue scaffolding since it is biocompatible and supports cell growth. Silk fibroin is enzymatically degradable and biodegradable, and this makes it useful for tissue engineering purposes. Silk fibroin possesses excellent biocompatibility and has been utilized in several applications in biomedicine, like bone tissue engineering, which calls for a certain amount of flexibility [66]. A comparison of biopolymers is represented below in Table **2**.

Table 2. Comparative analysis of biopolymers in bone regeneration: Osteogenic potential, biodegradability, biocompatibility, and mechanical properties.

Biopolymer Type	Osteogenic Potential [67]	Biodegradability [68, 69]	Biocompatibility [70]	Mechanical Properties [71]
PLA	High	Moderate	Excellent	High
PHBV	Moderate	High	Excellent	Moderate
Chitosan	High	High	Excellent	Low
Hydrogel Scaffolds	High	Variable	Excellent	Low to Moderate
Bioglass Composites	Very High	Moderate	Good	High

BIOPOLYMER-BASED BIOMATERIALS FOR BONE TISSUE ENGINEERING

Bone tissue engineering has been contemplated as a possible remedy to overcome some of the deficiencies of conventional bone grafting procedures, especially autologous grafts, that are associated with significant morbidity and need supplementary surgical procedures. Biopolymer-based scaffolds have been under focus because of their ability to simulate the extracellular matrix of bone tissue as well as to provide superior biocompatibility and controlled biodegradation characteristics. These scaffolds support cell adhesion, proliferation, and differentiation—crucial events in bone formation. The effectiveness of scaffolds in assisting bone regeneration depends to a large extent on their design parameters and fabrication techniques. Studies have shown that the inclusion of natural biopolymers such as gelatin, chitosan, and collagen in scaffold formulations improves both the structural properties and biological functions of such constructs. Preclinical studies [72, 73], have proven the high bone-repairing efficacy of composite scaffolds consisting of collagen, chitosan, and hydroxyapatite and have pointed out the positive effects of combining these materials. Similarly, research using 3D-printed scaffolds made up of polylactic acid (PLA) and hydroxyapatite (HA) proved improved bone growth in critical-sized defects [74], again emphasizing the contribution of composition to osteogenic performance. Scaffolds with well-defined porosity and architectural features can be produced using modern manufacturing methods such as electrospinning and 3D printing. These are critical parameters for enabling cellular migration and delivery of nutrition. Electrospun PLA/gelatin nanofibrous scaffolds are a novel approach to overcome the mechanical drawbacks of pure biopolymers while retaining good cell attachment properties [75]. Microfluidic technology has now been used for drug-loaded scaffold screening, increasing their suitability for targeted therapeutic use in bone regeneration. Evidence indicates that adding bioactive elements like graphene oxide and hydroxyapatite to biopolymer scaffolds greatly enhances their mechanical strength and osteoconductive properties. For example, experiments have proved that the addition of graphene oxide as an interface phase to polyether ketone (PEEK) scaffolds enhances their effectiveness in repairing bone defects [76]. Furthermore, research has shown that *in situ* hydroxyapatite formation over biopolymer particles significantly promotes the bioactivity of ensuing scaffolds, ensuring improved integration into adjacent bone tissue [77]. Scaffolds functionalized by growth hormones and microRNA (miRNA) inhibitors have also been investigated to further promote bone regeneration. Evidence depicts that scaffolds with miRNA inhibitors trigger precise transcription factors that promote osteoblast development and hence hasten bone healing procedures [78]. Their inclusion of stem cell-derived osteogenic cells offers an attractive option to further increase the process of bone regeneration since they can enhance support for the repair process [79].

Growth Factors, Cells, and other Bioactive Substances

Enhancing the osteogenic ability of these scaffolds, their incorporation is expected to enhance the effective repair of bones. The successful combination of biopolymers and bioactive substances has been found in several investigations to significantly increase the ability of the scaffold to stimulate bone healing. Osteogenesis largely relies on growth factors like BMP-2 (bone morphogenetic protein). Delivery of growth factors in scaffolds is of huge importance; mesenchymal stem cells (MSCs) on chitosan/alginate/hydroxyapatite scaffolds releasing BMP-2 demonstrated more healing of rat calvarial defects [80]. It further stressed the application of TGF-β1 in poly (lactic acid)/polycaprolactone scaffolds, which induced bone regeneration by promoting the increase in alkaline phosphatase (ALP) activity, a characteristic of osteoblast maturation [81].

The findings underscore the importance of incorporating growth factors to induce cellular responses that favor new bone formation. The efficacy of the scaffold also relies on the selection of proper biopolymers. Due to their similarity in structure with the natural bone matrix, naturally derived materials are desirable because they lower immunogenic response and improve cell adhesion and functionality [82]. The incorporation of bioactive ceramics, like hydroxyapatite, into chitosan scaffolds has been used to improve their mechanical and biological properties for advancing bone regeneration. Increased porosity makes the scaffold more interconnected to surrounding tissues and also expands its surface area to which cells can adhere [83]. The porosity and architecture of the scaffold are critical to encouraging vascularization, diffusion of nutrients, and cell migration. Scaffolds whose hole sizes are greater than 300 μm are required for osteoblast migration and new bone tissue ingrowth [84]. This is particularly important when working with large bone lesions, as the healing outcome can be significantly affected by proper scaffold porosity. It has been shown that stem cells added to scaffolds can enhance bone regeneration as well as growth factors. It is indicated in Fig. (**2**) that the employment of differentiated stem cells along with fibrin biopolymer scaffolds improved bone healing, which indicated that the live cell component of the scaffold could further advance osteogenic processes [85]. Besides, the co-administration of most growth factors is facilitated through the employment of hydrogels, which further advances the osteogenic capacity of the scaffolds [86].

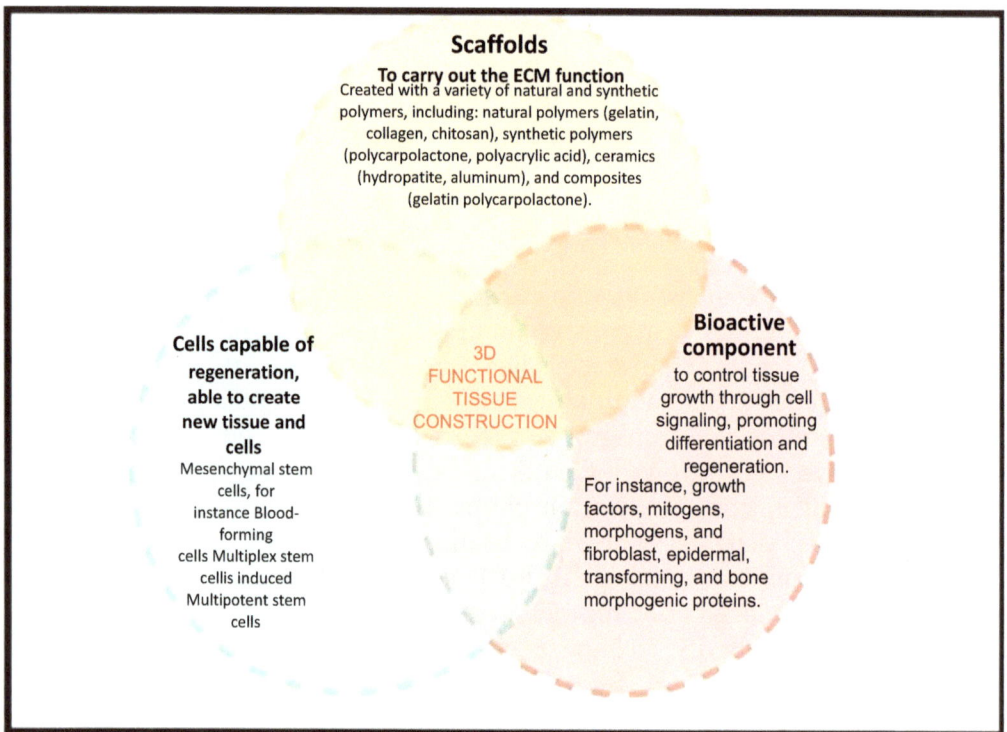

Scaffolds
To carry out the ECM function
Created with a variety of natural and synthetic polymers, including: natural polymers (gelatin, collagen, chitosan), synthetic polymers (polycarpolactone, polyacrylic acid), ceramics (hydropatite, aluminum), and composites (gelatin polycarpolactone).

Cells capable of regeneration, able to create new tissue and cells
Mesenchymal stem cells, for instance Blood-forming cells Multiplex stem cellis induced Multipotent stem cells

3D FUNCTIONAL TISSUE CONSTRUCTION

Bioactive component
to control tissue growth through cell signaling, promoting differentiation and regeneration. For instance, growth factors, mitogens, morphogens, and fibroblast, epidermal, transforming, and bone morphogenic proteins.

Fig. (2). Three dimensions in tissue engineering.

In vitro Testing and Characterization of Biopolymer Scaffolds

Biopolymer scaffolds must be tested and characterized *in vitro* to design biomaterials that function optimally within bone tissue engineering (BTE). With these processes, the scaffolds are ensured to possess the characteristics that facilitate cell proliferation, differentiation, and ultimately bone repair. Design elements of scaffolds critical to their functionality in biological settings, such as mechanical properties, degradation rates, and bioactivity, have been addressed in several studies. Stability and mechanical strength of the material are critical aspects of scaffold design. Polycaprolactone (PCL) scaffolds are reinforced with diopside nanopowder to ensure they can cope with dynamic remodeling under bone regeneration. They demonstrated that these nanocomposite fibrous scaffolds exhibited superior mechanical properties [87].

The mechanical properties of scaffolds are shaped by biodegradable poly D-L-lactide-co-glycolide. The scaffolds can be beneficial to bone regeneration because they induce calcium phosphate deposition and mesenchymal stem cell growth. Such findings stress the importance of scaffold composition optimization to achieve the desired mechanical properties. The architecture and porosity of scaffolds are critical parameters influencing the efficacy of the scaffolds. To allow nutrient delivery and cell migration, there is a need for interconnected porous structures with pore sizes of 50-300 μm [88]. The absence of connecting pores within the calcium phosphate/silk composite scaffolds lowered their mechanical properties but enhanced their osteoconductivity. The secret to an effective scaffold, therefore, is finding the optimal balance between porosity and mechanical stability. *In vitro* biocompatibility tests are important to determine the compatibility of scaffolds and cells [89]. The biocompatibility of chitosan/montmorillonite nanocomposite scaffolds indicated that they were non-toxic and could support cell attachment and growth [90], emphasizing how important it is to select proper materials when constructing a scaffold [91]. These findings underscore the necessity of thorough biocompatibility testing to ensure that scaffolds do not produce negative biological reactions. In addition, an important factor influencing the effectiveness of scaffolds in BTE is their degradation rate. A good scaffold will degrade at a rate corresponding to the rate at which tissue formation increases. Adipic acids are known to interact with natural polymers. They found that this interaction enhanced the mechanical and thermal stability of the scaffolds, which is important for maintaining structural integrity during degradation [92].

APPLICATIONS OF BIOPOLYMER BIOMATERIALS IN BONE REGENERATION

Bone Fracture Healing and Repair

Biodegradable materials can be structured to incorporate an interconnected porous architecture, which is critical for cell migration and nutrient transport. The ideal pore sizes in such materials are 50-300 μm. The addition of bioactive glass particles to gelatin-based scaffolds has also been found to improve the repair of osteoporotic bone defects, indicating the significance of composite materials in producing better mechanical and biological performance [93]. New advancements in 3D printing technology have enabled the production of complex scaffolds that can be customized to incorporate medications and growth hormones to enhance bone healing. For example, it has been established that the application of collagen composites and calcium phosphate in 3D printing enhances cell survival and stimulates bone repair [94]. In addition, it has been illustrated that the incorporation of lanthanum oxide nanoparticles into collagen-based hydrogels enhances angiogenesis and osteogenesis, which supports osseointegration and bone healing [95]. These developments reflect how biopolymer Scaffolds can be engineered to target the particular needs of bone regrowth. The other crucial factor influencing the effectiveness of biopolymer scaffolds in bone repair is the breakdown rate. Biodegradable scaffolds are preferred because they circumvent the necessity for subsequent surgeries to eliminate non-biodegradable devices. These include materials like polylactic-co-glycolic acid (PLGA). The degradation products, however, must be properly considered because they can potentially influence the healing process in general as well as local cellular responses [96]. For instance, while PLGA is highly acclaimed for its biocompatibility, concerns have also been raised about preventing the sites where bone grafts are applied from regenerating [97]. The design of the scaffold thus has to balance between how fast it is degraded and maintaining a need for constant mechanical stability during healing. In addition to this, the incorporation of bioactive compounds within biopolymer scaffolds enhances their osteogenic capacity immensely. For example, it has been shown that scaffolds that deliver growth factors such as BMP-7 and PDGF-B promote repair of the osteoporotic lesion by stimulating cellular processes associated with bone formation [98]. Some of the biopolymers utilized are indicated in Table **3**. Similar enhancement in bone regeneration has also been observed when stem cells are combined with fibrin biopolymer scaffolds, underlining the importance of creating a good environment for cell proliferation and differentiation.

Table 3. Biopolymer profiles and advantages.

S. No.	Biopolymers	Advantages
1.	Hydroxyapatite	Osteoconductive, mechanically stable, and nonimmunogenic [99].
2.	Starch	Biodegradable, biological renewable, abundant [100].
3.	Fibrin	ECM, which stimulates cartilage healing, makes it simple to functionalize [101].
4.	Alginate	Wound dressing, blood vessel tissue repair, drug and protein delivery, bone, cardiac, muscle, and liver tissue engineering [102].
5.	Gelatin	Encourages the growth, adhesion, and spread of chondrocytes [103].

Craniofacial Bone Regeneration

Platelet-rich fibrin (PRF) has been shown to enhance bone regrowth in craniofacial use by increasing osteogenic markers such as Runx2. PRF significantly accelerated cranial bone defect healing, leading to more substantial trabeculae and improved integration with the surrounding tissue [104]. Scaffolds that mimic intramembranous ossification are involved in the process of craniofacial bone formation. This procedure led to macrophage polarization, which promoted the healing of the bone [105].

Stem Cell Therapy and Biopolymer Scaffolds

The potential for efficacious regeneration of craniofacial bone is achieved by combining stem cell therapy with biopolymer scaffolds. The effectiveness of stem cell therapy in restoring craniofacial bone is randomized through a controlled experiment. The study sees the possibility for cellular therapies combined with biopolymer scaffolds to advance healing effects [106]. In addition, the application of fibrin biopolymers as scaffolds has been shown to facilitate the differentiation of mesenchymal stem cells to osteoblasts and induce bone formation by providing a suitable proliferation environment [107].

Innovative Approaches and Future Directions

New approaches have been designed for using chitosan and bioactive glasses to make biopolymer matrices. In recent studies, the influence of scaffolds prepared from bioactive glass, collagen, and cerium-doped chitosan on mesenchymal stem cell proliferation showed that the scaffolds may enhance osteogenic differentiation. A systematic review emphasized the biocompatibility and osteoinductive properties of chitosan [108], highlighting its effectiveness as a scaffold material for cranial bone.

BIOPOLYMER COMPOSITE MATERIALS FOR BONE REPLACEMENT

Biopolymer-ceramic Composites (*e.g.*, Collagen-hydroxyapatite)

Natural bone primarily consists of 69–80% of these two constituents, and thus, the significance of hydroxyapatite and collagen is established in the design of tissue engineering scaffolds [109]. It has been noted that combining biopolymers, for example, collagen, with ceramics, HA, improves the scaffolds' biological and mechanical behavior by enhancing cell attachment, growth, and differentiation [110 - 112]. Several techniques, such as electrospinning and 3D bioprinting, can be used to create biopolymer-ceramic composites. This technique yields scaffolds with controlled architecture and porosity [113]. As a proof of concept, it is demonstrated that the combination of collagen and bacterial cellulose could augment bone regeneration in vitro, suggesting that these composites might provide a setting that is conducive to osteoblast activity [114]. It has been shown that the implantation of bioactive ceramics, including HA, into polymer matrices can improve osteoconductivity and provide for the repair of bone defects. A minimally invasive technique of bone healing has recently been explored with the incorporation of nanocomposite hydrogels involving HA and other biopolymers. The gels are injectable into bone lesions with irregular shapes. Injectable systems deliver bioactive agents that can induce bone regeneration as well as structural reinforcement. Especially known for its antibacterial properties, the chitosan and gelatin blend with HA is suitable for use in situations where infection is likely to be high[[115]. In addition, by varying the production process and the ratio of biopolymer to ceramic, the composites' mechanical properties can be engineered [116]. For instance, the addition of Calcium phosphates to polymer matrices has been demonstrated to improve the mechanical stability and degradation rate of the scaffolds and align their properties with natural bone; such a flexibility is required for the scaffolds to sustain the *in vivo* mechanical loads and degrade at a rate that matches new bone formation [117]. The role of biodegradable polymers in bone regeneration is demonstrated in Fig. (**3**).

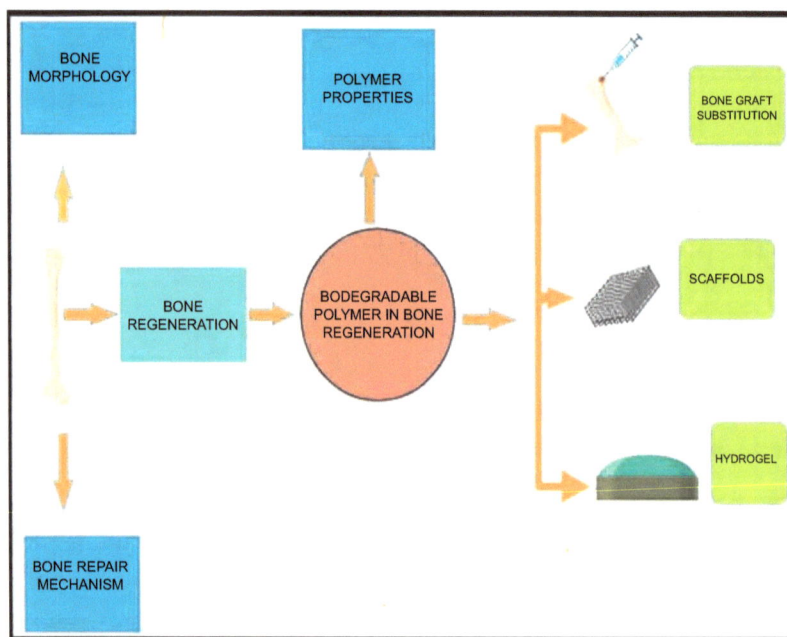

Fig. (3). Biodegradable polymer in bone regeneration.

Biopolymer-metal Composites for Orthopaedic Implants

Biopolymer-metal composites are becoming more and more known for their applications in orthopedic implants, combining the biological compatibility of biopolymers with the mechanical properties of metals. This combination is expected to maximize osseointegration, minimize stress shielding, and enhance the general performance of the implant. Titanium and its compounds, especially Ti-6Al-4V, are some of the most widely used metallic biomaterials for orthopedic applications due to their high mechanical properties, minimal toxicity, and good biocompatibility [118, 119]. Compared to all other metals, the elastic modulus of titanium alloys is closer to human bone and will decrease the stress-shielding situations where too much strain is taken by the implant and resulting in bone resorption (BMPs), on to titanium surfaces to further improve osseointegration [120]. Such coatings favor cell adhesion and proliferation, which are essential to the success of bone integration. Aside from titanium, biodegradable metals such as magnesium and zinc alloys are also of interest due to their ability to preclude secondary surgeries related to permanent implants [121, 122]. Magnesium-calcium (Mg-Ca) alloys, for example, are demonstrated to possess suitable mechanical characteristics in addition to the advantage of biodegradability, thereby reducing long-term complications. The inclusion of biopolymers into such metallic matrices may further enhance their biological functionality. For example, the mixture of biodegradable polymers and magnesium alloys might create composites that not only support mechanical loading but also promote tissue regeneration due to regulated degradation rates [123]. Furthermore, new titanium alloys with lower Young's moduli have been suggested to be developed so that they can more closely match the mechanical properties of bone to avoid the risk of stress shielding [124]. These advances are complemented by surface modification techniques, such as calcium phosphate coating, to enhance the biologic response of metal implants [125, 126]. These alterations are designed to enhance the osseointegration of the implant, thereby furthering its applications in the clinical setting. Adding biopolymers to metallic composites creates opportunities for functionalized mechanical characteristics and improved biocompatibility as well. For instance, the porous structures obtained with additive manufacturing may allow for bone ingrowth and an enhanced general integration of the implant with soft tissues [127, 128]. Besides meeting the mechanical needs, it also adds a biological feature to the performance of orthopedic implants.

Mechanical Properties and Osseointegration of Biopolymer Composites

Biopolymer composites, wherein natural polymers are blended with inorganic material, have the potential to improve mechanical strength while stimulating the biological interaction necessary for osseointegration. For example, incorporation of chitin with polylactic acid has proven to increase the tensile strength because of improved dispersion and crystallinity of composite materials [129]. Equivalently, the incorporation of calcium phosphate within collagen matrices has been reported to mimic the native extracellular matrix (ECM) of bone and, as such, enhance the properties of the scaffolds, both mechanical and biological [130]. The mechanical improvement is typically said to result from the efficient transfer of load from the polymer matrix to the inorganic fillers, which is seen in composites that incorporate halloysite nanotubes and bioactive glass [131, 132]. Osseointegration, the phenomenon whereby living bone tissue integrates into a firm bond with the surface of an implant, is largely influenced by the properties of biopolymer composite materials. This process, referring to the uneventful integration between bone and load-carrying implants, is a critical factor in the production and use of these biological composites. The incorporation of bioactive agents such as hydroxyapatite (HA) into biopolymer matrices has been reported to improve osteoconductivity and induce cellular activity required for bone regeneration. For instance, one study illustrated that scaffolds produced from biopolymer and HA had superior bioactivity and were able to integrate more successfully with the adjacent bone tissue. In addition, biopolymer multilayer coatings on metal implants have been reported to promote cell attachment and mineral deposition, essential for osseointegration [133]. Biopolymer surface composite properties also significantly influence their biological function. Surface roughness and wettability can influence protein adsorption, cell adhesion, and cell proliferation that follows [134]. For example, optimized surface properties in biopolymer scaffolds have been correlated with increased osteoblast-specific gene expression, further validating their osseointegration potential. Furthermore, the biodegradability of these composites is important as it allows the scaffold to be progressively replaced by new bone tissue, allowing for a successful integration process [135].

PRECLINICAL AND CLINICAL STUDIES

Animal Studies Assessing Biopolymer-based Bone Grafts

In the area of regenerative medicine, studies involving animals evaluating biopolymer-based bone grafts have attracted significant attention due to their capability of increasing bone integration and healing. Several studies have looked at the effectiveness of several biopolymer composites, frequently with growth factors or other bioactive agents, to enhance results in critical-sized bone defects. The osteoinductivity of deproteinized bovine bone was considerably enhanced by coprecipitated calcium phosphate granules with Bone Morphogenetic Protein 2 (BMP2) in one valuable study. According to this study, BMP2-coprecipitated calcium phosphate has been found to enhance the growth of new bone within critical-sized lesions, with the results being on par with those achieved by autologous bone grafting [136]. Likewise, the histomorphometric results of bovine demineralized bone grafts in critical-sized defects determined that these grafts were effective in facilitating the healing of bone, though autogenous grafts continued to be the gold standard [137]. The use of biopolymers like chitosan and fibrin in bone grafts has also been explored based on their mechanical and biological behavior. Chitosan-calcium phosphate composites highlight their potential as alternatives to bone grafting because they are biodegradable and osteoinductive [138]. The combination of fibrin and biphasic calcium phosphate forms a biopolymer, which provides an appropriate environment for the generation of bone precursor cells and neovascularization, indicating its potential for the bioengineering of tissues. Recent studies have centered on the biological enhancement of bone grafts using growth hormones, along with their mechanical attributes. For example, it has been noted that the Platelet-rich plasma (PRP) incorporated into autogenous bone grafts significantly enhanced the incorporation and assimilation of the grafts in the oral and maxillofacial field [139]. This was in agreement with the study, which showed that PRF incorporated with autogenous bone grafts resulted in enhanced clinical outcomes relative to autogenous bone alone [140]. Furthermore, studies have examined the influence of low-level laser treatment (LLLT) on bone healing in conjunction with biopolymer-based grafts. Autogenous bone grafts with a heterologous fibrin sealant might heal faster owing to LLLT, with a possible synergistic effect of potential use in clinical practice [141]. Likewise, recorded improved outcomes when LLLT was used for essential-sized defects given bovine bone grafts, suggesting that this treatment was likely to enhance the healing process [142].

Human Clinical Trials and Results

Bone regeneration research is an important aspect of regenerative medicine, especially for human clinical trials involving the application of biopolymers. Biopolymers, such as fibrin, and chitosan, along with PLGA, have been promising as scaffolds in bone tissue engineering because they can support scaffolds in bone tissue engineering because biological processes involved in bone repair, along with their biodegradability and biocompatibility [143].

Long-term Performance of biopolymer

• Clinical Results:

There is a diverse line of scaffolding technologies that have been explored for bone regeneration in orthopedics and dentistry which range from conventional bioactive materials to polymeric advanced and hybrid systems [144]. According to follow-up long-term studies, such materials support primary bone integration and formation that usually leads to complete healing in a specified timeframe.

• Growth Factors:

Growth factors, for instance, BMP-2, have been demonstrated by research to enhance osteogenic differentiation and create new bone tissue upon addition to biopolymer scaffolds, and at certain concentrations, BMP-2 greatly enhances the regenerative ability of scaffolds, which translates to increased strength and volume of bone [145].

• Hybrid Scaffolds:

The latest developments concern hybrid scaffolds that combine either natural or synthetic biopolymers with bioactive materials such as bioactive glass or calcium phosphates, which have demonstrated improvement in osteogenic differentiation, cell proliferation, and mechanical properties [146-147]. Based on long-term observations, such hybrid systems may better simulate the condition of bones to allow efficient healing in the long term.

Safety Factors to be Controlled

One of the main issues with biopolymer-based treatments is infection at the implant location. Scaffolds, in many cases, have porous structures that are prone to bacterial attachment and post-operative infection. Evidence has established the importance of incorporating antibacterial agents into scaffolds to reduce this risk and improve safety profiles [148].

Mechanical strength or tissue integration may be inadequate, leading to the failure of the implant. In load-bearing conditions, biopolymers such as PLA and PCL have exhibited varying degrees of mechanical performance. To detect any initial indication of implant failure, patients must be monitored regularly [149]. Even with high biocompatibility profiles of most biopolymers, inappropriate interactions can sometimes happen. Because some materials have been associated with inflammatory or foreign body reactions in long-term tests, additional studies are required to maximize material choice and formulation for fewer risks [150].

Extended Follow-up Information

Clinical trials of biopolymer-based therapy are typically successful in terms of patient recovery and satisfaction. For example, patients undergoing treatment for mandibular abnormalities with OCP-collagen scaffolds healed successfully and with minimal side effects. These patients maintain functional recovery with minimal side effects, as reported by long-term follow-ups.

Regenerative Capacity: Experiments also indicate that adding growth factors and bioactive chemicals to scaffold materials can enhance the regenerative capacity of biopolymer systems. For instance, it has been demonstrated that adding fibrin sealants with rhBMP-2 significantly speeds up bone regeneration of tibial lesions, proving that supplementary treatment can maximize clinical outcomes.

Emerging Technologies: The convergence of physical therapies is likely to advance the results of healing in clinical settings further, as evidenced by developments like photobiomodulation therapy and biopolymer scaffolds, which have been successful in augmenting specific bone regenerative processes.

Phosphorylation-enhanced chitosan hydrogels promoted osteoblasts' osteogenic differentiation and stimulated key pathways for signals that play a role in bone formation. This indicates employing a multimodal approach to enhance the outcome of bone regeneration. It has also been explored what mechanical properties of biopolymer-based bone cemented [151]. Suitable compressive module for low-load bearing applications yielded cement based on α-tricalcium phosphate that contained regenerating biomaterials. In the clinical environment, where mechanical stability is critical in the initial stages of recovery, this is particularly relevant [152]. Furthermore, fibrin biopolymer scaffolds can increase calcium phosphate deposition and mesenchymal stem cell differentiation, two pivotal steps in bone healing. It has been demonstrated that the regenerative potential of biopolymer systems can be further augmented by adding growth factors and bioactive molecules in addition to scaffold materials. The osteoconductive capability of a fibrin sealant in tibial bone defects is enhanced by mixing it with rhBMP-2, a recombinant human bone morphogenetic protein. As per the research, this mix significantly hastened bone healing, underlining the utility of combination therapies in regenerative medicine. Moreover, there has been work on applying photo-bio modulation therapy with biopolymer scaffolds [153]. Along with biphasic calcium phosphate and fibrin biopolymer, such a treatment focused on bone regenerating processes. This means biomaterial combination, together with physical treatment, might increase the results of healing during clinical applications. There are certain challenges in implementing biopolymers for regeneration in bones, particularly concerning the mechanical properties as well as the levels of degradation of biopolymers. Yet, recent advances in materials science have come up with hybrid scaffolds with the integration of biopolymers and inorganic materials such as hydroxyapatite, which enhance mechanical strength as well as trigger mineralization. These are requirements to validate scaffolds against physiological loads while allowing for bone healing [154].

FUTURE TRENDS AND CHALLENGES

Novel Biopolymers and Composite Biomaterials

Studies have shown the potential of phosphorylated chitosan hydrogels in stimulating osteogenic differentiation of osteoblasts. The presence of phosphorus groups within chitosan led to the formation of a mineral phase that looked like bone and greatly stimulated osteogenic differentiation through JNK and p38 activation. This implies that biopolymers can be engineered to enhance their biological activity, potentially making them more useful in bone regeneration processes. Hybrid biomaterials, or a mixture of biopolymers with inorganic components, have also been promising. For example, the application of alginate blended with nanohydroxyapatite has been investigated to fine-tune formulations for greater bone regeneration. The blend utilizes the osteoconductive nature of hydroxyapatite with the biodegradability and injectability of alginates, developing a scaffold that facilitates cell proliferation and osteogenesis. In addition, the use of bioactive glass ceramics incorporated with autogenous bone has also been demonstrated to enhance the structural organization of the organic bone matrix, and in turn, improve the efficacy of bone repair therapies [155].

Personalized and 3D Printed Biopolymer Implants

The development of 3D printing technology revolutionized the manufacturing of customized The advancement has transformed the production of customized biopolymer implants. 3D printing can fabricate scaffolds according to the specific mechanical and anatomical needs of each patient. By optimizing porosity and mechanical properties, hybrid scaffolds for enhanced, faster bone regeneration can be fabricated by combining fused deposition modeling with melt electro-wetting [156]. Such scaffolds may be designed to be similar to natural bones architecture, enhancing integration and healing. Furthermore, the use of micro-computed tomography (micro-CT) to observe the bone graft healing process in real-time has been a huge leap forward. This method examines the spatiotemporal features of patterns of fracture healing, enabling dynamic assessment of bone regeneration upon the application of different biomaterials [157]. Such an approach not only deepens our knowledge about the healing process but also promotes the tuning of biomaterials according to their *in vivo* performance. The integration of 3D printing and personalized medicine is also reflected by the fabrication of hydrogel scaffolds able to release growth factors or other therapeutic molecules under controlled conditions. Black phosphorus hydrogel scaffold may enhance bone regrowth by ensuring sustained phosphorus availability without calcium, which is essential for osteogenesis [158]. Through the avoidance of the shortcomings of conventional bone repair methods, the application of customized and 3D-printed biopolymer implants in bone regeneration is a breakthrough in tissue engineering. Because of their biocompatibility and adjustable properties, biopolymers are becoming more popular in the construction of scaffolds with the capacity to simulate naturally occurring extracellular matrix (ECM) naturally occurring and support the cellular activity necessary for bone healing. Biopolymer-based hydrogels have shown promise in bone regeneration applications, particularly due to their ability to enhance osteogenic differentiation and mineralization. For instance, incorporating chondroitin sulfate (CS) in hydrogels has been proven to enhance the production of hydroxyapatite (HA), which is essential for the material to be able to interact effectively with bone tissues. Moreover, it has been demonstrated that the addition of bioactive glass nanoparticles to gelatine/CS hydrogels increases mechanical properties and induces osteogenic differentiation, implying that the use of these materials to create scaffolds that stimulate bone regrowth is effective. The production of scaffolds has been utterly revolutionized by 3D printing technologies, which provide precise control over the design and composition of implants. The efficiency of 3D printed structures in bone regeneration has been established by the application of nano clay-based scaffolds, which have been shown to induce vascular ingrowth and form bone mineral tissue both *in vitro* and *in vivo* [159]. Added emphasis on the importance of biopolymers for the stimulation of bone regeneration is provided by the creation of GelMA (gelatin methacryloyl) bioinks, which enable the fabrication of complex structures with the ability to support stem cell osteogenic differentiation [160]. Composite scaffolds with enhanced mechanical strength and bioactivity have been synthesized through the blending of biopolymers and inorganic materials like calcium phosphates. A sample of the cooperative advantages of the materials in use here is improved directed bone regenerative processes from the co-combination of biphasic calcium phosphate with fibrin biopolymers [161]. Another sample proving the potential in bone tissue engineering of techniques reliant on biopolymer materials has been found as phosphorylated chitosan hydrogels, causing osteogenic differentiation through specific signal pathways. While encouraging preclinical data have been provided by preclinical studies, translating the 3D-printed scaffolds into practical applications for bone regeneration remains challenging. One of the main drawbacks of the study is the absence of systematic *in vivo* investigations that assess the efficacy and safety of these scaffolds. With advancing fabrication technologies, however, the gate is being opened to more effective tissue engineering techniques that can address these challenges, *e.g.*, the fabrication of high-fidelity nanocomposite bio-inks [162].

Future Technologies in Bone Tissue Engineering: Gene Editing and 3D Bioprinting

Gene editing technologies, particularly CRISPR/Cas9, have revolutionized regenerative medicine with enhanced bone repair. CRISPR/Cas9 makes it possible to target specific genes responsible for bone formation and healing by allowing precise genetic modification. For instance, studies have shown that osteogenic development in stem cells can be enhanced by fine-tuning signaling pathways through CRISPR [163]. By altering genes linked to osteoprogenitor cells, this targeted approach can increase the activation of bone regeneration stem cells and improve their efficacy in scaffold-based therapies [164]. Combined with the progress in bioprinting techniques, the CRISPR/Cas9 technology provides a great opportunity to create personalized therapies that are tailored to the needs of every patient in bone healing.

3D bioprinting methods have greatly improved the capability of biopolymer scaffolds for bone regeneration. The novel method allows for the generation of personalized, intricate structures closely resembling the physiological bone environment through the integration of synthetic and biological elements. Modern bioprinting technologies, such as direct ink writing and laser-assisted bioprinting, allow for the tight control of scaffold architecture, porosity, and mechanical properties necessary to support cellular attachment, growth, and differentiation [165]. Moreover, the seeding of growth factors and stem cells into scaffolds *via* these bioprinting methods also enhances the promise of successful bone regeneration [166]. The fundamental concept behind these novel technologies is the integration of biotechnology with engineering concepts.

The core idea of additive manufacturing, which creates intricate structures through the layer-by-layer depositing of material, is the basis for 3D bioprinting. This method allows customization based on the specific biological demands of bone tissue, including delivering nutrients and oxygen that are indispensable for vascularization [167]. In addition to this, extracellular matrix-simulating bioinks have emerged due to improved material science. These bioinks enhance cellular behavior and assist in the process of natural bone healing [168]. With the incorporation of bioactive components, such as nanoparticles, into the bio-inks, osteoconductivity and mechanical properties can be enhanced further, giving a more holistic approach to bone regeneration [169]. Recent research shows that pre-vascularized 3D printed PLA-hydroxyapatite scaffolds enriched with mineral cues and extracellular matrix features can stimulate both vascular differentiation and osteogenesis in engineered bone constructs integrated within these implants [170]. This is essential because tissue viability in a healthy state and delivery of oxygen and nutrients rely on healthy circulatory networks. Based on the evidence, scaffolds that enhance endothelial cell proliferation alongside osteogenic cells exhibit increased bone formation due to synergism between them [171]. In addition, the development of composite bio-inks containing multiple growth factors holds the potential to stimulate osteogenic differentiation and ensure proper scaffold degradation rates that match the bone regeneration schedules [172]. Despite these advancements, there remain some limitations to combining gene editing and bioprinting for bone regeneration. Combining the two on a clinical level is one of the principal challenges. Even as advances in the lab have phenomenal results, strict validation and regulatory acceptance are required before applying these results within clinical environments [173]. Several hurdles must be overcome due to the challenges of ensuring consistent and reproducible results in 3D bioprinting and the subtleties of proper utilization of CRISPR/Cas9. Moreover, it remains challenging to obtain uniform vascularization in printed structures; without proper vascular networks, osteogenic effectiveness and healing times can be compromised [174]. In the future, these technologies hold much promise. New research indicates a shift toward more integrated systems that employ tailored bioprinting techniques alongside CRISPR gene editing to augment bone repair. The cell microenvironment and scaffold function can be enhanced by newer advances in hybrid fabrication methods that combine bioprinting with other methods such as electrospinning or decellularized matrix infusion [175].

Commercialization and Translation to the Clinical

Orthopaedic R&D should focus on the commercialization and clinical application of biopolymer-based bone regeneration implants. Biopolymers are the best fit for use in bone regenerative surgery owing to their singular benefits, such as biocompatibility, biodegradability, and tailorability to a given application. Taking these laboratory results into the clinician's hands requires overcoming some hurdles, like the scalability of production, regulatory hurdles, and long-term safety and efficacy establishment. One of the biggest barriers to commercialization is the need for extensive proof of the capability of biopolymer implants to aid bone repair. Constant monitoring of implant load and condition of bone healing has been proven to yield informative data that can be used to maximize implant design and treatment regimen [176].

New products are first required to obtain regulatory clearance and meet applicable safety and effectiveness standards before they can be generally used in practice. Evidence-based strategies are crucial for this intention. Furthermore, the application of new materials, including ultra-high molecular weight polyethylene (UHMWPE), has been investigated for application in orthopedic devices. UHMWPE is a potential material for orthopedic devices such as joint replacements because of its superior mechanical properties, including high tensile strength and abrasion resistance [177].

Such implants can be made even more effective by fabricating biopolymer composites that incorporate UHMWPE and bioactive molecules to facilitate osseointegration and reduce the infection risk. In orthopedic surgery, 3D printing technology has been a breakthrough, enabling the customization of implants to the specific anatomical and pathological characteristics of each patient. Preoperative planning and the production of implants have also been highly advanced through the use of this technology, which has also been demonstrated to enhance surgical outcomes dramatically. Implant customization facilitated by 3D printing not only enhances the fit and function of the devices but also facilitates the inclusion of bioactive components that can contribute to bone integration and healing. The development of antimicrobial implant coatings is a key area of focus in the commercialization of biopolymer-based solutions [178]. Successful implants with hydroxyapatite covering and silver have been manufactured, for instance, to enhance bone fusion capacity and prevent postoperative infection [179]. Such improvements are critical as, in orthopedic implantation, infection remains the major reason for failure. Orthopedic implant longevity and success can be optimized by adding bioactive materials to the implant surface, which has the potential to significantly enhance osseointegration and reduce infection risk [180]. Additionally, the performance of implants can be enhanced even further by modifying their surface characteristics using biofunctionalization techniques. Titanium implants can be bio-functionalized to enhance osseointegration and minimize implant-associated infections [181].

CONCLUSION

This chapter explores the exciting area of biopolymers and how they can transform bone repair and regeneration. These exceptional biocompatibility and capacity of naturally occurring substances to sustain vital cellular processes have drawn a great deal of interest from the medical community. Biopolymers' versatility enables them to be molded into various forms that mimic the extracellular matrix, providing an ideal environment for bone growth and development. One of the primary advantages of biopolymers is their seamless integration with bodily tissues. Unlike synthetic materials that can trigger adverse reactions, biopolymers are generally well-tolerated and can degrade gradually, allowing the body's tissues to replace them. This characteristic makes them highly appealing for bone repair applications, where a scaffold that supports initial growth and is safely resorbed is desirable. This chapter explores various biopolymers, including collagen, chitosan, alginate, and hyaluronic acid, which have shown promise in bone regeneration. Each of these materials possesses unique properties, such as strength, flexibility, or the ability to bind growth factors. Researchers can design composite materials that are specifically suited to meet certain needs related to bone restoration by combining different biopolymers or altering their architectures. Modern uses of biopolymers in bone treatment have been thoroughly investigated. These include injectable hydrogels that can fill uneven voids and harden in place, nanofiber meshes that offer the perfect surface for cell attachment and proliferation, and 3D-printed scaffolds that precisely fit a patient's bone deformity. The use of biopolymers as delivery systems for growth factors, stem cells, or other bioactive substances that promote bone formation is also covered in this chapter. A review of the methods currently used in clinical settings is conducted to use biopolymers, including case studies and clinical trials that show their effectiveness. The chapter describes the regulatory environment for bone therapeutics based on biopolymers and the obstacles that need to be removed before these novel treatments may be used widely in clinical settings. The chapter concludes by discussing how biopolymers might change bone therapy in the future.

This involves creating smart materials that can react dynamically to the healing process, combining biopolymers with cutting-edge technologies like gene editing or 3D bioprinting, and creating personalized treatments catered to the individual needs of each patient. The biopolymer-based bone therapeutics, include production scaling up for clinical usage, mechanical strength assurance, and rate control. They talk about current studies that aim to get beyond these obstacles and enhance the functionality of biopolymer scaffolds. To progress biopolymer-based bone therapeutics, the subject is interdisciplinary, and the chapter ends by stressing the cooperation of materials scientists, biologists, engineers, and physicians. It highlights the necessity for continued study and development in this field because of the huge potential to improve patient outcomes in orthopedic surgery, dental implantology, and other fields where bone regeneration is crucial. All things considered, the chapter provides readers with a comprehensive grasp of the current and prospective uses of biopolymers in bone regeneration and repair, along with the chance to learn more about this intriguing and rapidly evolving field of biomedical research.

REFERENCES

[1]   Paschalis EP, Tatakis DN, Robins S, *et al.* Lathyrism-induced alterations in collagen cross-links influence the mechanical properties of bone material without affecting the mineral. Bone 2011; 49(6): 1232-41.
[http://dx.doi.org/10.1016/j.bone.2011.08.027] [PMID: 21920485]

[2]   Liu Y, Luo D, Wang T. Hierarchical structures of bone and bioinspired bone tissue engineering. Small 2016; 12(34): 4611-32.
[http://dx.doi.org/10.1002/smll.201600626] [PMID: 27322951]

[3]   Aido M, Kerschnitzki M, Hoerth R, *et al.* Relationship between nanoscale mineral properties and

calcein labeling in mineralizing bone surfaces. Connect Tissue Res 2014; 55(sup1): 15-7.
[http://dx.doi.org/10.3109/03008207.2014.923869]

[4] McNerny EMB, Gong B, Morris MD, Kohn DH. Bone fracture toughness and strength correlate with collagen cross-link maturity in a dose-controlled lathyrism mouse model. J Bone Miner Res 2015; 30(3): 455-64.
[http://dx.doi.org/10.1002/jbmr.2356] [PMID: 25213475]

[5] Wang X, Bank RA, Tekoppele JM, Agrawal CM. The role of collagen in determining bone mechanical properties. J Orthop Res 2001; 19(6): 1021-6.
[http://dx.doi.org/10.1016/S0736-0266(01)00047-X] [PMID: 11781000]

[6] Pathi SP, Kowalczewski C, Tadipatri R, Fischbach C. A novel 3-D mineralized tumor model to study breast cancer bone metastasis. PLoS One 2010; 5(1): e8849.
[http://dx.doi.org/10.1371/journal.pone.0008849] [PMID: 20107512]

[7] Sanchez-Rodriguez E, Benavides-Reyes C, Torres C, *et al.* Changes with age (from 0 to 37 D) in tibiae bone mineralization, chemical composition and structural organization in broiler chickens. Poult Sci 2019; 98(11): 5215-25.
[http://dx.doi.org/10.3382/ps/pez363] [PMID: 31265108]

[8] He F, Chiou AE, Loh HC, *et al.* Multiscale characterization of the mineral phase at skeletal sites of breast cancer metastasis. Proc Natl Acad Sci USA 2017; 114(40): 10542-7.
[http://dx.doi.org/10.1073/pnas.1708161114] [PMID: 28923958]

[9] Kim SE, Lee E, Jang K, Shim KM, Kang SS. Evaluation of porcine hybrid bone block for bone grafting in dentistry. In Vivo 2018; 32(6): 1419-26.

[10] Li S-D, Chen Y-B, Qiu L-G, Qin M-Q. G-CSF indirectly induces apoptosis of osteoblasts during hematopoietic stem cell mobilization. Clin Transl Sci 2017; 10(4): 287-91.
[http://dx.doi.org/10.1111/cts.12467] [PMID: 28556597]

[11] Otero JE, Chen T, Zhang K, Abu-Amer Y. Constitutively active canonical NF-κB pathway induces severe bone loss in mice. PLoS One 2012; 7(6): e38694.
[http://dx.doi.org/10.1371/journal.pone.0038694] [PMID: 22685599]

[12] Hong H, Shi Z, Qiao P, *et al.* Interleukin-3 plays dual roles in osteoclastogenesis by promoting the development of osteoclast progenitors but inhibiting the osteoclastogenic process. Biochem Biophys Res Commun 2013; 440(4): 545-50.
[http://dx.doi.org/10.1016/j.bbrc.2013.09.098] [PMID: 24103757]

[13] Rantlha M, Sagar T, Kruger MC, Coetzee M, Deepak V. Ellagic acid inhibits RANKL-induced osteoclast differentiation by suppressing the p38 MAP kinase pathway. Arch Pharm Res 2017; 40(1): 79-87.
[http://dx.doi.org/10.1007/s12272-016-0790-0] [PMID: 27384064]

[14] Kowada T, Kikuta J, Kubo A, *et al. In vivo* fluorescence imaging of bone-resorbing osteoclasts. J Am Chem Soc 2011; 133(44): 17772-6.
[http://dx.doi.org/10.1021/ja2064582] [PMID: 21939210]

[15] Liu Y, Wang Z, Ma C, *et al.* Dracorhodin perchlorate inhibits osteoclastogenesis through repressing RANKL-stimulated NFATc1 activity. J Cell Mol Med 2020; 24(6): 3303-13.
[http://dx.doi.org/10.1111/jcmm.15003] [PMID: 31965715]

[16] Jansen IDC, Vermeer JAF, Bloemen V, Stap J, Everts V. Osteoclast fusion and fission. Calcif Tissue Int 2012; 90(6): 515-22.
[http://dx.doi.org/10.1007/s00223-012-9600-y] [PMID: 22527205]

[17] Qin A, Cheng TS, Lin Z, *et al.* Versatile roles of V-ATPases accessory subunit Ac45 in osteoclast formation and function. PLoS One 2011; 6(11): e27155.
[http://dx.doi.org/10.1371/journal.pone.0027155] [PMID: 22087256]

[18] Chiu YH, Ritchlin CT. DC-STAMP: a key regulator in osteoclast differentiation. J Cell Physiol 2016;

231(11): 2402-7.
[http://dx.doi.org/10.1002/jcp.25389] [PMID: 27018136]

[19]   Siddiqui JA, Partridge NC. Physiological bone remodeling: systemic regulation and growth factor involvement. Physiology (Bethesda) 2016; 31(3): 233-45.
[http://dx.doi.org/10.1152/physiol.00061.2014] [PMID: 27053737]

[20]   Liang M, Yin X, Zhang S, *et al.* Osteoclast-derived small extracellular vesicles induce osteogenic differentiation *via* inhibiting ARHGAP1. Mol Ther Nucleic Acids 2021; 23: 1191-203.
[http://dx.doi.org/10.1016/j.omtn.2021.01.031] [PMID: 33664997]

[21]   Whitlock JM, de Castro LF, Collins MT, Chernomordik LV, Boyce AM. An inducible explant model for dissecting osteoclast-osteoblast coordination in health and disease. bioRxiv 2022.
[http://dx.doi.org/10.1101/2022.10.27.514052]

[22]   Sun W, Zhao C, Li Y, *et al.* Osteoclast-derived microRNA-containing exosomes selectively inhibit osteoblast activity. Cell Discov 2016; 2(1): 16015.
[http://dx.doi.org/10.1038/celldisc.2016.15] [PMID: 27462462]

[23]   Sims NA, Martin TJ. Coupling the activities of bone formation and resorption: a multitude of signals within the basic multicellular unit. Bonekey Rep 2014; 3: 481.
[http://dx.doi.org/10.1038/bonekey.2013.215] [PMID: 24466412]

[24]   Zhou YM, Yang YY, Jing YX, *et al.* BMP9 reduces bone loss in ovariectomized mice by dual regulation of bone remodeling. J Bone Miner Res 2020; 35(5): 978-93.
[http://dx.doi.org/10.1002/jbmr.3957] [PMID: 31914211]

[25]   Zhang H, Zhang M, Zhai D, *et al.* Polyhedron-like biomaterials for innervated and vascularized bone regeneration. Adv Mater 2023; 35(42): 2302716.
[http://dx.doi.org/10.1002/adma.202302716] [PMID: 37434296]

[26]   Amarasekara DS, Kim S, Rho J. Regulation of osteoblast differentiation by cytokine networks. Int J Mol Sci 2021; 22(6): 2851.
[http://dx.doi.org/10.3390/ijms22062851] [PMID: 33799644]

[27]   Yuan FL, Wu Q, Miao ZN, *et al.* Osteoclast-derived extracellular vesicles: novel regulators of osteoclastogenesis and osteoclast–osteoblasts communication in bone remodeling. Front Physiol 2018; 9: 628.
[http://dx.doi.org/10.3389/fphys.2018.00628] [PMID: 29910740]

[28]   Okamoto M, Murai J, Imai Y, *et al.* Conditional deletion of *Bmpr1a* in differentiated osteoclasts increases osteoblastic bone formation, increasing volume of remodeling bone in mice. J Bone Miner Res 2011; 26(10): 2511-22.
[http://dx.doi.org/10.1002/jbmr.477] [PMID: 21786321]

[29]   To TT, Witten PE, Renn J, Bhattacharya D, Huysseune A, Winkler C. Rankl-induced osteoclastogenesis leads to loss of mineralization in a medaka osteoporosis model. Development 2012; 139(1): 141-50.
[http://dx.doi.org/10.1242/dev.071035] [PMID: 22096076]

[30]   Bernhardt A, Thieme S, Domaschke H, Springer A, Rösen-Wolff A, Gelinsky M. Crosstalk of osteoblast and osteoclast precursors on mineralized collagen—towards an *in vitro* model for bone remodeling. J Biomed Mater Res A 2010; 95A(3): 848-56.
[http://dx.doi.org/10.1002/jbm.a.32856] [PMID: 20824694]

[31]   Xie Y, Chen Y, Zhang L, Ge W, Tang P. The roles of bone-derived exosomes and exosomal micro RNA s in regulating bone remodelling. J Cell Mol Med 2017; 21(5): 1033-41.
[http://dx.doi.org/10.1111/jcmm.13039] [PMID: 27878944]

[32]   Sravani Sala , Sala S, Vadaga AK. Importance of biopolymers in pharmaceutical and medical fields. J Pharm Insights Res 2024; 2(3): 115-22.
[http://dx.doi.org/10.69613/spw06q40]

[33]   Gautieri A, Vesentini S, Redaelli A, Buehler MJ. Hierarchical structure and nanomechanics of collagen microfibrils from the atomistic scale up. Nano Lett 2011; 11(2): 757-66.
[http://dx.doi.org/10.1021/nl103943u] [PMID: 21207932]

[34]   Stuart K, Panitch A. Influence of chondroitin sulfate on collagen gel structure and mechanical properties at physiologically relevant levels. Biopolymers 2008; 89(10): 841-51.
[http://dx.doi.org/10.1002/bip.21024] [PMID: 18488988]

[35]   Xin X, Borzacchiello A, Netti PA, Ambrosio L, Nicolais L. Hyaluronic-acid-based semi-interpenetrating materials. J Biomater Sci Polym Ed 2004; 15(9): 1223-36.
[http://dx.doi.org/10.1163/1568562041753025] [PMID: 15503636]

[36]   Fernandes-Cunha GM, Chen KM, Chen F, *et al. In situ*-forming collagen hydrogel crosslinked *via* multi-functional PEG as a matrix therapy for corneal defects. Sci Rep 2020; 10(1): 16671.
[http://dx.doi.org/10.1038/s41598-020-72978-5] [PMID: 33028837]

[37]   Nagaraj A, Etxeberria AE, Naffa R, Zidan G, Seyfoddin A. 3D-printed hybrid collagen/GelMA hydrogels for tissue engineering applications. Biology (Basel) 2022; 11(11): 1561.
[http://dx.doi.org/10.3390/biology11111561] [PMID: 36358262]

[38]   Yang JM, Lin HT, Wu TH, Chen CC. Wettability and antibacterial assessment of chitosan containing radiation-induced graft nonwoven fabric of polypropylene- *g* -acrylic acid. J Appl Polym Sci 2003; 90(5): 1331-6.
[http://dx.doi.org/10.1002/app.12787]

[39]   Bhatt R, Jaffe M. Biopolymers in medical implants. In: Naran AS, Boddu SHS, Eds. Excipient Applications in Formulation Design and Drug Delivery. 2015; pp. 311-48.

[40]   Mistry K, Grinberg N. Separation of peptides and proteins by capillary electrochromatography. J Liq Chromatogr Relat Technol 2004; 27(7-9): 1179-202.
[http://dx.doi.org/10.1081/JLC-120030601]

[41]   Persikov AV, Ramshaw JAM, Brodsky B. Prediction of collagen stability from amino acid sequence. J Biol Chem 2005; 280(19): 19343-9.
[http://dx.doi.org/10.1074/jbc.M501657200] [PMID: 15753081]

[42]   Zarrintaj P, Seidi F, Youssefi Azarfam M, *et al.* Biopolymer-based composites for tissue engineering applications: A basis for future opportunities. Compos, Part B Eng 2023; 258: 110701.
[http://dx.doi.org/10.1016/j.compositesb.2023.110701]

[43]   Krishani M, Shin WY, Suhaimi H, Sambudi NS. Development of scaffolds from bio-based natural materials for tissue regeneration applications: a review. Gels 2023; 9(2): 100.
[http://dx.doi.org/10.3390/gels9020100] [PMID: 36826270]

[44]   Chana RS, Wheeler DC, Thomas GJ, Williams JD, Davies M. Low-density lipoprotein stimulates mesangial cell proteoglycan and hyaluronan synthesis. Nephrol Dial Transplant 2000; 15(2): 167-72.
[http://dx.doi.org/10.1093/ndt/15.2.167] [PMID: 10648661]

[45]   Schanté CE, Zuber G, Herlin C, Vandamme TF. Chemical modifications of hyaluronic acid for the synthesis of derivatives for a broad range of biomedical applications. Carbohydr Polym 2011; 85(3): 469-89.
[http://dx.doi.org/10.1016/j.carbpol.2011.03.019]

[46]   Jabeen N, Atif M. Polysaccharides based biopolymers for biomedical applications: A review. Polym Adv Technol 2024; 35(1): e6203.
[http://dx.doi.org/10.1002/pat.6203]

[47]   Bessalah S, Faraz A, Dbara M, *et al.* Antibacterial, anti-biofilm, and anti-inflammatory properties of gelatin–chitosan–moringa-biopolymer-based wound dressings towards *Staphylococcus aureus* and *Escherichia coli.*. Pharmaceuticals (Basel) 2024; 17(5): 545.
[http://dx.doi.org/10.3390/ph17050545] [PMID: 38794116]

[48]   Ndlovu SP, Ngece K, Alven S, Aderibigbe BA. Gelatin-based hybrid scaffolds: promising wound dressings. Polymers (Basel) 2021; 13(17): 2959.
[http://dx.doi.org/10.3390/polym13172959] [PMID: 34502997]

[49]   Gopi S, Amalraj A, Thomas S. Effective drug delivery system of biopolymers based on nanomaterials and hydrogels-a review. Drug Des 2016; 5(2): 2169-0138.
[http://dx.doi.org/10.4172/2169-0138.1000129]

[50]   Opriş O, Mormile C, Lung I, Stegarescu A, Soran ML, Soran A. An overview of biopolymers for drug delivery applications. Appl Sci (Basel) 2024; 14(4): 1383.
[http://dx.doi.org/10.3390/app14041383]

[51]   Ranganadhareddy A, Chandrsekhar C. Polyhydroxyalkanoates, the biopolymers of microbial origin - a review. J Biochem Technol 2022; 13(3-2022): 1-6.
[http://dx.doi.org/10.51847/3qf2Wvuzl2]

[52]   Jadoun S, Riaz U, Budhiraja V. Biodegradable conducting polymeric materials for biomedical applications: a review. Med Devices Sens 2021; 4(1): e10141.
[http://dx.doi.org/10.1002/mds3.10141]

[53]   Mohan TP, Devchand K, Kanny K. Barrier and biodegradable properties of corn starch-derived biopolymer film filled with nanoclay fillers. J Plast Film Sheeting 2017; 33(3): 309-36.
[http://dx.doi.org/10.1177/8756087916682553]

[54]   Sánchez-Cid P, Jiménez-Rosado M, Alonso-González M, Romero A, Perez-Puyana V. Applied rheology as tool for the assessment of chitosan hydrogels for regenerative medicine. Polymers (Basel) 2021; 13(13): 2189.
[http://dx.doi.org/10.3390/polym13132189] [PMID: 34209385]

[55]   Jirvankar P, Agrawal S, Chambhare N, Agrawal R. Harnessing biopolymer gels for theranostic applications: imaging agent integration and real-time monitoring of drug delivery. Gels 2024; 10(8): 535.
[http://dx.doi.org/10.3390/gels10080535] [PMID: 39195064]

[56]   Taran M, Etemadi S, Safaei M, Chacon EL, Bertolo MRV, Plepis AMG, *et al.* Collagen-chitosa--hydroxyapatite composite scaffolds for bone repair in ovariectomized rats. Sci Rep 2023; 13(1): 28.
[PMID: 36593236]

[57]   Lyons JG, Plantz MA, Hsu WK, Hsu EL, Minardi S. Nanostructured biomaterials for bone regeneration. Front Bioeng Biotechnol 2020; 8: 922.
[http://dx.doi.org/10.3389/fbioe.2020.00922] [PMID: 32974298]

[58]   Farjaminejad S, Farjaminejad R, Garcia-Godoy F. Nanoparticles in bone regeneration: a narrative review of current advances and future directions in tissue engineering. J Funct Biomater 2024; 15(9): 241.
[http://dx.doi.org/10.3390/jfb15090241] [PMID: 39330217]

[59]   Zhang G, Zhen C, Yang J, *et al.* Recent advances of nanoparticles on bone tissue engineering and bone cells. Nanoscale Adv 2024; 6(8): 1957-73.
[http://dx.doi.org/10.1039/D3NA00851G] [PMID: 38633036]

[60]   Gong T, Xie J, Liao J, Zhang T, Lin S, Lin Y. Nanomaterials and bone regeneration. Bone Res 2015; 3(1): 15029.
[http://dx.doi.org/10.1038/boneres.2015.29] [PMID: 26558141]

[61]   Idumah CI. Progress in polymer nanocomposites for bone regeneration and engineering. Polym Polymer Compos 2021; 29(5): 509-27.
[http://dx.doi.org/10.1177/0967391120913658]

[62]   Hajiali H, Ouyang L, Llopis-Hernandez V, Dobre O, Rose FRAJ. Review of emerging nanotechnology in bone regeneration: progress, challenges, and perspectives. Nanoscale 2021; 13(23): 10266-80.
[http://dx.doi.org/10.1039/D1NR01371H] [PMID: 34085085]

[63]   Brassolatti P, Bossini PS, Andrade ALM, *et al.* Comparison of two different biomaterials in the bone regeneration (15, 30 and 60 days) of critical defects in rats. Acta Cir Bras 2021; 36(6): e360605.
[http://dx.doi.org/10.1590/acb360605] [PMID: 34287608]

[64]   Alkaron W, Almansoori A, Balázsi C, Balázsi K. A critical review of natural and synthetic polymer-based biological apatite composites for bone tissue engineering. J Compos Sci 2024; 8(12): 523.
[http://dx.doi.org/10.3390/jcs8120523]

[65]   Jahan K, Tabrizian M. Composite biopolymers for bone regeneration enhancement in bony defects. Biomater Sci 2016; 4(1): 25-39.
[http://dx.doi.org/10.1039/C5BM00163C] [PMID: 26317131]

[66]   Kashirina A, Yao Y, Liu Y, Leng J. Biopolymers as bone substitutes: a review. Biomater Sci 2019; 7(10): 3961-83.
[http://dx.doi.org/10.1039/C9BM00664H] [PMID: 31364613]

[67]   2.Zeghoud S, Hemmami H, Alhamad AA, Segueni A, Dahmri M, Guedouda N, Zahira M, Amor IB. Biopolymers for enhancement of bone regeneration. IJS Global Health 2024; 7(2): e0303.

[68]   Girón J, Kerstner E, Medeiros T, *et al.* Biomaterials for bone regeneration: an orthopedic and dentistry overview. Braz J Med Biol Res 2021; 54(9): e11055.
[http://dx.doi.org/10.1590/1414-431x2021e11055]

[69]   Alkhursani SA, Ghobashy MM, Al-Gahtany SA, *et al.* Application of nano-inspired scaffolds-based biopolymer hydrogel for bone and periodontal tissue regeneration. Polymers (Basel) 2022; 14(18): 3791.
[http://dx.doi.org/10.3390/polym14183791] [PMID: 36145936]

[70]   Xing Y, Qiu L, Liu D, Dai S, Sheu CL. The role of smart polymeric biomaterials in bone regeneration: a review. Front Bioeng Biotechnol 2023; 11: 1240861.
[http://dx.doi.org/10.3389/fbioe.2023.1240861] [PMID: 37662432]

[71]   Fahmy MD, Jazayeri HE, Razavi M, Masri R, Tayebi L. Three-dimensional bioprinting materials with potential application in preprosthetic surgery. J Prosthodont 2016; 25(4): 310-8.
[http://dx.doi.org/10.1111/jopr.12431] [PMID: 26855004]

[72]   Chacon EL, Bertolo MRV, de Guzzi Plepis AM, *et al.* Collagen-chitosan-hydroxyapatite composite scaffolds for bone repair in ovariectomized rats. Sci Rep 2023; 13(1): 28.
[http://dx.doi.org/10.1038/s41598-022-24424-x] [PMID: 36593236]

[73]   Shuai C, Yang W, Feng P, Peng S, Pan H. Accelerated degradation of HAP/PLLA bone scaffold by PGA blending facilitates bioactivity and osteoconductivity. Bioact Mater 2021; 6(2): 490-502.
[http://dx.doi.org/10.1016/j.bioactmat.2020.09.001] [PMID: 32995675]

[74]   Zhang H, Mao X, Du Z, *et al.* Three dimensional printed macroporous polylactic acid/hydroxyapatite composite scaffolds for promoting bone formation in a critical-size rat calvarial defect model. Sci Technol Adv Mater 2016; 17(1): 136-48.
[http://dx.doi.org/10.1080/14686996.2016.1145532] [PMID: 27877865]

[75]   Ming L, Zhipeng Y, Fei Y, *et al.* Microfluidic-based screening of resveratrol and drug-loading PLA/gelatine nano-scaffold for the repair of cartilage defect. Artif Cells Nanomed Biotechnol 2018; 46(sup1): 336-46.

[76]   Peng S, Feng P, Wu P, *et al.* Graphene oxide as an interface phase between polyetheretherketone and hydroxyapatite for tissue engineering scaffolds. Sci Rep 2017; 7(1): 46604.
[http://dx.doi.org/10.1038/srep46604] [PMID: 28425470]

[77]   Feng P, Peng S, Shuai C, *et al. In situ* generation of hydroxyapatite on biopolymer particles for fabrication of bone scaffolds owning bioactivity. ACS Appl Mater Interfaces 2020; 12(41): 46743-55.
[http://dx.doi.org/10.1021/acsami.0c13768] [PMID: 32940994]

[78]   Ou L, Lan Y, Feng Z, *et al.* Functionalization of SF/HAP scaffold with GO-PEI-miRNA inhibitor

complexes to enhance bone regeneration through activating transcription factor 4. Theranostics 2019; 9(15): 4525-41.
[http://dx.doi.org/10.7150/thno.34676] [PMID: 31285777]

[79]   Creste CFZ, Orsi PR, Landim-Alvarenga FC, *et al.* Highly effective fibrin biopolymer scaffold for stem cells upgrading bone regeneration. Materials (Basel) 2020; 13(12): 2747.
[http://dx.doi.org/10.3390/ma13122747] [PMID: 32560388]

[80]   He X, Liu Y, Yuan X, Lu L. Enhanced healing of rat calvarial defects with MSCs loaded on BMP-2 releasing chitosan/alginate/hydroxyapatite scaffolds. PLoS One 2014; 9(8): e104061.
[http://dx.doi.org/10.1371/journal.pone.0104061] [PMID: 25084008]

[81]   Cheng CH, Shie MY, Lai YH, Foo NP, Lee MJ, Yao CH. Fabrication of 3D printed poly (lactic acid)/polycaprolactone scaffolds using TGF-β1 for promoting bone regeneration. Polymers (Basel) 2021; 13(21): 3731.
[http://dx.doi.org/10.3390/polym13213731] [PMID: 34771286]

[82]   Sanjaya IG, Maliawan S. Chitosan as bone scaffold for craniofacial bone regeneration: a systematic review. Open Access Maced J Med Sci 2022; 10(F): 773-9.
[http://dx.doi.org/10.3889/oamjms.2022.10684]

[83]   Casagrande S, Tiribuzi R, Cassetti E, *et al.* Biodegradable composite porous poly (dl-lactide--o-glycolide) scaffold supports mesenchymal stem cell differentiation and calcium phosphate deposition. Artif Cells Nanomed Biotechnol 2018; 46(sup1): 219-29.

[84]   Du J, Gan S, Bian Q, *et al.* Preparation and characterization of porous hydroxyapatite/β-cyclodextrin-based polyurethane composite scaffolds for bone tissue engineering. J Biomater Appl 2018; 33(3): 402-9.
[http://dx.doi.org/10.1177/0885328218797545] [PMID: 30223737]

[85]   Zorzella Creste CF, Orsi PR, Landim-Alvarenga FC, *et al.* Improvement of bone regeneration using fibrin biopolymer combined with differentiated stem cells. bioRxiv 2019; 608166.
[http://dx.doi.org/10.1101/608166]

[86]   Dyondi D, Webster TJ, Banerjee R. A nanoparticulate injectable hydrogel as a tissue engineering scaffold for multiple growth factor delivery for bone regeneration. Int J Nanomedicine 2013; 8: 47-59.
[PMID: 23293519]

[87]   Hosseini Y, Emadi R, Kharaziha M, Doostmohammadi A. Reinforcement of electrospun poly(ε-caprolactone) scaffold using diopside nanopowder to promote biological and physical properties. J Appl Polym Sci 2017; 134(6): app.44433.
[http://dx.doi.org/10.1002/app.44433]

[88]   Quemeneur F, Tourne-Peteilh C, Drouet C, *et al.* Foam-based bionanocomposite scaffold for bone tissue engineering. Key Eng Mater 2017; 758: 145-9.
[http://dx.doi.org/10.4028/www.scientific.net/KEM.758.145]

[89]   Zhang Y, Wu C, Friis T, Xiao Y. The osteogenic properties of CaP/silk composite scaffolds. Biomaterials 2010; 31(10): 2848-56.
[http://dx.doi.org/10.1016/j.biomaterials.2009.12.049] [PMID: 20071025]

[90]   Dastjerdi R, Sharafi M, Kabiri K, Mivehi L, Samadikuchaksaraei A. An acid-free water-born quaternized chitosan/montmorillonite loaded into an innovative ultra-fine bead-free water-born nanocomposite nanofibrous scaffold; *in vitro* and *in vivo* approaches. Biomed Mater 2017; 12(4): 045014.
[http://dx.doi.org/10.1088/1748-605X/aa7608] [PMID: 28561741]

[91]   Kumar S, Madhav NVS, Verma A, Pathak K. A smart approach for delivery of nanosized phenytoin using biomaterial isolated from *Fragaria ananassa*. Int J Pharm Investig 2020; 10(3): 305-11.
[http://dx.doi.org/10.5530/ijpi.2020.3.55]

[92]   Mitra T, Sailakshmi G, Gnanamani A, Mandal AB. Adipic acid interaction enhances the mechanical

and thermal stability of natural polymers. J Appl Polym Sci 2012; 125(S2): E490-500.
[http://dx.doi.org/10.1002/app.36957]

[93]   Diba M, Camargo WA, Brindisi M, *et al*. Composite colloidal gels made of bisphosphonate-functionalized gelatin and bioactive glass particles for regeneration of osteoporotic bone defects. Adv Funct Mater 2017; 27(45): 1703438.
[http://dx.doi.org/10.1002/adfm.201703438]

[94]   Inzana JA, Olvera D, Fuller SM, *et al*. 3D printing of composite calcium phosphate and collagen scaffolds for bone regeneration. Biomaterials 2014; 35(13): 4026-34.
[http://dx.doi.org/10.1016/j.biomaterials.2014.01.064] [PMID: 24529628]

[95]   Vijayan V, Sreekumar S, Ahina KM, Lakra R, Kiran MS. Lanthanum Oxide Nanoparticles Reinforced Collagen Ҝ-Carrageenan Hydroxyapatite Biocomposite as Angio-Osteogenic Biomaterial for *In Vivo* Osseointegration and Bone Repair. Adv Biol 2023; 7(8): 2300039.
[http://dx.doi.org/10.1002/adbi.202300039] [PMID: 37080950]

[96]   Huang SL, Wen B, Bian WG, Yan HW. Reconstruction of comminuted long-bone fracture using CF/CPC scaffolds manufactured by rapid prototyping. Med Sci Monit 2012; 18(11): BR435-40.
[http://dx.doi.org/10.12659/MSM.883536] [PMID: 23111734]

[97]   Koga T, Kumazawa S, Okimura Y, Zaitsu Y, Umeshita K, Asahina I. Evaluation of poly lactic-co-glycolic acid-coated β-tricalcium phosphate bone substitute as a graft material for ridge preservation after tooth extraction in dog mandible: a comparative study with conventional β-tricalcium phosphate granules. Materials (Basel) 2020; 13(16): 3452.
[http://dx.doi.org/10.3390/ma13163452] [PMID: 32764407]

[98]   Zhang Y, Cheng N, Miron R, Shi B, Cheng X. Delivery of PDGF-B and BMP-7 by mesoporous bioglass/silk fibrin scaffolds for the repair of osteoporotic defects. Biomaterials 2012; 33(28): 6698-708.
[http://dx.doi.org/10.1016/j.biomaterials.2012.06.021] [PMID: 22763224]

[99]   Fahmy MD, Jazayeri HE, Razavi M, Masri R, Tayebi L. Three-dimensional bioprinting materials with potential application in preprosthetic surgery. J Prosthodont 2016; 25(4): 310-8.
[http://dx.doi.org/10.1111/jopr.12431] [PMID: 26855004]

[100]   Filippi M, Born G, Chaaban M, Scherberich A. Natural polymeric scaffolds in bone regeneration. Front Bioeng Biotechnol 2020; 8: 474.
[http://dx.doi.org/10.3389/fbioe.2020.00474] [PMID: 32509754]

[101]   Pramanik S, Kharche S, More N, Ranglani D, Singh G, Kapusetti G. Natural biopolymers for bone tissue engineering: a brief review. Engineered Regeneration 2023; 4(2): 193-204.
[http://dx.doi.org/10.1016/j.engreg.2022.12.002]

[102]   Almeida HV, Eswaramoorthy R, Cunniffe GM, Buckley CT, O'Brien FJ, Kelly DJ. Fibrin hydrogels functionalized with particulated cartilage extracellular matrix and incorporating freshly isolated stromal cells as an injectable for cartilage regeneration. Acta Biomater 2016; 36: 55-62.
[http://dx.doi.org/10.1016/j.actbio.2016.03.008]

[103]   Agheb M, Dinari M, Rafienia M, Salehi H. Novel electrospun nanofibers of modified gelatin-tyrosine in cartilage tissue engineering. Mater Sci Eng C 2017; 71: 240-51.
[http://dx.doi.org/10.1016/j.msec.2016.10.003] [PMID: 27987704]

[104]   Li Q, Reed D, Min L, *et al*. Lyophilized platelet-rich fibrin (PRF) promotes craniofacial bone regeneration through Runx2. Int J Mol Sci 2014; 15(5): 8509-25.
[http://dx.doi.org/10.3390/ijms15058509] [PMID: 24830554]

[105]   Tang Y, Chen Y, Huang L, Gao F, Sun H, Huang C. Intramembranous ossification imitation scaffold with the function of macrophage polarization for promoting critical bone defect repair. ACS Appl Bio Mater 2020; 3(6): 3569-81.
[http://dx.doi.org/10.1021/acsabm.0c00233] [PMID: 35025227]

[106] Kaigler D, Pagni G, Park CH, *et al.* Stem cell therapy for craniofacial bone regeneration: a randomized, controlled feasibility trial. Cell Transplant 2013; 22(5): 767-77.
[http://dx.doi.org/10.3727/096368912X652968] [PMID: 22776413]

[107] Della Coletta BB, Jacob TB, Moreira LAC, *et al.* Photobiomodulation therapy on the guided bone regeneration process in defects filled by biphasic calcium phosphate associated with fibrin biopolymer. Molecules 2021; 26(4): 847.
[http://dx.doi.org/10.3390/molecules26040847] [PMID: 33562825]

[108] Hammouda HF, Farag MM, El Deftar MMF, Abdel-Gabbar M, Mohamed BM. Effect of Ce-doped bioactive glass/collagen/chitosan nanocomposite scaffolds on the cell morphology and proliferation of rabbit's bone marrow mesenchymal stem cells-derived osteogenic cells. J Genet Eng Biotechnol 2022; 20(1): 33.
[http://dx.doi.org/10.1186/s43141-022-00302-x] [PMID: 35192077]

[109] Mishchenko O, Yanovska A, Kosinov O, *et al.* Synthetic calcium–phosphate materials for bone grafting. Polymers (Basel) 2023; 15(18): 3822.
[http://dx.doi.org/10.3390/polym15183822] [PMID: 37765676]

[110] Chen YY, Li HL, Chen CC, Jiang CP. Biofabrication of biopolymer and biocomposite scaffolds for bone tissue engineering. Key Eng Mater 2012; 523-524: 374-9.
[http://dx.doi.org/10.4028/www.scientific.net/KEM.523-524.374]

[111] Bendtsen ST, Wei M. Synthesis and characterization of a novel injectable alginate–collagen–hydroxyapatite hydrogel for bone tissue regeneration. J Mater Chem B Mater Biol Med 2015; 3(15): 3081-90.
[http://dx.doi.org/10.1039/C5TB00072F] [PMID: 32262508]

[112] Pina S, Oliveira JM, Reis RL. Natural-based nanocomposites for bone tissue engineering and regenerative medicine: a review. Adv Mater 2015; 27(7): 1143-69.
[http://dx.doi.org/10.1002/adma.201403354] [PMID: 25580589]

[113] Phogat K, Ghosh SB, Bandyopadhyay-Ghosh S. Recent advances on injectable nanocomposite hydrogels towards bone tissue rehabilitation. J Appl Polym Sci 2023; 140(4): e53362.
[http://dx.doi.org/10.1002/app.53362]

[114] Saska S, Teixeira LN, Tambasco de Oliveira P, *et al.* Bacterial cellulose-collagen nanocomposite for bone tissue engineering. J Mater Chem 2012; 22(41): 22102-12.
[http://dx.doi.org/10.1039/c2jm33762b]

[115] Tripathi A, Saravanan S, Pattnaik S, Moorthi A, Partridge NC, Selvamurugan N. Bio-composite scaffolds containing chitosan/nano-hydroxyapatite/nano-copper–zinc for bone tissue engineering. Int J Biol Macromol 2012; 50(1): 294-9.
[http://dx.doi.org/10.1016/j.ijbiomac.2011.11.013] [PMID: 22123094]

[116] Ismail R, Munanda R, Rusiyanto , *et al.* Design, manufacturing and characterization of biodegradable bone screw from PLA prepared by Fused Deposition Modelling (FDM) 3D printing technique. J Adv Res Fluid Mech Therm Sci 2023; 103(2): 205-15.
[http://dx.doi.org/10.37934/arfmts.103.2.205215]

[117] Tang X, Mao L, Liu J, *et al.* Fabrication, characterization and cellular biocompatibility of porous biphasic calcium phosphate bioceramic scaffolds with different pore sizes. Ceram Int 2016; 42(14): 15311-8.
[http://dx.doi.org/10.1016/j.ceramint.2016.06.172]

[118] Gomes CC, Moreira LM, Santos VJSV, *et al.* Assessment of the genetic risks of a metallic alloy used in medical implants. Genet Mol Biol 2011; 34(1): 116-21.
[http://dx.doi.org/10.1590/S1415-47572010005000118] [PMID: 21637553]

[119] He W, Chuang A, Cao Z, Liaw PK. Biocompatibility study of zirconium-based bulk metallic glasses for orthopedic applications. Metall Mater Trans, A Phys Metall Mater Sci 2010; 41(7): 1726-34.

[http://dx.doi.org/10.1007/s11661-009-0150-5]

[120] Gherasim O, Grumezescu AM, Grumezescu V, *et al.* Bioactive coatings loaded with osteogenic protein for metallic implants. Polymers (Basel) 2021; 13(24): 4303.
[http://dx.doi.org/10.3390/polym13244303] [PMID: 34960852]

[121] Choi S, Kwon J, Suk K, *et al.* The clinical use of osteobiologic and metallic biomaterials in orthopedic surgery: the present and the future. Materials (Basel) 2023; 16(10): 3633.
[http://dx.doi.org/10.3390/ma16103633] [PMID: 37241260]

[122] Salahshoor M, Guo Y. Biodegradable orthopedic magnesium-calcium (MgCa) alloys, processing, and corrosion performance. Materials (Basel) 2012; 5(1): 135-55.
[http://dx.doi.org/10.3390/ma5010135] [PMID: 28817036]

[123] Jia B, Yang H, Zhang Z, *et al.* Biodegradable Zn–Sr alloy for bone regeneration in rat femoral condyle defect model: *In vitro* and *in vivo* studies. Bioact Mater 2021; 6(6): 1588-604.
[http://dx.doi.org/10.1016/j.bioactmat.2020.11.007] [PMID: 33294736]

[124] Oriňaková R, Gorejová R, Čákyová V, *et al.* Biodegradable zinc-based materials with a polymer coating designed for biomedical applications. J Appl Polym Sci 2024; 141(2): e54773.
[http://dx.doi.org/10.1002/app.54773]

[125] Bărbînță AC, Luca D, Strugaru SI, *et al.* New titanium alloys potentially used for metal-ceramic applications in medicine. Key Eng Mater 2013; 587: 287-92.
[http://dx.doi.org/10.4028/www.scientific.net/KEM.587.287]

[126] Su Y, Cockerill I, Zheng Y, Tang L, Qin YX, Zhu D. Biofunctionalization of metallic implants by calcium phosphate coatings. Bioact Mater 2019; 4: 196-206.
[http://dx.doi.org/10.1016/j.bioactmat.2019.05.001] [PMID: 31193406]

[127] Kirmanidou Y, Sidira M, Drosou ME, *et al.* New Ti-alloys and surface modifications to improve the mechanical properties and the biological response to orthopedic and dental implants: A review. BioMed Res Int 2016; 2016(1): 1-21.
[http://dx.doi.org/10.1155/2016/2908570] [PMID: 26885506]

[128] Babaie E, Bhaduri SB. Fabrication aspects of porous biomaterials in orthopedic applications: A review. ACS Biomater Sci Eng 2018; 4(1): 1-39.
[http://dx.doi.org/10.1021/acsbiomaterials.7b00615] [PMID: 33418675]

[129] Lewallen EA, Riester SM, Bonin CA, *et al.* Biological strategies for improved osseointegration and osteoinduction of porous metal orthopedic implants. Tissue Eng Part B Rev 2015; 21(2): 218-30.
[http://dx.doi.org/10.1089/ten.teb.2014.0333] [PMID: 25348836]

[130] Nasrin R, Biswas S, Rashid TU, *et al.* Preparation of Chitin-PLA laminated composite for implantable application. Bioact Mater 2017; 2(4): 199-207.
[http://dx.doi.org/10.1016/j.bioactmat.2017.09.003] [PMID: 29744430]

[131] Baheiraei N, Nourani MR, Mortazavi SMJ, *et al.* Development of a bioactive porous collagen/β-tricalcium phosphate bone graft assisting rapid vascularization for bone tissue engineering applications. J Biomed Mater Res A 2018; 106(1): 73-85.
[http://dx.doi.org/10.1002/jbm.a.36207] [PMID: 28879686]

[132] Eryildiz M, Altan M. Fabrication of polylactic acid/halloysite nanotube scaffolds by foam injection molding for tissue engineering. Polym Compos 2020; 41(2): 757-67.
[http://dx.doi.org/10.1002/pc.25406]

[133] Akram M, Arshad N, Braem A. Nanoparticle-modified coatings by electrophoretic deposition for corrosion protection of magnesium. Surf Eng 2022; 38(4): 430-9.
[http://dx.doi.org/10.1080/02670844.2022.2095151]

[134] Uzulmez B, Demirsoy Z, Can O, Gulseren G. Bioinspired multi-layer biopolymer-based dental implant coating for enhanced osseointegration. Macromol Biosci 2023; 23(7): 2300057.
[http://dx.doi.org/10.1002/mabi.202300057] [PMID: 37097091]

[135] Zakharova VA, Kildeeva NR. Biopolymer matrices based on chitosan and fibroin: A review focused on methods for studying surface properties. Polysaccharides 2021; 2(1): 154-67.
[http://dx.doi.org/10.3390/polysaccharides2010011]

[136] Aslam Khan MU, Abd Razak SI, Al Arjan WS, *et al.* Recent advances in biopolymeric composite materials for tissue engineering and regenerative medicines: a review. Molecules 2021; 26(3): 619.
[http://dx.doi.org/10.3390/molecules26030619] [PMID: 33504080]

[137] Liu T, Zheng Y, Wu G, Wismeijer D, Pathak JL, Liu Y. BMP2-coprecipitated calcium phosphate granules enhance osteoinductivity of deproteinized bovine bone, and bone formation during critical-sized bone defect healing. Sci Rep 2017; 7(1): 41800.
[http://dx.doi.org/10.1038/srep41800] [PMID: 28139726]

[138] Hocaoğlu TP, Gençoğlan S, Arslan M, Benlïdayi ME, Kürkçü M. Histomorphometric assessment of the impact of bovine demineralized bone graft on bone healing *versus* autogenous, allogeneic and synthetic grafts in experimentally- induced critical size bone defects in rats. Cumhur Dent J 2018; 21(4): 387-95.
[http://dx.doi.org/10.7126/cumudj.475498]

[139] van de Graaf GMM, De Zoppa ALV, Moreira RC, Maestrelli SC, Marques RFC, Campos MGN. Morphological and mechanical characterization of chitosan-calcium phosphate composites for potential application as bone-graft substitutes. Res Biomed Eng 2015; 31(4): 334-42.
[http://dx.doi.org/10.1590/2446-4740.0786]

[140] Kumar KAJ, Rao JB, Pavan Kumar B, Mohan AP, Patil K, Parimala K. A prospective study involving the use of platelet rich plasma in enhancing the uptake of bone grafts in the oral and maxillofacial region. J Maxillofac Oral Surg 2013; 12(4): 387-94.
[http://dx.doi.org/10.1007/s12663-012-0466-3] [PMID: 24431876]

[141] Findik Y, Kökdere NN, Baykul T. The use of platelet-rich fibrin (PRF) and PRF-mixed particulated autogenous bone graft in the treatment of bone defects: An experimental and histomorphometrical study. Dent Res J (Isfahan) 2015; 12(5): 418-24.
[http://dx.doi.org/10.4103/1735-3327.166188] [PMID: 26604954]

[142] de Oliveira Gonçalves JB, Buchaim DV, de Souza Bueno CR, *et al.* Effects of low-level laser therapy on autogenous bone graft stabilized with a new heterologous fibrin sealant. J Photochem Photobiol B 2016; 162: 663-8.
[http://dx.doi.org/10.1016/j.jphotobiol.2016.07.023] [PMID: 27497370]

[143] Bosco AF, Faleiros PL, Carmona LR, *et al.* Effects of low-level laser therapy on bone healing of critical-size defects treated with bovine bone graft. J Photochem Photobiol B 2016; 163: 303-10.
[http://dx.doi.org/10.1016/j.jphotobiol.2016.08.040] [PMID: 27611453]

[144] Todd EA, Mirsky NA, Silva BL, *et al.* Functional scaffolds for bone tissue regeneration: a comprehensive review of materials, methods, and future directions. J Funct Biomater 2024; 15(10): 280.
[http://dx.doi.org/10.3390/jfb15100280]

[145] Łuczak JW, Palusińska M, Matak D, *et al.* The future of bone repair: emerging technologies and biomaterials in bone regeneration. Int J Mol Sci 2024; 25(23): 12766.
[http://dx.doi.org/10.3390/ijms252312766] [PMID: 39684476]

[146] Degli Esposti M, Changizi M, Salvatori R, *et al.* Comparative study on bioactive filler/biopolymer scaffolds for potential application in supporting bone tissue regeneration. ACS Appl Polym Mater 2022; 4(6): 4306-18.
[http://dx.doi.org/10.1021/acsapm.2c00270]

[147] Percival KM, Paul V, Husseini GA. Recent advancements in bone tissue engineering: integrating smart scaffold technologies and bio-responsive systems for enhanced regeneration. Int J Mol Sci 2024; 25(11): 6012.
[http://dx.doi.org/10.3390/ijms25116012]

[148] Farjaminejad S, Farjaminejad R, Hasani M, *et al.* Advances and challenges in polymer-based scaffolds for bone tissue engineering: a path towards personalized regenerative medicine. Polymers (Basel) 2024; 16(23): 3303.
[http://dx.doi.org/10.3390/polym16233303] [PMID: 39684048]

[149] Tupe A, Patole V, Ingavle G, *et al.* Recent advances in biomaterial-based scaffolds for guided bone tissue engineering: challenges and future directions. Polym Adv Technol 2024; 35(11): e6619.
[http://dx.doi.org/10.1002/pat.6619]

[150] Szwed-Georgiou A, Płociński P, Kupikowska-Stobba B, *et al.* Bioactive materials for bone regeneration: biomolecules and delivery systems. ACS Biomater Sci Eng 2023; 9(9): 5222-54.
[http://dx.doi.org/10.1021/acsbiomaterials.3c00609] [PMID: 37585562]

[151] Kawai T, Suzuki O, Matsui K, Tanuma Y, Takahashi T, Kamakura S. Octacalcium phosphate collagen composite facilitates bone regeneration of large mandibular bone defect in humans. J Tissue Eng Regen Med 2017; 11(5): 1641-7.
[http://dx.doi.org/10.1002/term.2110] [PMID: 26612731]

[152] Liu L, Miao Y, Shi X, Gao H, Wang Y. Phosphorylated chitosan hydrogels inducing osteogenic differentiation of osteoblasts *via* JNK and p38 signaling pathways. ACS Biomater Sci Eng 2020; 6(3): 1500-9.
[http://dx.doi.org/10.1021/acsbiomaterials.9b01374] [PMID: 33455392]

[153] Harrison R, Criss ZK, Feller L, *et al.* Mechanical properties of α-tricalcium phosphate-based bone cements incorporating regenerative biomaterials for filling bone defects exposed to low mechanical loads. J Biomed Mater Res B Appl Biomater 2016; 104(1): 149-57.
[http://dx.doi.org/10.1002/jbm.b.33362] [PMID: 25677680]

[154] Machado GC, Maher CG, Ferreira PH, *et al.* Efficacy and safety of paracetamol for spinal pain and osteoarthritis: systematic review and meta-analysis of randomised placebo controlled trials. BMJ 2015; 350
[http://dx.doi.org/10.1136/bmj.h1225]

[155] Zhou L, Fan L, Zhang FM, *et al.* Hybrid gelatin/oxidized chondroitin sulfate hydrogels incorporating bioactive glass nanoparticles with enhanced mechanical properties, mineralization, and osteogenic differentiation. Bioact Mater 2021; 6(3): 890-904.
[http://dx.doi.org/10.1016/j.bioactmat.2020.09.012] [PMID: 33073063]

[156] Barros J, Ferraz MP, Azeredo J, Fernandes MH, Gomes PS, Monteiro FJ. Alginate-nanohydroxyapatite hydrogel system: Optimizing the formulation for enhanced bone regeneration. Mater Sci Eng C 2019; 105: 109985.
[http://dx.doi.org/10.1016/j.msec.2019.109985] [PMID: 31546404]

[157] Biguetti CC, Cavalla F, Tim CR, *et al.* Bioactive glass-ceramic bone repair associated or not with autogenous bone: a study of organic bone matrix organization in a rabbit critical-sized calvarial model. Clin Oral Investig 2019; 23(1): 413-21.
[http://dx.doi.org/10.1007/s00784-018-2450-x] [PMID: 29700614]

[158] Eichholz KF, Pitacco P, Burdis R, *et al.* Integrating melt electrowriting and fused deposition modeling to fabricate hybrid scaffolds supportive of accelerated bone regeneration. Adv Healthc Mater 2024; 13(3): 2302057.
[http://dx.doi.org/10.1002/adhm.202302057] [PMID: 37933556]

[159] Wehrle E, Tourolle né Betts DC, Kuhn GA, *et al.* Spatio-temporal characterization of fracture healing patterns and assessment of biomaterials by time-lapsed *in vivo* micro-computed tomography. Sci Rep 2021; 11(1): 8660.
[http://dx.doi.org/10.1038/s41598-021-87788-6] [PMID: 33883593]

[160] Huang K, Wu J, Gu Z. Black phosphorus hydrogel scaffolds enhance bone regeneration *via* a sustained supply of calcium-free phosphorus. ACS Appl Mater Interfaces 2019; 11(3): 2908-16.
[http://dx.doi.org/10.1021/acsami.8b21179] [PMID: 30596421]

[161] Cidonio G, Glinka M, Kim YH, *et al.* Nanoclay-based 3D printed scaffolds promote vascular ingrowth *ex vivo* and generate bone mineral tissue *in vitro* and *in vivo*. Biofabrication 2020; 12(3): 035010.
[http://dx.doi.org/10.1088/1758-5090/ab8753] [PMID: 32259804]

[162] Tavares MT, Gaspar VM, Monteiro MV, S Farinha JP, Baleizão C, Mano JF. GelMA/bioactive silica nanocomposite bioinks for stem cell osteogenic differentiation. Biofabrication 2021; 13(3): 035012.
[http://dx.doi.org/10.1088/1758-5090/abdc86] [PMID: 33455952]

[163] Fu B, Shen J, Chen Y, *et al.* Narrative review of gene modification: applications in three-dimensional (3D) bioprinting. Annals of translational medicine. 2021; 9(19): 1502.

[164] Mei Q, Rao J, Bei HP, Liu Y, Zhao X. 3D bioprinting photo-crosslinkable hydrogels for bone and cartilage repair. Int J Bioprint 2021; 7(3): 367.
[http://dx.doi.org/10.18063/ijb.v7i3.367]

[165] Xing F, Xiang Z, Rommens PM, Ritz U. 3D bioprinting for vascularized tissue-engineered bone fabrication. Materials (Basel) 2020; 13(10): 2278.
[http://dx.doi.org/10.3390/ma13102278] [PMID: 32429135]

[166] Anada T, Pan CC, Stahl AM, *et al.* Vascularized bone-mimetic hydrogel constructs by 3D bioprinting to promote osteogenesis and angiogenesis. Int J Mol Sci 2019; 20(5): 1096.

[167] Arif ZU, Khalid MY, Noroozi R, Hossain M, Shi HH, Tariq A, Ramakrishna S, Umer R. Additive manufacturing of sustainable biomaterials for biomedical applications. Asian Journal of Pharmaceutical Sciences. 2023; 18(3): 100812.
[http://dx.doi.org/10.1016/j.ajps.2023.100812]

[168] Genova T, Roato I, Carossa M, Motta C, Cavagnetto D, Mussano F. Advances on bone substitutes through 3D bioprinting. Int J Mol Sci 2020; 21(19): 7012.
[http://dx.doi.org/10.3390/ijms21197012] [PMID: 32977633]

[169] Bakhtiary N, Liu C, Ghorbani F. Bioactive inks development for osteochondral tissue engineering: A mini-review. Gels 2021; 7(4): 274.
[http://dx.doi.org/10.3390/gels7040274] [PMID: 34940334]

[170] Bo T, Pascucci E, Capuani S, *et al.* 3D bioprinted mesenchymal stem cell laden scaffold enhances subcutaneous vascularization for delivery of cell therapy. Biomed Microdevices 2024; 26(3): 29.
[http://dx.doi.org/10.1007/s10544-024-00713]

[171] Daly AC, Pitacco P, Nulty J, Cunniffe GM, Kelly DJ. 3D printed microchannel networks to direct vascularisation during endochondral bone repair. Biomaterials 2018; 162: 34-46.
[http://dx.doi.org/10.1016/j.biomaterials.2018.01.057] [PMID: 29432987]

[172] Lapomarda A, Cerqueni G, Geven MA, *et al.* Physicochemical characterization of pectin-gelatin biomaterial formulations for 3D bioprinting. Macromol Biosci 2021; 21(9): 2100168.
[http://dx.doi.org/10.1002/mabi.202100168] [PMID: 34173326]

[173] Stanco D, Urbán P, Tirendi S, Ciardelli G, Barrero J. 3D bioprinting for orthopaedic applications: Current advances, challenges and regulatory considerations. Bioprinting 2020; 20: e00103.
[http://dx.doi.org/10.1016/j.bprint.2020.e00103] [PMID: 34853818]

[174] McMillan A, McMillan N, Gupta N, Kanotra SP, Salem AK. 3D bioprinting in otolaryngology: a review. Adv Healthc Mater 2023; 12(19): 2203268.
[http://dx.doi.org/10.1002/adhm.202203268] [PMID: 36921327]

[175] Budharaju H, Sundaramurthi D, Sethuraman S. Biofabrication & cryopreservation of tissue engineered constructs for on-demand applications. Biofabrication 2024; 16(4): 042008.
[http://dx.doi.org/10.1088/1758-5090/ad7906] [PMID: 39258414]

[176] Cidonio G, Alcala-Orozco CR, Lim KS, *et al.* Osteogenic and angiogenic tissue formation in high fidelity nanocomposite Laponite-gelatin bioinks. Biofabrication 2019; 11(3): 035027.
[http://dx.doi.org/10.1088/1758-5090/ab19fd] [PMID: 30991370]

[177] Windolf M, Varjas V, Gehweiler D, *et al.* Continuous implant load monitoring to assess bone healing status—evidence from animal testing. Medicina (Kaunas) 2022; 58(7): 858.
[http://dx.doi.org/10.3390/medicina58070858] [PMID: 35888576]

[178] Wahed SB, Dunstan CR, Boughton PA, Ruys AJ. Biofunctionalization of UHMWPE polymer and its application in Anterior Cruciate Ligament reconstruction. Preprints 2020, 2020080071.
[http://dx.doi.org/10.20944/preprints202008.0071.v1]

[179] Morimoto T, Hirata H, Eto S, *et al.* Development of silver-containing hydroxyapatite-coated antimicrobial implants for orthopaedic and spinal surgery. Medicina (Kaunas) 2022; 58(4): 519.
[http://dx.doi.org/10.3390/medicina58040519] [PMID: 35454358]

[180] Beck S, Sehl C, Voortmann S, *et al.* Sphingosine is able to prevent and eliminate *Staphylococcus epidermidis* biofilm formation on different orthopedic implant materials *in vitro*. J Mol Med (Berl) 2020; 98(2): 209-19.
[http://dx.doi.org/10.1007/s00109-019-01858-x] [PMID: 31863153]

[181] Sui J, Liu S, Chen M, Zhang H. Surface bio-functionalization of anti-bacterial titanium implants: a review. Coatings 2022; 12(8): 1125.
[http://dx.doi.org/10.3390/coatings12081125]

CHAPTER 11

Mechanisms of Immunotherapeutic Biopolymers in Autoimmune Disease

Deepak Kumar[1], **Piyush Anand**[1] and **Shashi Kant Singh**[1,*]

[1] *Faculty of Pharmaceutical Sciences, Mahayogi Gorakhnath University Gorakhpur, Uttar Pradesh 273007, India*

Abstract: Autoimmune disorders result from dysregulated immune responses directed against the body's tissues; immunotherapeutic biopolymers are becoming increasingly important as transformational agents in this regard. These biopolymers, which comprise organic materials like peptides and polysaccharides, function in different ways to help the immune system's restoration to equilibrium. A noteworthy method pertains to the control of immune cell function, wherein biopolymers augment the functionalities of regulatory T cells while inhibiting the generation of pro-inflammatory cytokines. Furthermore, certain biopolymers, such as hyaluronic acid and chitosan, have anti-inflammatory qualities that are vital in lowering tissue damage and inflammation linked to autoimmune diseases. By influencing the gut microbiota, which has been linked to the etiology of several autoimmune disorders, these biopolymers can also balance the immune system. The adaptability of immunotherapeutic biopolymers is further demonstrated by their capacity to target particular pathways, such as the NF-κB signaling cascade and cytokine production. Recent developments in nanotechnology have made it possible to create delivery systems based on biopolymers that improve the bioavailability and effectiveness of medicinal drugs. Not only does this novel method increase therapeutic specificity, but it also reduces systemic adverse effects that are frequently linked to traditional medications. Personalized and efficient therapy techniques might transform the management of autoimmune diseases as research advances and immunotherapeutic biopolymers are incorporated into clinical practice.

Keywords: Antigen presentation, Autoimmune diseases, Gut microbiota, Immunological regulation, Immunotherapeutic biopolymers, Inflammation, Regulatory T cells.

INTRODUCTION

Autoimmune diseases can be triggered by genetics, bacterial or viral infections, or other factors, including the body's immune cells activating abnormally and

* **Corresponding author Shashi Kant Singh:** Faculty of Pharmaceutical Sciences, Mahayogi Gorakhnath University Gorakhpur, Uttar Pradesh 273007, India; E-mail: Shashikantsingh59@gmail.com

destroying host tissues or organs. An estimated 24 million Americans suffer from more than 100 autoimmune disorders, with 80% of those affected being women, according to data made available online by the American Autoimmune Disorders Association. In addition, five to ten percent of Americans suffer from one or more autoimmune disorders. Patients frequently have abnormal T lymphocyte activation and autoantibodies that impact specific organs. Type I diabetes (pancreas) includes pernicious anemia (stomach), Hashimoto's thyroiditis (thyroid gland), and Addison's disease (adrenal glands). Additional organs and tissues may be impacted by rheumatoid arthritis, dermatomyositis, Systemic Lupus Erythematosus (SLE), and other conditions of a similar kind. Autoimmune diseases frequently recur with long delays. During clinical diagnosis, the majority of patients frequently exhibit tissue damage and persistent problems. Currently, anti-inflammatory, nonspecific, broad-spectrum, or immunosuppressive medications (such as tacrolimus, corticosteroids, or cyclosporine) are used to treat autoimmune diseases. These therapies mostly lessen the body's inflammatory cell proliferation and immune response, which can lessen clinical symptoms but not the disease's underlying cause or its consequences. Furthermore, using cytotoxic and immunosuppressive medications often over an extended period will weaken the body's defense against infection and cancer [1]. The objective has been to restore immunological equilibrium, and in the last several years, research has concentrated on creating treatments that may precisely reduce immunity without compromising healthy immune function. Because of their vital roles in preserving and enhancing immunological tolerance, immunoregulatory tolerogenic dendritic cells (Tol DCs), as opposed to earlier immunosuppressive therapies, have garnered a lot of attention in the treatment of autoimmune diseases. Currently, the *in vitro* harvesting of autologous tolerogenic DC is costly and has a risk of transfusion-associated failure. Another outcome of the *in vitro* culture of tolerogenic DCs is non-specific immunosuppression. However, disease-specific autoantigens must be loaded into these DCs for them to produce autoimmune cells that are specific to the targets [2].

Key Features and Types

Immunotherapeutic biopolymers are a cutting-edge approach to treating autoimmune diseases because of their unique capacity to successfully alter immune responses. These biopolymers are designed to interact with the immune system in a way that reduces aberrant immune activity, which is characteristic of autoimmune diseases and increases tolerance [3]. Biocompatibility is one of the most important characteristics of immunotherapeutic biopolymers. This feature guarantees that the substances utilized do not cause negative responses when ingested. Biopolymer synthesis and characterization are also important factors for establishing their biocompatibility, efficacy, and stability in drugs. Different techniques of synthesis, such as chemical polymerization, enzymatic polymerization, and microbial fermentation, are utilized based on the biopolymer type. For instance, chitosan is generally derived through the deacetylation of chitin, whereas hyaluronic acid is biosynthesized through bacterial fermentation by utilizing Streptococcus species. Characterization of biopolymers is needed to determine their purity, molecular weight, and structural integrity, which further determines their functionality in immunotherapeutic applications. Gel Permeation Chromatography (GPC) is typically used for the determination of molecular weight distribution, while Fourier-Transform Infrared Spectroscopy (FTIR) and Nuclear Magnetic Resonance (NMR) spectroscopy confirm structural composition. These biopolymers' surface topography and nano-scale interactions are also elucidated by Atomic Force Microscopy (AFM) and Scanning Electron Microscopy (SEM). High-Performance Liquid Chromatography (HPLC) identifies purity, where contaminants are kept as low as possible, and batches can be reliably identical. Biodegradability as well as potential for drug-loading are important physical properties that guide biopolymer performance. Thermogravimetric Analysis (TGA) and Differential Scanning Calorimetry (DSC) are most frequently applied in assessing thermal stability, which dictates storage and formulation stability. All these sophisticated characterization methods together enhance our knowledge of biopolymer-based drug delivery systems and how they can potentially be used for the treatment of autoimmune diseases. Maintaining patient safety and improving the treatment's overall efficacy depends heavily on biocompatibility [4]. The regulated release aspect is also crucial. It is possible to construct immunotherapeutic biopolymers such that their therapeutic chemicals are released in a controlled way. This regulated release is crucial for maintaining therapeutic levels of the medicine over an extended period, which can contribute to improved patient outcomes by guaranteeing constant exposure to the treatment [5].

These biopolymers also possess the important quality of versatility. They may be engineered to carry nucleic acids, proteins, and peptides, among other medicinal substances. This flexibility enables customized treatment plans that may be modified under the requirements of each unique patient and the particular autoimmune disorder.

Types of Biopolymers Used in Immunotherapy

At present, several immunotherapeutic biopolymers are being investigated for their potential to treat autoimmune diseases. One of the most promising types of biopolymers is *nanoparticles*. To improve treatment precision, they can target certain immune cells and encapsulate therapeutic molecules. Nanoparticles can greatly increase therapeutic effectiveness by delivering medications directly to the areas of immune dysregulation [6].

Hydrogels are another type of biopolymer that has attracted attention in the immunotherapy field. They can act as a scaffold to promote cell migration or be loaded with drugs to modify immunological responses. Hydrogels can also mimic the extracellular matrix to promote cell adhesion and proliferation, which are two processes necessary for tissue regeneration and repair.

The purpose of immunostimulatory polymers is to improve the immune response. These polymers work as agonists to stimulate the immune system to identify and react to certain targets by binding to Pattern Recognition Receptors (PRRs). Fighting autoimmune diseases can be greatly aided by this method of immune system activation [7].

Mechanisms of Action

Immunotherapeutic biopolymers work through a variety of different methods. Among which, tolerance induction is one main mechanism. These biopolymers can teach the immune system to perceive some autoantigens as harmless by providing them. Retraining is essential for lowering autoantibody production and stopping more tissue damage brought on by autoimmune disorders [8].

Furthermore, by affecting the activity of different immune cells, immunotherapeutic biopolymers can alter immunological responses. They may, for example, encourage the development of regulatory T cells, which are essential for inhibiting unwarranted immunological reactions. Restoring immune system equilibrium and lessening the symptoms of autoimmune diseases depend on this regulation.

To conclude, immunotherapeutic biopolymers provide a viable approach to the management of autoimmune disorders. Their special qualities, like biocompatibility, controlled release, adaptability, and the capacity to alter immune responses and create tolerance, place them in a strong position to support the ongoing management and treatment of chronic diseases [9].

IMMUNOTHERAPY IN AUTOIMMUNE DISEASES

Overview of Immunotherapy

Immunotherapy, which uses the body's immune system to combat diseases, especially autoimmune disorders and cancer, has completely changed the medical world. Immunotherapy works based on improving or modifying the immune response to help the body identify and get rid of pathogenic cells or to help autoimmune disorders return to normal. An overview of immunotherapy, including its varieties, uses, problems, and prospects, is presented in this chapter. The infographic representation of autoimmune diseases and excessive inflammation is depicted in Fig. (1).

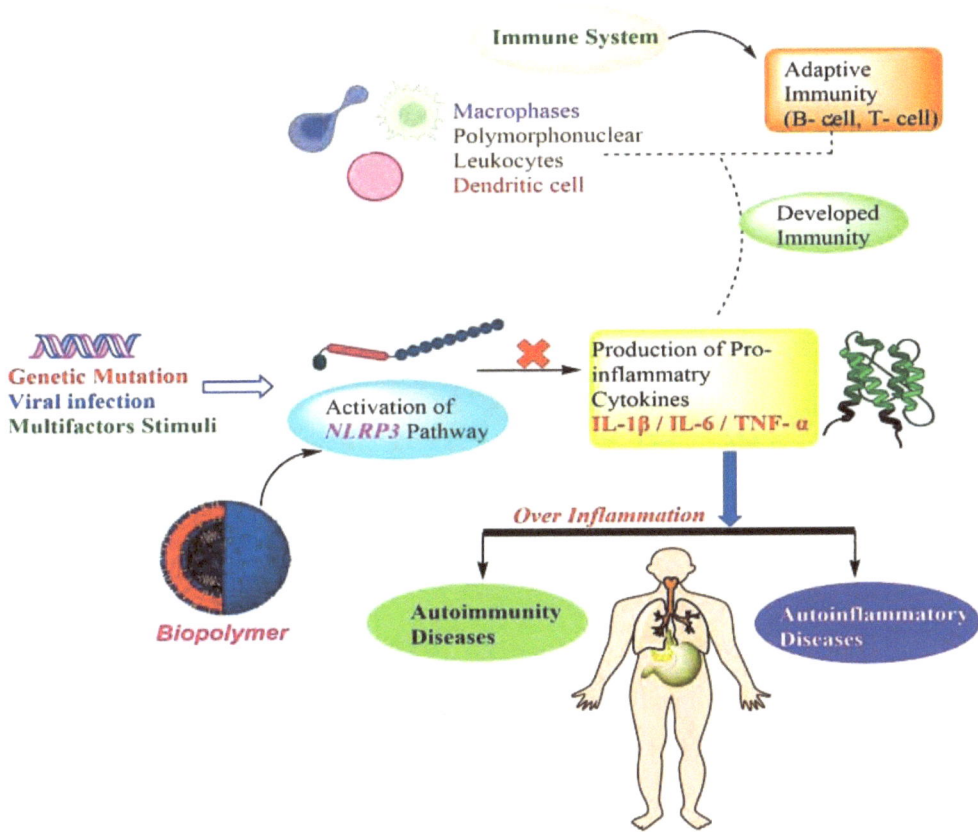

Fig. (1). The infographic representation of autoimmune illnesses and excessive inflammation.

Types of Immunotherapies

Immunotherapy includes a range of methods, such as: (See Table **1** for detailed evidence on natural and artificial biopolymers, their therapeutic roles, and future applications).

Monoclonal Antibodies

These artificially created antibodies are designed to specifically target antigens on immunological or cancer cells. They can enlist immune effector cells, prevent immunological checkpoints, or cause direct cell death [11].

Cytokine Therapy

It involves the use of signaling proteins called cytokines to control immune responses. Interleukins and interferons are examples of therapeutic cytokines that can stimulate immune cell activity, support anti-tumor immunity, and control autoimmune reactions [12].

Vaccinations

Therapeutic vaccinations aim to provoke an immune response against certain antigens linked to autoimmune disorders or malignancies. The immune system may be trained by these vaccinations to identify and target certain cells [13].

Cell-based Therapies

In this method, immune cells, like T cells or dendritic cells, are increased or altered outside of the body and then reinfused into the patient. One notable example is CAR T-cell therapy, which has demonstrated potential in the management of certain hematological cancers [14].

Gene Therapy

Genetic changes can boost the immune response. This includes techniques that alter the expression of immune system-related genes or insert genes encoding therapeutic proteins [15].

IMMUNOTHERAPY'S WORKING MECHANISMS

Activation of Immune Cells

Immunotherapy can promote the growth and activation of important immune cells, including Natural Killer (NK) and T cells. These activated cells can modify autoreactive immune responses in autoimmune diseases or specifically target and kill cancer cells.

Targeting Immune Checkpoints

Regulating mechanisms known as immunological checkpoints keep the immune system in control and stop overreactions. To prevent cancer cells from spreading and to influence the immune system in cases of autoimmune diseases, immunotherapy (Table **1**), known as checkpoint inhibitors, blocks these pathways [16].

Enhancing Antigen Presentation

The way that autoantigens or tumors appear to the immune system can be enhanced by immunotherapy. This improvement makes it easier to identify damaged cells, which strengthens the immune system's defenses against autoimmune disorders, self-tissues, or tumors [17].

Inducing Immune Memory

To help the immune system identify and combat cancer cells or abnormal self-tissues upon preexposure, certain immunotherapies try to develop long-lasting immunological memory. For patients to remain in remission and avoid their sickness from returning, this is essential.

Table 1. Natural and synthetic biopolymers used in autoimmune disease, their treatment and future directions.

Biopolymers	Autoimmune Diseases	Treatment	Future Direction	Reference
Starch	Autoimmune hypoglycemia (Hirata's disease)	Hydrothermally improved slow-release maize starch has been used as a therapy for autoimmune hypoglycemia when dietary restrictions are not enough to control blood glucose levels. It was discovered that insulin autoantibodies were a sign of autoimmune hypoglycemia. Dietary restriction alone was not enough to control blood sugar levels in this patient with autoimmune hypoglycemia. However, therapy with hydrothermally improved slow-release maize starch resulted in sustained euglycemia.	1. Starch blending with compostable biopolymers. 2. Nanocomposite films 3. Biomedical implants 4. Cosmetic formulation	[18]

Biopolymers	Autoimmune Diseases	Treatment	Future Direction	Reference
Chitosan	Rheumatoid arthritis, Sclerosis, Type 1 Diabetes	Leflunomide for rheumatoid arthritis can be taken orally *via* chitosan-coated clove oil-based nanoemulsions.	1. Heavy metal removal 2. Wastewater purification 3. Chitosan nanoparticles for medical and industrial applications 4. Battery electrodes and supercapacitors	[19]
Polycaprolactone	Rheumatoid arthritis	Poly-ε-caprolactone nanoparticles for long-term intra-articular immune regulation in ajuvant-induced arthritis.	1. PCL Scaffolds formation 2. Biomedical engineering 3. Water treatment 4. In the sustainable packaging of material.	[20]
Polyamides (Pyrrole-imidazole)	Autoimmune Diseases	Pyrrole-imidazole polyamide is used to target IL-23 for the treatment of autoimmune diseases. Dendritic cells and macrophages have a high efficiency of entry for polyamide into their nuclei. Furthermore, it particularly suppressed the expression of IL-23 and stopped c-Rel from binding to the IL-23p19 promoter *in vivo*.	1. Biomaterials 2. Cell signaling 3. Biohybrid systems	[21]
Silica	Rheumatoid arthritis, Silicosis, Sclerosis, Nephritis, Vasculitis	Exposure to silica has been connected with silicosis (pulmonary fibrosis), rheumatoid arthritis (Caplan's disease), systemic lupus erythematosus, and ANCA-induced vasculitis/nephritis.	1. Silica-based conjugates 2. Silica-based biosensors 3. Silica-based bioprinting.	[22]

(Table 1) cont.....

Biopolymers	Autoimmune Diseases	Treatment	Future Direction	Reference
Polylactic-coglycolic acid (PLGA)	Autoimmune diseases (Inflammatory diseases)	Several colloidal systems are used to treat inflammatory diseases besides PLGA-NPS. Exosomes, gold NPs, and lipid NPs are often used as substitutes for polymeric systems. Utilizing unique characteristics, every nanoscale system optimizes its anti-inflammatory efficacy. Exosomes are of particular importance among them because, depending on the cells from which they originate, they exhibit an innate immunomodulatory action.	1. AI-powered diagnostics 2. Biomedical engineering	[23]
Hyaluronic acid	Rheumatoid arthritis, Osteoarthritis	In rheumatic diseases, HA plays a critical pathophysiological role, particularly concerning joint function and health. This decrease in HA levels is explained by an imbalance between the synthesis and breakdown of HA, which is caused by an enhancement in enzymatic activity called hyaluronidases. OA-related cartilage deterioration, synovitis, and discomfort are all influenced by dysregulated HA metabolism, which includes increased synthesis and breakdown. The disease development is further exacerbated by the changed biomechanical properties of HA, which impact inflammation, chondrocyte activity, and joint lubrication.	1. HA-based biomaterials 2. Cell signaling	[24]

BIOPOLYMERS AS IMMUNOTHERAPEUTIC AGENTS

Immunotherapy has made remarkable strides in the last several years, especially in the customized treatment of cancer, autoimmune diseases, and infectious diseases. Biopolymers, which are naturally occurring macromolecules with the potential to modulate immune responses and qualities like biocompatibility and biodegradability, have emerged as crucial players in the development of novel immunotherapeutic techniques. These materials, derived from biological sources like proteins, polysaccharides, and nucleic acids, have shown great promise in increasing the efficacy of immunotherapy by functioning as carriers of substances that modulate the immune system or as active participants in immunological regulation [25].

Different Biopolymers for Immunotherapy Uses

Biopolymers are divided into several categories (Table 2) based on their chemical composition and place of origin. Each category has special qualities that can be used for immunotherapeutic applications.

Polysaccharides

The immunomodulatory qualities of polysaccharides, such as hyaluronic acid, alginate, and chitosan, have been the subject of several investigations. Because it stimulates both innate and adaptive immune responses, chitosan, which is derived from chitin, is a prospective option for vaccine delivery methods. An anionic polymer found naturally in the environment, alginate has been shown to enhance immunological detection and stability when used to encapsulate antigens [26].

Proteins

The potential application of proteins, including collagen, silk fibroin, and fibrin, in immunotherapy has been investigated. It has been demonstrated that collagen, a structural protein present in the extracellular matrix, promotes cell adhesion and proliferation while also acting as a scaffold for immune cells. Due to its biocompatibility, silk fibroin has been investigated for its ability to transfer immune-regulating chemicals, such as cytokines [27].

Nucleic Acids

Biopolymers based on DNA and RNA are becoming essential components of gene-based immunotherapy. To promote a focused immune response, antigens or immuno-modulatory proteins are encoded using plasmid DNA and messenger RNA (mRNA). Recent progress with mRNA vaccines against COVID-19 has shown that nucleic acid-based biopolymers are effective in eliciting robust and targeted immune responses [28].

IMMUNE MODULATION MECHANISMS ASSOCIATED WITH BIOPOLYMERS

There are several methods in which biopolymers might engage with the immune system, contingent upon their structure and mode of dispersion. Because biopolymers may either increase or diminish the immune response, they are useful tools in the treatment of several diseases [29].

Enhancement of Antigen Presentation

The capacity of biopolymers to improve antigen presentation is one of their most important functions in immunotherapy. Antigens can be given directly to Antigen-presenting Cells (APCs), such as dendritic cells and macrophages, and kept safe from destruction by encasing them in a biopolymer matrix. By increasing the effectiveness of antigen presentation, this tailored administration raises T cell activation [30, 31].

Table 2. Properties of biopolymers, their impact on the immune system, their use in biomedicine, and Nano conversion.

Biopolymers	Properties	Effect on the Immune System	Biomedical Applications	Nano Conversion
Cellulose	Strength, low cost, low density, outstanding biocompatibility, and excellent durability are among its many qualities	1. By stimulating macrophages, cellulose nanocrystals can cause a pro-inflammatory response. 2. Dendritic cells may develop tolerogenic qualities as a result of cellulose nanofibers.	Tissue engineering, regenerative medicine, wound healing, treatment of autoimmune diseases	Cellulose Nanofibers (CNFs), Cellulose Nanocrystals (CNCs)
Alginates	Biocompatible, biodegradable, low-cost, flexible and cell gel transfer	1. Alginate can stimulate macrophage-like cells, which in turn can cause proinflammatory cytokines to be released. 2. Dendritic cell activation is more successful with particulate alginate than with high-viscosity alginate.	Wound dressings, tissue engineering, dental applications, 3D bioprinting, vaccine adjuvants, skin substitutes	Nanofibers, nanogels, nanocapsules, nanocomposites
Chitosan	Biocompatible, biodegradable, non-toxic	1. Chitosan can activate both humoral and cell-mediated immune responses. 2. Immunomodulatory activity.	Tissue engineering, wound healing, drug delivery, cancer treatment, in autoimmune disease	Nanofibers, nanogels, nanotubes, nanocomposites

Biopolymers	Properties	Effect on the Immune System	Biomedical Applications	Nano Conversion
Glycosaminoglycans	Biodegradable and biocompatible	1. Chemokines and cytokines interact with GAGs to help deliver inflammatory chemicals to receptors. This results in airway infiltration and immune cell migration. 2. Leukocyte GAGs aid in the concentration of chemokines on the cell surface, improving receptor activation.	Tissue engineering, bone and cartilage regeneration, and wound healing, act as biomarkers	Nanoparticles, nanopores, nanocomposites
Lignin	Biodegradable, biocompatible, binder, accelerator	1. Lignin can increase the release of cytokines and nitric oxide, and differentiate and activate CD8 T cells and CD14 monocytes. 2. The gut immune system can be modulated by lignin-carbohydrate complexes.	Wound healing, bone tissue engineering, biosensors and bioimaging	Nanocomposite, nanocrystals, nanotubes, nanoporous materials
Chondroitin sulfate	Biocompatible, biodegradable, highly viscous	1. Psoriasis and other autoimmune disorders may benefit from CS. 2. CS can stimulate immunity by acting as an immunostimulatory. *In vitro* tests, CS boosted RAW264.7 cell production of ROS, pinocytosis, and phagocytosis. Moreover, CS can raise macrophage secretion of TNF-α, NO, IL-10, and IL-6.	Wound healing, autoimmune diseases treatment, repairment of bones	Prodrug nanoparticles

(Table 2) cont.....

Biopolymers	Properties	Effect on the Immune System	Biomedical Applications	Nano Conversion
Pectin	Biocompatible, safe, mucoadhesive, inert, gel-forming ability	1. Pectin interacts directly with the intestinal barrier and with immunological receptors, including Toll-like Receptors (TLR), to decrease inflammation. 2. Pectin can either stimulate or suppress macrophage and dendritic cell responses.	Drug delivery, wound healing, tissue engineering	Nanoparticles, nanocomposites, and encapsulated particles
Casein	Biocompatible, biodegradable, non-toxic, bioaccessible	1. Casein stimulates the production of antibodies and lymphocytes, which strengthens the immune system. 2. By decreasing the production of IL-4 and increasing the release of IgG, IgA, and IL-17A, β-casein can strengthen the immune system.	Tissue engineering, tooth remineralization, biosensing and bioimaging	Nanoparticles, nanocomposites
Xanthan gum	Pseudoplastic, soluble, stable, polyelectrolyte, thixotropic	1. Xanthan gum increases the immunogenicity 2. Antineoplastic and adjuvant effects	Tissue engineering and wound management	Nanoparticle, nanocomposite, nano gels
Fucoidan	Biodegradable, biocompatible	1. Natural killer cell activity, phagocytosis, and immune cell proliferation can all be induced by fucoidan. It can also boost the synthesis of serum antibodies, interleukin-2, and interferon-γ. 2. By controlling immunological responses, fucoidan can enhance allergic reactions.	Tissue regeneration, bone regeneration, immunomodulation, scaffold formation, wound healing	Nanoparticles, nanotubes, nano gels, nanocarriers

Activation of Specific Immune Cells

Biopolymers can activate particular immune cells. Polysaccharide-based nanoparticles, for instance, have been utilized to transport adjuvants or cytokines that activate natural killer cells, macrophages, or dendritic cells. When used in conjunction with cancer immunotherapy, this activation may result in a better immune response, as the aim is to modulate the body's tendency to identify and eradicate tumor cells [32].

Cytokine Delivery

The regulated delivery of cytokines through biopolymers has demonstrated significant potential in the field of immunotherapy. Signaling molecules called cytokines are essential for coordinating immune responses. On the other hand, systemic cytokine injection frequently causes serious harm. Cytokines have been delivered locally to appropriate sites through the use of biopolymers, especially hydrogels, which minimize adverse effects while preserving therapeutic efficacy [33].

Induction of Immune Tolerance

To stop the immune system from targeting self-tissues in autoimmune disorders and organ transplants, immunological tolerance must be induced. Tolerogenic substances that enhance immunological tolerance by modifying the function of regulatory T cells (Tregs) have been delivered *via* biopolymers. For example, antigens may now be delivered using polysaccharide-based nanoparticles in a way that promotes Treg induction and keeps transplant recipients from developing autoimmune or rejection [34].

BIOPOLYMER APPLICATIONS IN IMMUNOTHERAPY

There are many different medicinal uses for biopolymers as immunotherapeutic agents, including the treatment of cancer, the development of vaccines, and the control of autoimmune disorders. The graphical demonstration of a diverse variety of uses for biopolymers, showcasing their potential to transform several industries from food packaging to medicinal equipment, is depicted in Fig. (**2**) .

Cancer Immunotherapy

The application of biopolymers has significantly advanced cancer immunotherapy, especially in the creation of drug and gene delivery nanocarriers. Biodegradable polymer nanoparticles, for example, have been used to carry immune checkpoint inhibitors or tumor antigens, which improve the immune system's ability to identify and eradicate cancer cells. Furthermore, therapeutic drugs have been directly delivered to the tumor microenvironment using hydrogels derived from biopolymers, eliciting a prolonged and targeted immune response [35, 36].

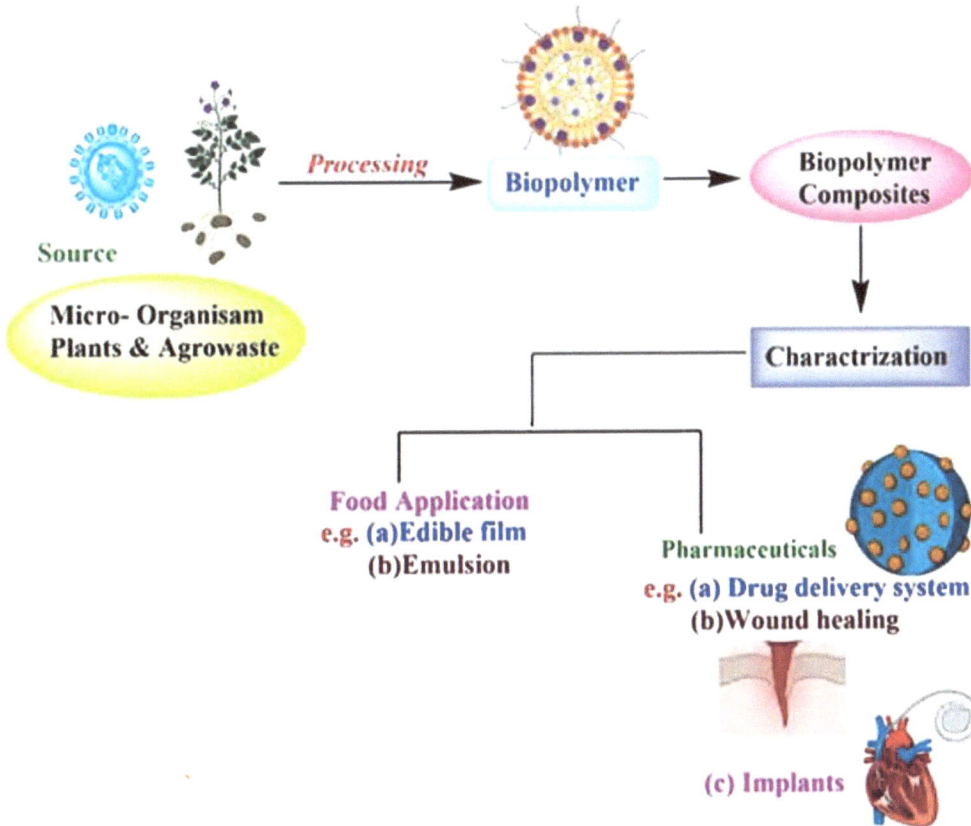

Fig. (2). The diverse variety of uses for biopolymers.

Vaccine Development

Biopolymers have been essential in the creation of innovative vaccination platforms, such as those against cancer and infectious diseases. Adjuvants based on polysaccharides, for instance, strengthen the immune response to vaccinations by promoting APCs and aiding in the identification of antigens. To administer mRNA vaccines, like the ones used in the COVID-19 pandemic, biopolymer nanoparticles have also been produced. This shows how well these materials work to induce protective immune responses [37, 38].

Autoimmune Diseases

Autoimmune diseases like Multiple Sclerosis (MS) and rheumatoid arthritis are caused by the immune system mistakenly attacking healthy tissues. Immunosuppressive medications or tolerogenic substances that alter the immune response and encourage immunological tolerance have been delivered *via* biopolymers. For instance, antigens and Treg responses are delivered *via* chitosan-based nanoparticles, which reduce inflammation and stop tissue damage [39, 40].

Organ Transplantation

Biopolymers have demonstrated potential in this field by delivering immunosuppressive drugs in a regulated and localized way, which poses a significant challenge in the prevention of transplant rejection. Immunosuppressants have been applied to transplant sites slowly using hydrogels produced from naturally occurring biopolymers, reducing systemic adverse effects and increasing graft survival rates [41 - 43].

BIOLOGICAL BASIS OF AUTOIMMUNE DISEASES

The immune system is necessary to keep the body tolerant of its cells and tissues while defending it against dangerous infections, including bacteria, viruses, and fungi. It functions by the use of an intricate system of tissues, cells, and organs that identify and eradicate external intruders. The innate immune system, which provides fast, non-specific defense, and the immune system that adapts, which creates a focused, lasting response, are the two basic systems that mediate the immune system's response [44].

Immune System Function and Dysfunction

The natural immune system, which employs external obstacles like the skin and mucus membranes in addition to specialized immune cells, such as cells called dendritic cells and macrophages, is the body's first line of defense. These cells can recognize molecular patterns associated with pathogens (PAMPs) and initiate inflammatory responses due to the presence of receptors that recognize patterns (PRRs) [45]. T and B cells make up the highly specialized and antigen-specific adaptive immune system. Due to the immunological memory this system offers, successive exposures to the same virus might elicit stronger and faster reactions [46].

Autoimmunity and Immune System Dysfunction

The immune system may mistakenly target the body's tissues, cells, or organs if this delicate balance is thrown off. Autoimmune diseases stem from this. These self-directed attacks are mostly caused by a decline in immunological tolerance and the breakdown of regulatory mechanisms that frequently prevent autoimmunity. There is strong evidence that genetic predisposition, environmental factors (such as poisons or viruses), and hormone impacts are significant contributors to this breakdown, even if the exact causes are still unclear [47]. During development, autoreactive T and B cells can evade central tolerance and trigger autoimmune reactions. Autoreactive cells should normally be destroyed or rendered inactive in the bone marrow or thymus, but malfunctions in this process might allow the cells to persist. After reaching the periphery, these cells can be stimulated by immune response-inducing molecules mimicking bystander activation, or epitope-spreading pathways [48, 49].

Autoimmunity Mechanisms

The immune system's failure to differentiate between self and non-self leads to an attack on the body's tissues, which is the intricate phenomenon known as autoimmunity. Knowing the fundamental causes of autoimmunity is crucial to creating successful treatment plans. Genetic predisposition is one of the main pathways, wherein certain alleles of the Major Histocompatibility Complex (MHC) genes enhance a person's sensitivity to autoimmune disorders. These genetic variables may modify an individual's immune response, increasing their risk of developing autoimmune diseases. The development of autoimmunity is also significantly influenced by environmental factors. Viruses and bacteria, in particular, can cause or worsen autoimmune reactions by tricking the immune system into thinking that self-antigens are pathogen-derived antigens through a process known as molecular mimicry [50]. Furthermore, exposure to specific toxins, chemicals, or dietary components can disrupt immune regulation, which further contributes to autoimmunity. Dysregulation of immune tolerance is another important mechanism. Autoimmune patients may also have impaired regulatory T cells (Tregs), which normally suppress inappropriate immune responses. This impairment results in a failure to control autoreactive T cells, which allows them to proliferate and attack healthy tissues. Additionally, autoimmune patients may have inappropriate B cell activation, which produces autoantibodies that target the body's cells, causing inflammation and tissue damage. Finally, epigenetic modifications, which alter gene expression without changing the DNA sequence, also contribute to autoimmunity [51, 52].

Ultimately, it is now evident that immune regulation is significantly influenced by the gut microbiome. An imbalance in the bacteria in the gut, or dysbiosis, has been connected to many autoimmune diseases and may impact immune system performance. In summary, a variety of environmental, genetic, and immunological variables that are all implicated in the multiple pathways of autoimmunity influence the etiology of these complex disorders. Understanding these procedures is crucial for creating treatment plans and improving patient outcomes [53, 54]. Fig. (**3**) illustrates how malfunctioning mitochondria can leak RNA and DNA, which can trigger immune responses that worsen autoimmune diseases.

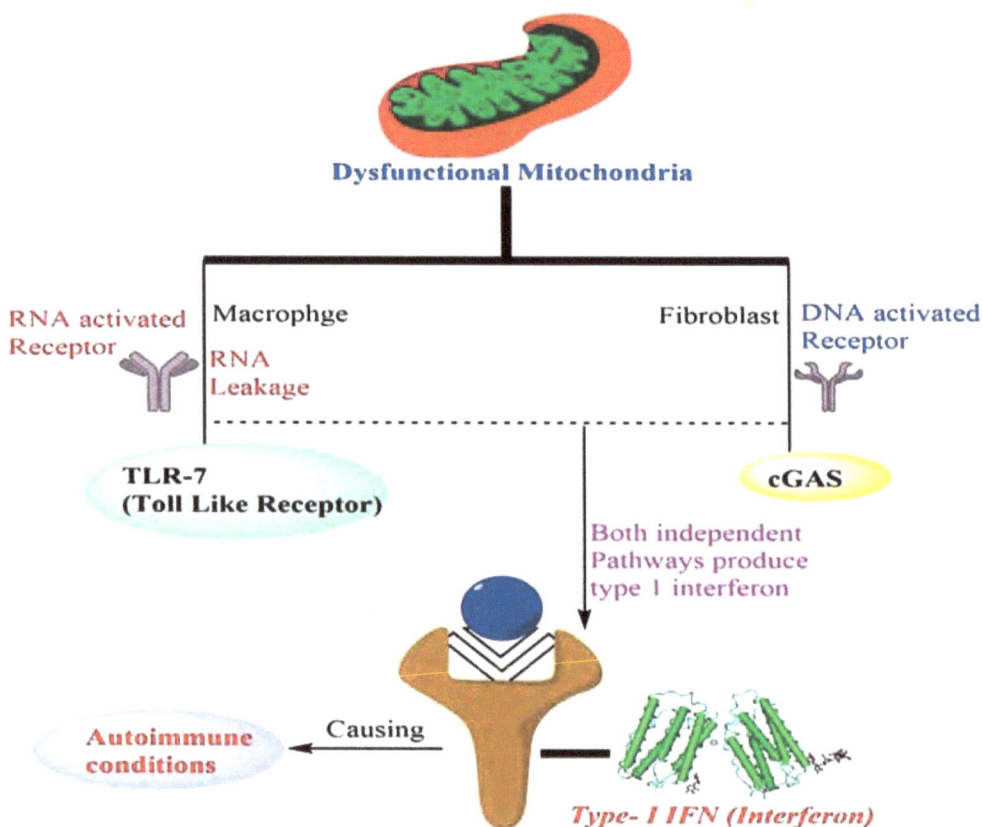

Fig. (3). Malfunctioning mitochondria can leak RNA and DNA.

CLINICAL APPLICATIONS AND TRIALS

Immunotherapeutic biopolymers are garnering interest as novel therapeutic approaches for autoimmune disorders, which arise from the immune system's inadvertent targeting of the body's tissues. These biopolymers, which come from natural sources, can be designed to improve the way medicinal substances are delivered, control immunological reactions, and encourage tolerance to self-antigens [55]. They can be used to treat the complicated issues associated with autoimmune disorders because of their special qualities.

CURRENT TREATMENTS

Immunosuppressive medications that try to lessen the hyperactive immune response are usually part of the conventional treatment for autoimmune diseases. DMARDs, corticosteroids, and biologics that target certain immune pathways are common therapies. Although these therapies can be useful in controlling symptoms and stopping the course of the disease, they frequently exhibit serious adverse effects, including an increased risk of infections, persistent side effects, and the possibility of developing drug resistance. Immunotherapeutic biopolymers enhance the effectiveness and safety of already available treatments, providing a supplementary strategy [56]. Anti-inflammatory medications, for instance, can be specifically delivered to inflammatory tissues by encapsulating them in biopolymer-based nanoparticles. By minimizing systemic exposure and reducing adverse effects, this targeted strategy may improve patient outcomes [57].

Limitations of Current Treatments

Immunotherapeutic biopolymers may have advantages, but there are also certain drawbacks. Variability in patient reactions to therapy is a significant difficulty that can be attributed to environmental triggers, disease heterogeneity, and hereditary variables. Furthermore, because the long-term safety profile of biopolymer-based therapeutics is yet unknown due to their potential for unexpected effects and interactions with the immune system, further research is needed to completely understand these interactions. The challenges of manufacturing and scaling up are also very high. Regulatory approval and clinical application of biopolymer formulations depend on the consistent quality and effectiveness of these formulations, which can only be achieved through extensive clinical trials that prove the safety and efficacy of the therapy [58, 69].

Clinical Trials and Upcoming Therapy

The application of immunotherapeutic biopolymers in different autoimmune diseases is being investigated in ongoing clinical studies. Numerous clinical trials have investigated the safety and effectiveness of biopolymer-based immunotherapies for autoimmune diseases. Below are key studies from the field of recent advancement that present study design, sample size, and outcome [60].

Phase II Clinical Trial on Chitosan Nanoparticles in Rheumatoid Arthritis (RA)

The efficacy and safety of methotrexate in chitosan nanoparticles compared to placebo in 120 patients with moderate-to-severe rheumatoid arthritis were assessed in a randomized, double-blind, placebo-controlled Phase II clinical trial. The main outcome measures were a decrease in Disease Activity Score (DAS28), a clinically validated index for the measurement of RA severity, and the inflammatory biomarkers, levels of C-Reactive Protein (CRP) and Erythrocyte Sedimentation Rate (ESR). Chitosan Nanoparticles in Rheumatoid Arthritis (RA) – Phase II Clinical Trial A randomized, placebo-controlled, double-blind Phase II clinical trial was carried out to compare the efficacy and safety of chitosan nanoparticles-encapsulated methotrexate in moderate-to-severe rheumatoid arthritis patients. In addition, levels of inflammatory biomarkers, such as CRP and ESR, were significantly decreased, supporting the increased anti-inflammatory action of the chitosan-based treatment [61]. Notably, chitosan nanoparticle-treated patients experienced fewer side effects, namely decreased gastrointestinal toxicity, which is a frequent correlate of conventional methotrexate treatment. These findings point to the therapeutic promise of drug delivery *via* nanoparticles to better treat RA through improved efficacy and reduced systemic toxicity [62].

Hyaluronic Acid-based Immunotherapy for Multiple Sclerosis (MS) – Phase I/II Clinical Trial

An open-label, dose-escalation Phase I/II trial was performed to explore the safety and therapeutic value of hyaluronic acid-based dendritic cell immunotherapy in Relapsing-remitting Multiple Sclerosis (RRMS). The research involved 60 MS patients who received monthly hyaluronic acid-loaded dendritic cell therapy for six months. The major outcomes were a diminution of Annualized Relapse Rate (ARR), "an important clinical indicator of MS disease activity", and MRI-based lesion progression analysis to measure neuroinflammation. Secondary outcome measures involved neurological function change as measured through Expanded Disability Status Scale (EDSS) scores and the regulation of pivotal pro-inflammatory cytokines, *i.e.*, IL-17, IFN-γ, and TNF-α, implicated in MS disease mechanisms. The study indicated a 40% reduction in ARR, equivalent to a remarkable reduction in disease relapse incidence. MRI also showed fewer new or enlarging lesions, demonstrating the potential neuroprotective effect. Patients treated with hyaluronic acid-based immunotherapy also had lower levels of inflammatory cytokines, which validates its immunomodulatory effect [63] These results suggest that dendritic cell therapy with the support of biopolymer has the potential to be a promising immune-modulating treatment for MS patients, with a new therapeutic strategy being capable of yielding better clinical result. For example, preliminary studies on the delivery of anti-inflammatory drugs in rheumatoid arthritis using chitosan-based nanoparticles have demonstrated enhanced joint function and decreased inflammation. In a similar vein, alginate hydrogels are being investigated for their potential to deliver biologics with prolonged release in diseases, such as multiple sclerosis. As research advances, the management of autoimmune diseases may be improved through the incorporation of immunotherapeutic biopolymers into standard treatment regimens. These biopolymers have the potential to greatly enhance patient outcomes and quality of life by addressing the shortcomings of existing therapies and offering targeted, efficient treatment options [64].

Biopolymer-based Therapies in Clinical Trials

Biopolymer-based treatments give a viable new avenue in medical therapy, notably in the domains of the use of tissue engineering, regeneration medicine, and delivery of drugs. These treatments make use of polymers that occur naturally, including alginate, chitosan, and hyaluronic acid, to form scaffolds or carriers that can boost the effectiveness of medications. The biocompatibility of biopolymers reduces the possibility of negative responses when ingested, which is one of their main benefits [65].

The possible applications of biopolymer-based systems are being studied in many contexts through ongoing clinical studies. To reduce systemic side effects and increase the number of therapeutic medications at the tumor site, researchers are investigating the use of biopolymer nanoparticles for specific drug delivery to cancerous cells. Additionally, biopolymers are being employed in regenerative medicine to encourage tissue repair and regeneration, which might offer relief from conditions including cartilage degeneration and wound healing. Although biopolymer-based treatments show great promise, there are still obstacles in getting them from the lab to the patient's bedside. It is imperative to solve issues, including manufacturing scalability, regulatory obstacles, and the requirement for comprehensive clinical testing to guarantee safety and efficacy. However, the ongoing studies represent a change in direction toward novel therapeutic approaches that capitalize on the special qualities of biopolymers, opening the door for improvements in patient outcomes and care [66].

FUTURE DIRECTIONS AND CHALLENGES

Emerging Technologies

Recent technical developments are expected to have a major impact on the development of immunotherapeutic biopolymers. Optimizing medication delivery systems through the application of nanotechnology is a major area of interest. One way to improve the bioavailability and targeting of therapeutic drugs is by engineering nanoparticles to encapsulate them and release them at specific locations of immune dysregulation. In the future, scientists may be able to create multipurpose nanocarriers that can transport several drugs at once, enabling more thorough immune response regulation. Moreover, new technologies like bioprinting and 3D printing may make it easier to create personalized biopolymer scaffolds that support tissue regeneration in injured regions, in addition to acting as drug transporters. Another cutting-edge strategy is the direct correction of immune-related genetic defects using gene-editing technologies like CRISPR when combined with biopolymer systems [67]. The combination of CRISPR-Cas9 gene editing for biopolymers has developed as a breakthrough strategy for autoimmune disease management through genetic correction and targeted immunomodulation. Conventional viral vectors employed for CRISPR delivery are frequently associated with off-target mutagenesis and side effects of immunogenicity, thereby making delivery systems based on biopolymers a prime option. Current research has shown the effectiveness of chitosan-encapsulated CRISPR-Cas9 systems in knocking out pro-inflammatory cytokine genes like TNF-α and IL-6, successfully reducing joint inflammation in rheumatoid arthritis models. Likewise, hyaluronic acid-based CRISPR-loaded nanoparticles have been engineered to silence autoreactive T-cell receptors in models of multiple sclerosis, resulting in long-term remission of disease. PLGA nanoparticles were also investigated to deliver CRISPR-Cas9 Ribonucleoproteins (RNPs) of low cytotoxicity and high precision in editing the gene [68]. Biopolymer-delivered CRISPR's foremost advantages in its biodegradability, lower toxicity, and delivery control, ensuring that gene expression can be persistently modified without the long-term drawbacks of viral vectors. Future efforts are likely to address improving efficiency in delivery, genome-targeting specificity, and optimal safety profiles for therapeutic purposes. The use of gene editing together with biopolymer-mediated immunotherapies may offer avenues for customized and long-lasting disease-modifying treatments in autoimmune diseases [69]. This would allow for the development of highly individualized therapeutics for autoimmune disorders.

Personalized Medicine Approaches

Personalized medicine is becoming more important in the treatment of autoimmune diseases, and biopolymers are well-positioned to play a significant role in this evolution. Treatments that are intended for everyone may not always work because of the immune system's high degree of individualization when responding to sickness. By tailoring biopolymer-based treatments to the specific genetic, epigenetic, and environmental traits of each patient, it may be possible to develop more targeted and effective drugs. One of the promising directions in this field is the development of biopolymers that may be engineered to respond to specific biomarkers associated with an individual's health state. It may be possible to design treatments that alter the immune system's response to autoantigens more successfully by selecting biopolymers based on patient-specific immune cell profiles [70]. This method may decrease negative effects and improve long-term outcomes. Moreover, advancements in personalized medicine might enable the customization of dosage forms, guaranteeing that every patient gets the precise amount of therapeutic agent required to fulfill their specific requirements [71, 72].

CONCLUSION

One potential area for the therapy of autoimmune diseases is studying the possibility of immunotherapeutic biopolymers. These biopolymers provide a unique method of regulating immune responses since they are produced from natural macromolecules, such as proteins, polysaccharides, and nucleic acids. Contrary to conventional immunosuppressive treatments, which frequently have serious side effects and do not deal with the underlying causes of autoimmune diseases, biopolymers offer a more specialized and suitable treatment. They can be used as a flexible tool to manage conditions, such as rheumatoid arthritis, multiple sclerosis, and type 1 diabetes, because of their capacity to create immunological tolerance, regulate immune cell activity, and reduce inflammation. To maximize therapeutic results, current research and clinical trials are concentrated on improving these biopolymers' safety and efficacy through the use of cutting-edge delivery methods. It is anticipated that the incorporation of cutting-edge technologies, such as customized medicine and nanotechnology, would enhance these therapies and provide more effective and individualized healthcare options. Nonetheless, there are still issues with regulatory barriers, manufacturing scalability, and the requirement for thorough clinical testing to guarantee efficacy and safety. There is a great deal of promise for immunotherapeutic biopolymers to transform the way autoimmune diseases are being treated. Improving patient outcomes and lowering the burden of autoimmune disorders are anticipated as a result of their capacity to offer focused, effective, and customized treatment choices, which represents substantial progress in the area of immunotherapy. To fully achieve their potential and get over current obstacles, further research and development in this field are required.

ABBREVIATIONS

APCs	Antigen Presenting Cells
CS	Chondroitin Sulfate
CNFs	Cellulose Nanofibers
CNCs	Cellulose Nanocrytals

DCs	Dendritic Cells
GAGs	Glycosaminoglycans
HA	Hyaluronic Acid
NCs	Nanocarriers
NP	Nanoparticles
NK	Natural Killer cells
PRR	Pattern Recognition Pattern
PAMPs	Pathogen – Associated Molecular Pattern
SLE	Systemic Lupus Erythematous

ACKNOWLEDGEMENTS

The authors are thankful to Mahayogi Gorakhnath University, Gorakhpur, for providing the necessary facilities to execute this manuscript.

CONFLICTS OF INTEREST/COMPETING INTERESTS

The authors declare no competing interests.

AUTHORS' CONTRIBUTIONS

Conceptualization: All the authors contributed equally to the development of this book's chapter framework and concept. This collaborative effort ensured that a well-rounded perspective was preserved throughout the study.

REFERENCES

[1] Desai MK, Brinton RD. Autoimmune disease in women: endocrine transition and risk across the lifespan. Front Endocrinol 2019; 10: 265.
[http://dx.doi.org/10.3389/fendo.2019.00265]

[2] Domogalla MP, Rostan PV, Raker VK, Steinbrink K. Tolerance through education: how tolerogenic dendritic cells shape immunity. Front Immunol 2017; 8: 1764.
[http://dx.doi.org/10.3389/fimmu.2017.01764] [PMID: 29375543]

[3] Mitarotonda R, Giorgi E, Eufrasio-da-Silva T, *et al.* Immunotherapeutic nanoparticles: From autoimmune disease control to the development of vaccines. Biomater Adv 2022; 135: 212726.
[http://dx.doi.org/10.1016/j.bioadv.2022.212726] [PMID: 35475005]

[4] Huzum B, Puha B, Necoara R, *et al.* Biocompatibility assessment of biomaterials used in orthopedic devices: An overview (Review). Exp Ther Med 2021; 22(5): 1315.
[http://dx.doi.org/10.3892/etm.2021.10750] [PMID: 34630669]

[5] Thang NH, Chien TB, Cuong DX. Polymer-based hydrogels applied in drug delivery: An overview. Gels 2023; 9(7): 523.
[http://dx.doi.org/10.3390/gels9070523] [PMID: 37504402]

[6] Desai N, Rana D, Salave S, *et al.* Chitosan: a potential biopolymer in drug delivery and biomedical applications. Pharmaceutics 2023; 15(4): 1313.
[http://dx.doi.org/10.3390/pharmaceutics15041313] [PMID: 37111795]

[7]   Mishra S, Shah H, Patel A, Tripathi SM, Malviya R, Prajapati BG. Applications of bioengineered polymer in the field of nano-based drug delivery. ACS Omega 2024; 9(1): 81-96.
        [http://dx.doi.org/10.1021/acsomega.3c07356] [PMID: 38222544]

[8]   Carey ST, Bridgeman C, Jewell CM. Biomaterial strategies for selective immune tolerance: advances and gaps. Adv Sci 2023; 10(8): 2205105.
        [http://dx.doi.org/10.1002/advs.202205105]

[9]   Mukherjee AG, Wanjari UR, Namachivayam A, *et al.* Role of immune cells and receptors in cancer treatment: an immunotherapeutic approach. Vaccines (Basel) 2022; 10(9): 1493.
        [http://dx.doi.org/10.3390/vaccines10091493] [PMID: 36146572]

[10]   Fountzilas E, Lampaki S, Koliou GA, *et al.* Real-world safety and efficacy data of immunotherapy in patients with cancer and autoimmune disease: the experience of the Hellenic Cooperative Oncology Group. Cancer Immunol Immunother 2022; 71(2): 327-37.
        [PMID: 34164709]

[11]   Muluh TA, Lu X, Zhang Y, *et al.* Combined immunotherapy and targeted therapies for cancer treatment: recent advances and future perspectives. Curr Cancer Drug Targets 2023; 23(4): 251-64.
        [http://dx.doi.org/10.2174/1568009623666221020104603] [PMID: 36278447]

[12]   Tayal V, Kalra BS. Cytokines and anti-cytokines as therapeutics — An update. Eur J Pharmacol 2008; 579(1-3): 1-12.
        [http://dx.doi.org/10.1016/j.ejphar.2007.10.049] [PMID: 18021769]

[13]   Guimarães LE, Baker B, Perricone C, Shoenfeld Y. Vaccines, adjuvants and autoimmunity. Pharmacol Res 2015; 100: 190-209.
        [http://dx.doi.org/10.1016/j.phrs.2015.08.003] [PMID: 26275795]

[14]   Jogalekar MP, Rajendran RL, Khan F, Dmello C, Gangadaran P, Ahn BC. CAR T-Cell-based gene therapy for cancers: new perspectives, challenges, and clinical developments. Front Immunol 2022; 13: 925985.
        [http://dx.doi.org/10.3389/fimmu.2022.925985] [PMID: 35936003]

[15]   Freitas MV, Frâncio L, Haleva L, Matte UDS. Protection is not always a good thing: the immune system's impact on gene therapy. Genet Mol Biol 2022; 45(3 Suppl 1): e20220046.
        [http://dx.doi.org/10.1590/1678-4685-GMB-2022-0046]

[16]   Chu J, Gao F, Yan M, *et al.* Natural killer cells: a promising immunotherapy for cancer. J Transl Med 2022; 20(1): 240.
        [http://dx.doi.org/10.1186/s12967-022-03437-0]

[17]   Caspi RR. Immunotherapy of autoimmunity and cancer: the penalty for success. Nat Rev Immunol 2008; 8(12): 970-6.
        [http://dx.doi.org/10.1038/nri2438] [PMID: 19008897]

[18]   Lechner K, Aulinger B, Brand S, Waldmann E, Parhofer KG. Hydrothermally modified slow release corn starch: a potential new therapeutic option for treating hypoglycemia in autoimmune hypoglycemia (Hirata's disease). Eur J Clin Nutr 2015; 69(12): 1369-70.
        [http://dx.doi.org/10.1038/ejcn.2015.151] [PMID: 26373963]

[19]   Gadhave RV, Das A, Mahanwar PA, Gadekar PT. Starch based bio-plastics: the future of sustainable packaging Open J Polym Chem 2018; 8(2)).
        [http://dx.doi.org/10.4236/ojpchem.2018.82003]

[20]   Woodruff MA, Hutmacher DW. The return of a forgotten polymer—Polycaprolactone in the 21st century. Prog Polym Sci 2010; 35(10): 1217-56.
        [http://dx.doi.org/10.1016/j.progpolymsci.2010.04.002]

[21]   Fireman EM, Fireman Klein E. Association between silicosis and autoimmune disease. Curr Opin Allergy Clin Immunol 2024; 24(2): 45-50.
        [http://dx.doi.org/10.1097/ACI.0000000000000966] [PMID: 38277164]

[22] Gándara Z, Rubio N, Castillo RR. Delivery of therapeutic biopolymers employing silica-based nanosystems. Pharmaceutics 2023; 15(2): 351.
[http://dx.doi.org/10.3390/pharmaceutics15020351] [PMID: 36839672]

[23] Makadia HK, Siegel SJ. Poly lactic-co-glycolic acid (PLGA) as biodegradable controlled drug delivery carrier. Polymers (Basel) 2011; 3(3): 1377-97.
[http://dx.doi.org/10.3390/polym3031377] [PMID: 22577513]

[24] Yasin A, Ren Y, Li J, Sheng Y, Cao C, Zhang K. Advances in hyaluronic acid for biomedical applications. Front Bioeng Biotechnol 2022; 10: 910290.
[http://dx.doi.org/10.3389/fbioe.2022.910290] [PMID: 35860333]

[25] Sprott H, Fleck C. Hyaluronic acid in rheumatology. Pharmaceutics 2023; 15(9): 2247.
[http://dx.doi.org/10.3390/pharmaceutics15092247] [PMID: 37765216]

[26] Leach DG, Young S, Hartgerink JD. Advances in immunotherapy delivery from implantable and injectable biomaterials. Acta Biomater 2019; 88: 15-31.
[http://dx.doi.org/10.1016/j.actbio.2019.02.016] [PMID: 30771535]

[27] Riley RS, June CH, Langer R, Mitchell MJ. Delivery technologies for cancer immunotherapy. Nat Rev Drug Discov 2019; 18(3): 175-96.
[http://dx.doi.org/10.1038/s41573-018-0006-z] [PMID: 30622344]

[28] Murphy EJ, Fehrenbach GW, Abidin IZ, *et al.* Polysaccharides—naturally occurring immune modulators. Polymers (Basel) 2023; 15(10): 2373.
[http://dx.doi.org/10.3390/polym15102373] [PMID: 37242947]

[29] Baranwal J, Barse B, Fais A, Delogu GL, Kumar A. Biopolymer: A sustainable material for food and medical applications. Polymers (Basel) 2022; 14(5): 983.
[http://dx.doi.org/10.3390/polym14050983] [PMID: 35267803]

[30] Li D, Wang Y, Zhu S, Hu X, Liang R. Recombinant fibrous protein biomaterials meet skin tissue engineering. Front Bioeng Biotechnol 2024; 12: 1411550.
[http://dx.doi.org/10.3389/fbioe.2024.1411550] [PMID: 39205856]

[31] Tripathi AS, Zaki MEA, Al-Hussain SA, *et al.* Material matters: exploring the interplay between natural biomaterials and host immune system. Front Immunol 2023; 14: 1269960.
[http://dx.doi.org/10.3389/fimmu.2023.1269960] [PMID: 37936689]

[32] Leitner WW, Ying H, Restifo NP. DNA and RNA-based vaccines: principles, progress and prospects. Vaccine 1999; 18(9-10): 765-77.
[http://dx.doi.org/10.1016/S0264-410X(99)00271-6] [PMID: 10580187]

[33] Chehelgerdi M, Chehelgerdi M. The use of RNA-based treatments in the field of cancer immunotherapy. Mol Cancer 2023; 22(1): 106.
[http://dx.doi.org/10.1186/s12943-023-01807-w] [PMID: 37420174]

[34] Börjesson M, Westman G. Crystalline nanocellulose-preparation, modification, and properties. Cellulose - Fundamental Aspects and Current Trends 2015; 7
[http://dx.doi.org/10.5772/61899]

[35] Čolić M, Tomić S, Bekić M. Immunological aspects of nanocellulose. Immunol Lett 2020; 222: 80-9.
[http://dx.doi.org/10.1016/j.imlet.2020.04.004] [PMID: 32278785]

[36] Pourmadadi M, Rahmani E, Shamsabadipour A, *et al.* Novel carboxymethyl cellulose based nanocomposite: A promising biomaterial for biomedical applications. Process Biochem 2023; 130: 211-26.
[http://dx.doi.org/10.1016/j.procbio.2023.03.033]

[37] Abka-khajouei R, Tounsi L, Shahabi N, Patel AK, Abdelkafi S, Michaud P. Structures, properties and applications of alginates. Mar Drugs 2022; 20(6): 364.
[http://dx.doi.org/10.3390/md20060364] [PMID: 35736167]

[38]   Lee KY, Mooney DJ. Alginate: Properties and biomedical applications. Prog Polym Sci 2012; 37(1): 106-26.
[http://dx.doi.org/10.1016/j.progpolymsci.2011.06.003] [PMID: 22125349]

[39]   Choukaife H, Doolaanea AA, Alfatama M. Alginate nanoformulation: Influence of process and selected variables. Pharmaceuticals (Basel) 2020; 13(11): 335.
[http://dx.doi.org/10.3390/ph13110335] [PMID: 33114120]

[40]   Zaharoff DA, Rogers CJ, Hance KW, Schlom J, Greiner JW. Chitosan solution enhances both humoral and cell-mediated immune responses to subcutaneous vaccination. Vaccine 2007; 25(11): 2085-94.
[http://dx.doi.org/10.1016/j.vaccine.2006.11.034] [PMID: 17258843]

[41]   Fong D, Hoemann CD. Chitosan immunomodulatory properties: perspectives on the impact of structural properties and dosage. Future Sci OA 2018; 4(1): FSO225.
[http://dx.doi.org/10.4155/fsoa-2017-0064] [PMID: 29255618]

[42]   Caird R, Williamson M, Yusuf A, *et al.* Targeting of glycosaminoglycans in genetic and inflammatory airway disease. Int J Mol Sci 2022; 23(12): 6400.
[http://dx.doi.org/10.3390/ijms23126400] [PMID: 35742845]

[43]   Schöbel L, Boccaccini AR. A review of glycosaminoglycan-modified electrically conductive polymers for biomedical applications. Acta Biomater 2023; 169: 45-65.
[http://dx.doi.org/10.1016/j.actbio.2023.07.054] [PMID: 37532132]

[44]   Vassie JA, Whitelock JM, Lord MS. Glycosaminoglycan functionalized nanoparticles exploit glycosaminoglycan functions. Glycosaminoglycans. Methods Mol Biol 2015; 1229: 557-65.
[http://dx.doi.org/10.1007/978-1-4939-1714-3_44]

[45]   Kim J, Bang J, Park S, *et al.* Enhanced barrier properties of biodegradable PBAT/acetylated lignin films. Sustain Mater Technol 2023; 37: e00686.
[http://dx.doi.org/10.1016/j.susmat.2023.e00686]

[46]   Ma QH. Lignin biosynthesis and its diversified roles in disease resistance. Genes (Basel) 2024; 15(3): 295.
[http://dx.doi.org/10.3390/genes15030295] [PMID: 38540353]

[47]   Sharma R, Kuche K, Thakor P, *et al.* Chondroitin Sulfate: Emerging biomaterial for biopharmaceutical purpose and tissue engineering. Carbohydr Polym 2022; 286: 119305.
[http://dx.doi.org/10.1016/j.carbpol.2022.119305] [PMID: 35337491]

[48]   Zhou J, Nagarkatti P, Zhong Y, Nagarkatti M. Immune modulation by chondroitin sulfate and its degraded disaccharide product in the development of an experimental model of multiple sclerosis. J Neuroimmunol 2010; 223(1-2): 55-64.
[http://dx.doi.org/10.1016/j.jneuroim.2010.04.002] [PMID: 20434781]

[49]   Lim JJ, Hammoudi TM, Bratt-Leal AM, *et al.* Development of nano- and microscale chondroitin sulfate particles for controlled growth factor delivery. Acta Biomater 2011; 7(3): 986-95.
[http://dx.doi.org/10.1016/j.actbio.2010.10.009] [PMID: 20965281]

[50]   Lara-Espinoza C, Carvajal-Millán E, Balandrán-Quintana R, López-Franco Y, Rascón-Chu A. Pectin and pectin-based composite materials: Beyond food texture. Molecules 2018; 23(4): 942.
[http://dx.doi.org/10.3390/molecules23040942] [PMID: 29670040]

[51]   Beukema M, Faas MM, de Vos P. The effects of different dietary fiber pectin structures on the gastrointestinal immune barrier: impact *via* gut microbiota and direct effects on immune cells. Exp Mol Med 2020; 52(9): 1364-76.
[http://dx.doi.org/10.1038/s12276-020-0449-2] [PMID: 32908213]

[52]   Gandhi S, Roy I. Drug delivery applications of casein nanostructures: A minireview. J Drug Deliv Sci Technol 2021; 66: 102843.
[http://dx.doi.org/10.1016/j.jddst.2021.102843]

[53]   Hou K, Wu ZX, Chen XY, *et al.* Microbiota in health and diseases. Signal Transduct Target Ther 2022; 7(1): 135.
[http://dx.doi.org/10.1038/s41392-022-00974-4]

[54]   Zhao M A, Chu J, Feng S, *et al.* Immunological mechanisms of inflammatory diseases caused by gut microbiota dysbiosis: a review. Biomed Pharmacother 2023; 164: 114985.
[http://dx.doi.org/10.1016/j.biopha.2023.114985]

[55]   Chaturvedi S, Kulshrestha S, Bhardwaj K, Jangir R. A review on properties and applications of xanthan gum. Microb Polym Appl Ecol Perspect 2021; 87-107.
[http://dx.doi.org/10.1007/978-981-16-0045-6_4]

[56]   Silveira M, Vargas S, Mendonça M, *et al.* Xanthan gum enhances humoral immune response elicited by a DNA vaccine against leptospirosis in mice. BMC Proc. 8: 1-2.
[http://dx.doi.org/10.1186/1753-6561-8-S4-P153]

[57]   Tariq A, Bhawani SA, Alotaibi KM. Xanthan gum-based nanocomposites for tissue engineering. Polysaccharide-Based Nanocomposites for Gene Delivery and Tissue Engineering. Woodhead Publishing. 2021; pp. 191-206.
[http://dx.doi.org/10.1016/B978-0-12-821230-1.00009-8]

[58]   Yeh C, Shih CJ, Liu TC, Chiou Y. Effects of oligo-fucoidan on the immune response, inflammatory status and pulmonary function in patients with asthma: a randomized, double-blind, placebo-controlled trial. Sci Rep 2022; 12(1): 18150.
[http://dx.doi.org/10.1038/s41598-022-21527-3] [PMID: 36307493]

[59]   Chollet L, Saboural P, Chauvierre C, Villemin JN, Letourneur D, Chaubet F. Fucoidans in nanomedicine. Mar Drugs 2016; 14(8): 145.
[http://dx.doi.org/10.3390/md14080145] [PMID: 27483292]

[60]   Gheorghita R, Anchidin-Norocel L, Filip R, Dimian M, Covasa M. Applications of biopolymers for drugs and probiotics delivery. Polymers 2021; 13: 2729.
[http://dx.doi.org/10.3390/polym13162729]

[61]   Minami T, Nakanishi Y, Izumi M, Harada T, Hara N. Enhancement of antigen-presenting capacity and antitumor immunity of dendritic cells pulsed with autologous tumor-derived RNA in mice. J Immunother 2003; 26(5): 420-31.
[http://dx.doi.org/10.1097/00002371-200309000-00005] [PMID: 12973031]

[62]   Gardner A, de Mingo Pulido Á, Ruffell B. Dendritic cells and their role in immunotherapy. Front Immunol 2020; 11: 924.
[http://dx.doi.org/10.3389/fimmu.2020.00924] [PMID: 32508825]

[63]   Grego EA, Siddoway AC, Uz M, *et al.* Polymeric nanoparticle-based vaccine adjuvants and delivery vehicles. Nanoparticles Rational Vaccine Des 2020; 29-76.
[http://dx.doi.org/10.1007/82_2020_226]

[64]   Gholami A, Mohkam M, Soleimanian S, Sadraeian M, Lauto A. Bacterial nanotechnology as a paradigm in targeted cancer therapeutic delivery and immunotherapy. Microsyst Nanoeng 2024; 10(1): 113.
[http://dx.doi.org/10.1038/s41378-024-00743-z] [PMID: 39166136]

[65]   Yu Z, Shen X, Yu H, Tu H, Chittasupho C, Zhao Y. Smart polymeric nanoparticles in cancer immunotherapy. Pharmaceutics 2023; 15(3): 775.
[http://dx.doi.org/10.3390/pharmaceutics15030775] [PMID: 36986636]

[66]   Zeng Y, Xiang Y, Sheng R, *et al.* Polysaccharide-based nanomedicines for cancer immunotherapy: A review. Bioact Mater 2021; 6(10): 3358-82.
[http://dx.doi.org/10.1016/j.bioactmat.2021.03.008] [PMID: 33817416]

[67]   Luo X, Miller SD, Shea LD. Immune tolerance for autoimmune disease and cell transplantation. Annu Rev Biomed Eng 2016; 18(1): 181-205.

[http://dx.doi.org/10.1146/annurev-bioeng-110315-020137] [PMID: 26928211]

[68]   Cifuentes-Rius A, Desai A, Yuen D, Johnston APR, Voelcker NH. Inducing immune tolerance with dendritic cell-targeting nanomedicines. Nat Nanotechnol 2021; 16(1): 37-46.
[http://dx.doi.org/10.1038/s41565-020-00810-2] [PMID: 33349685]

[69]   Sambi, Manpreet *et al.* Current challenges in cancer immunotherapy: Multimodal approaches to improve efficacy and patient response Rates. Journal of oncology 2019; 4508794.
[http://dx.doi.org/10.1155/2019/4508794] [PMID: 30941175]

[70]   Cordeiro AS, Alonso MJ, de la Fuente M. Nanoengineering of vaccines using natural polysaccharides. Biotechnol Adv 2015; 33(6): 1279-93.
[http://dx.doi.org/10.1016/j.biotechadv.2015.05.010] [PMID: 26049133]

[71]   Passeri L, Marta F, Bassi V, Gregori S. Tolerogenic dendritic cell-based approaches in autoimmunity. Int J Mol Sci 2021; 22(16): 8415.
[http://dx.doi.org/10.3390/ijms22168415] [PMID: 34445143]

[72]   Kim A, Xie F, Abed OA, Moon JJ. Vaccines for immune tolerance against autoimmune disease. Adv Drug Deliv Rev 2023; 203: 115140.
[http://dx.doi.org/10.1016/j.addr.2023.115140] [PMID: 37980949]

SUBJECT INDEX

A

Acemannan 6, 7
Adjuvants Therapeutic 1
Agonist 124
Albumin 142, 188
Aloe vera 5, 6
Amino acids 3, 10, 127
Antagonist 124
Antibodies 3
Anticoagulant properties 15, 16
Antimicrobial properties 5, 6, 208
Antithrombin 16
Apoptosis 142
Autoimmune disorders 148
Ayurveda 9

B

Biocompatibility 1, 2, 4, 7, 8, 9, 189
Biodegradability 1, 2, 4, 7, 8, 9, 13, 14, 139, 188
Biomedical research/engineering 1, 9
Biomolecules 1
Biopolymers 1, 2, 3, 4, 5, 9, 13, 14, 189, 236, 237, 262, 280, 288, 296, 300
Biotechnology 1, 13
Bovine Serum Albumin (BSA) 142

C

Cancer 3, 4, 139, 148, 167, 174, 188, 262, 264
Carbon Nanotubes (CNTs) 296
Catalysis 3
Cell proliferation 7
Cellulose 4, 189, 288
Chemotherapeutics (Biopolymer-based) 167
Chinese Medicine 5, 6, 7
Chitin 7
Chitosan 4, 5, 7, 8, 189, 264, 280, 288, 296

Collagen 7, 139, 296
Controlled drug release 4, 139
Cosmetic applications 4
Cryo-electron microscopy 13
Customizability/Interactive Capabilities 2

D

Delivery methods 148
Diabetes management 3, 14
DNA (Deoxyribonucleic Acid) 1, 3, 4, 11, 12, 13, 15
Drug 1, 2, 4, 8, 9, 13, 14, 15, 139, 148, 167, 188, 236, 237, 262
　delivery system 1, 2, 4, 8, 9, 13, 14, 15, 139, 148, 167, 188, 236, 237, 262
　formulations 4

E

Egyptian Medicine 6
Efficacy and safety 148
Electrospinning 288, 290
Emil Fischer 11, 12
Enzyme-
　replacement therapy 3
　Enzyme-substrate interaction 11
Enzymes 3, 11, 139
Excipient 4

F

Fibroin 7, 296, 300

G

Gene therapy 1, 3
Gelatin 139, 188, 280, 296
Genetic 3, 13
　blueprint 3

www.ingramcontent.com/pod-product-compliance
Lightning Source LLC
Chambersburg PA
CBHW050800220326
41598CB00006B/76